Heinz Rennenberg and Mark A. Adams (Eds.)

Nitrogen and Phosphorus Nutrition of Trees and Forests

MDPI

This book is a reprint of the Special Issue that appeared in the online, open access journal, *Forests* (ISSN 1999-4907) from 2014–2015 (available at: http://www.mdpi.com/journal/forests/special_issues/N_P).

Guest Editors
Heinz Rennenberg
Institute of Forest Sciences
University of Freiburg
Germany

Mark A. Adams
Faculty of Agriculture and Environment
University of Sydney
Australia

Editorial Office
MDPI AG
Klybeckstrasse 64
Basel, Switzerland

Publisher
Shu-Kun Lin

Managing Editor
Echo Zhang

1. Edition 2016

MDPI • Basel • Beijing • Wuhan

ISBN 978-3-03842-185-6 (Hbk)
ISBN 978-3-03842-186-3 (PDF)

Table of Contents

Heinz Rennenberg and Michael Dannenmann
Nitrogen Nutrition of Trees in Temperate Forests—The Significance of Nitrogen Availability
in the Pedosphere and Atmosphere
Reprinted from: *Forests* **2015**, *6*(8), 2820-2835

**Richard Hrivnák, Michal Slezák, Benjamín Jarčuška, Ivan Jarolímek and
Judita Kochjarová**
Native and Alien Plant Species Richness Response to Soil Nitrogen and Phosphorus in
Temperate Floodplain and Swamp Forests
Reprinted from: *Forests* **2015**, *6*(10), 3501-3513

Tae Kyung Yoon, Nam Jin Noh, Haegeun Chung, A-Ram Yang and Yowhan Son
Soil Nitrogen Transformations and Availability in Upland Pine and Bottomland Alder Forests
Reprinted from: *Forests* **2015**, *6*(9), 2941-2958

Timothy J. Albaugh, Thomas R. Fox, H. Lee Allen and Rafael A. Rubilar
Juvenile Southern Pine Response to Fertilization Is Influenced by Soil Drainage and Texture
Reprinted from: *Forests* **2015**, *6*(8), 2799-2819

Ania Kobylinski and Arthur L. Fredeen
Importance of Arboreal Cyanolichen Abundance to Nitrogen Cycling in Sub-Boreal Spruce
and Fir Forests of Central British Columbia, Canada
Reprinted from: *Forests* **2015**, *6*(8), 2588-2607

List of Contributors

Timothy J. Albaugh: Virginia Tech Department of Forest Resources and Environmental Conservation, 228 Cheatham Hall, Blacksburg, VA 24061, USA.

H. Lee Allen: ProFor Consulting, Cary, NC 27511, USA.

T. Andrew Black: Faculty of Land and Food Systems, University of British Columbia, Vancouver, BC V6T 1Z4, Canada.

Michael A. Blazier: Hill Farm Research Station, Louisiana State University Agricultural Center, Homer, LA 71040, USA.

Yang Cao: State Key Laboratory of Soil Erosion and Dryland Farming on the Loess Plateau, Northwest A & F University, Yangling 712100, China; Institute of Soil and Water Conservation, Chinese Academy of Sciences and Ministry of Water Resources, Yangling 712100, China.

Mingliang Che: State Key Laboratory of Resources and Environmental Information System, Institute of Geographic Sciences and Nature Resources Research, University of Chinese Academy of Sciences, Beijing 100101, China.

Yunming Chen: State Key Laboratory of Soil Erosion and Dryland Farming on the Loess Plateau, Northwest A & F University, Yangling 712100, China; Institute of Soil and Water Conservation, Chinese Academy of Sciences and Ministry of Water Resources, Yangling 712100, China.

Baozhang Chen: School of Environment Science and Spatial Informatics, China University of Mining and Technology, Xuzhou 221116, China; State Key Laboratory of Resources and Environmental Information System, Institute of Geographic Sciences and Nature Resources Research, University of Chinese Academy of Sciences, Beijing 100101, China.

Haegeun Chung: Department of Environmental Engineering, Konkuk University, Seoul 05029, Korea.

Dao Ngoc Chuong: Co-Innovation Center for Sustainable Forestry in Southern China, Nanjing Forestry University, Nanjing 210037, China.

Ryan Coleman: Hill Farm Research Station, Louisiana State University Agricultural Center, Homer, LA 71040, USA.

Michael Dannenmann: Institute for Meteorology and Climate Research, Atmospheric Environmental Research (IMK-IFU), Karlsruhe Institute of Technology (KIT), Kreuzeckbahnstrasse 19, Garmisch-Partenkirchen 82467, Germany.

Shiping Deng: Department of Plant and Soil Sciences, Oklahoma State University, Stillwater, OK 74048, USA.

Sharon L. Doty: School of Environmental and Forest Sciences, College of the Environment, University of Washington, Seattle, WA 98195, USA.

Xianming Dou: School of Resources and Earth Sciences; School of Environment Science and Spatial Informatics, China University of Mining and Technology, Xuzhou 221116, China.

Gregory J. Ettl: School of Environmental and Forest Sciences, College of the Environment, University of Washington, Seattle, WA 98195, USA.

Shengzuo Fang: Co-Innovation Center for Sustainable Forestry in Southern China, Nanjing Forestry University, Nanjing 210037, China.

Thomas R. Fox: Virginia Tech Department of Forest Resources and Environmental Conservation, 228 Cheatham Hall, Blacksburg, VA 24061, USA.

Arthur L. Fredeen: NRES Institute.

Fredrik From: Department of Forest Genetics and Plant Physiology, Umeå Plant Science Centre, Swedish University of Agricultural Sciences, Skogsmarksgränd 1, 90183 Umeå, Sweden.

Ah Reum Han: Department of Forest Sciences, Seoul National University, Seoul 151-921, Korea.

Richard Hrivnák: Institute of Botany, Slovak Academy of Sciences, Dúbravská cesta 9, SK-845 23 Bratislava, Slovakia.

Gazali Issah: Western Applied Research Corporation (WARC), P.O. Box 89 Scott, SK S0K 4A0, Canada.

Benjamín Jarčuška: Institute of Forest Ecology, Štúrova 2, SK-960 53 Zvolen, Slovakia.

Ivan Jarolímek: Institute of Botany, Slovak Academy of Sciences, Dúbravská cesta 9, SK-845 23 Bratislava, Slovakia.

Rachhpal S. Jassal: Faculty of Land and Food Systems, University of British Columbia, Vancouver, BC V6T 1Z4, Canada.

Mi-Ae Jeong: Department of Landscape Architecture; Department of Forest Sciences, Seoul National University, Seoul 151-921, Korea.

Jaeyeob Jeong: Centre for Environmental Risk Assessment and Remediation, University of South Australia, Adelaide, SA 5095, Australia.

Shyam L. Kandel: School of Environmental and Forest Sciences, College of the Environment, University of Washington, Seattle, WA 98195, USA.

Zareen Khan: School of Environmental and Forest Sciences, College of the Environment, University of Washington, Seattle, WA 98195, USA.

Choonsig Kim: Department of Forest Resources, Gyeongnam National University of Science and Technology, Jinju 660-758, Korea.

Anthony A. Kimaro: World Agroforestry Centre, ICRAF-Tanzania Programme, Dar-es-Salaam, Tanzania.

J. Diane Knight: Department of Soil Science, University of Saskatchewan, 51 Campus Drive, Saskatoon, SK S7N 5A8, Canada.

Ania Kobylinski: Natural Resources and Environmental Studies (NRES) Graduate Program, University of Northern British Columbia (UNBC), 3333 University Way, Prince George, BC V2N 4Z9, Canada.

Judita Kochjarová: Institute of Botany, Slovak Academy of Sciences, Dúbravská cesta 9, SK-845 23 Bratislava, Slovakia; Botanical Garden, Comenius University, SK-038 15 Blatnica, Slovakia.

John Kort: Agroforestry Development Centre, Agriculture and Agri-Food Canada (retired), P.O. Box 940, Indian Head, SK S0G 2K0, Canada.

Ho-Seop Ma: Department of Forest Environmental Resources, Gyeongsang National University, Jinju 660-701, Korea.

Nam Jin Noh: River Basin Research Center, Gifu University, Gifu 501-1193, Japan.

Annika Nordin: Department of Forest Genetics and Plant Physiology, Umeå Plant Science Centre, Swedish University of Agricultural Sciences, Skogsmarksgränd 1, 90183 Umeå, Sweden.

Yunmi Park: Division of Special-purpose Trees, Korea Forest Research Institute, Suwon 441-847, Korea.

Pil Sun Park: Department of Environment and Forest Resources, Chungnam National University, Daejeon 305-764, Korea; Department of Forest Sciences, Seoul National University, Seoul 151-921, Korea.

Jae-Hyun Park: Department of Forest Resources, Gyeongnam National University of Science and Technology, Jinju 660-758, Korea.

Daniela N. Ramos: School of Environmental and Forest Sciences, College of the Environment, University of Washington, Seattle, WA 98195, USA.

Heinz Rennenberg: Institute of Forest Sciences, University of Freiburg, Georges-Koehler-Allee 53/54, Freiburg 79110, Germany.

Rafael A. Rubilar: Cooperativa de Productividad Forestal. Facultad de Ciencias Forestales, Universidad de Concepción. Victoria 631, Casilla 160-C, Concepción, Chile.

D. Andrew Scott: Southern Research Station, USDA Forest Service, P.O. Box 1927, Normal, AL 35762, USA.

Michal Slezák: Institute of Botany, Slovak Academy of Sciences, Dúbravská cesta 9, SK-845 23 Bratislava, Slovakia; Department of Biology and Ecology, Faculty of Education, Catholic University, Hrabovská cesta 1, SK-034 01 Ružomberok, Slovakia; Department of Phytology, Faculty of Forestry, Technical University in Zvolen, T. G. Masaryka 24, SK-960 53 Zvolen, Slovakia.

Jaeeun Sohng: Department of Forest Sciences, Seoul National University, Seoul 151-921, Korea; School of Forestry & Environmental Studies, Yale University, New Haven, CT 06511, USA.

Yowhan Son: Department of Environmental Science and Ecological Engineering, Graduate School, Korea University, Seoul 02841, Korea; Department of Biological and Environmental Science, Qatar University, Doha P.O. Box 2713, Qatar.

Joachim Strengbom: Department of Ecology, Swedish University of Agricultural Sciences, Box 7044, 75007 Uppsala, Sweden.

Luozhong Tang: Co-Innovation Center for Sustainable Forestry in Southern China, Nanjing Forestry University, Nanjing 210037, China.

Ye Tian: Co-Innovation Center for Sustainable Forestry in Southern China, Nanjing Forestry University, Nanjing 210037, China.

Yafei Yan: Co-Innovation Center for Sustainable Forestry in Southern China, Nanjing Forestry University, Nanjing 210037, China; College of Forestry, Henan University of Science and Technology, Luoyang 471023, China.

A-Ram Yang: Forest Practice Research Center, Korea Forest Research Institute, Pocheon 11186, Korea.

Tae Kyung Yoon: Department of Environmental Science and Engineering, Ewha Womans University, Seoul 03760, Korea.

About the Guest Editors

Heinz Rennenberg received his PhD in 1977 from the University of Cologne, Germany working on sulfur metabolism of tobacco tissue cultures. After completing a Postdoc at the MSU-DOE Plant Research Laboratory, East Lansing, USA, he returned to Cologne were he got his habilitation in 1984 for his work on biosphere -atmosphere exchange of trace gases and an Associate Professor position in 1985. In 1986 he moved to the Institute of Atmospheric Environmental Research in Garmisch-Partenkirchen, Germany, and became Vice-Director of this institute in 1988. Since 1992 he is full professor of Tree Physiology at the University of Freiburg, Germany. His main fields of research include: Regulation of sulfur and nitrogen nutrition of plants, stress physiology and biosphere -atmosphere exchange processes (from molecular to whole plant). Since 2004 he is a member of Leopoldina, German Academy of Sciences. From 2010 to 2012 Heinz Rennenberg was President of the Federation of European Societies of Plant Biology (FESPB).

Mark Adams received his B.Sc. honours and PhD from the University of Melbourne. He is currently Professor of Biogeochemistry and Director of the Centre for Carbon, Water and Food. Mark Adams has held professorial appointments at the University of Western Australia, the University of Melbourne, and the University of New South Wales. He has been the recipient of a range of fellowships and awards in Australia, France, New Zealand, and Germany. In addition to tropical and temperate Australian forests, woodlands and grasslands, Mark Adams has conducted research in Pakistan, Papua New Guinea and Kenya, as well as in Europe and the USA. In 2006 he finished a six-year term as a member of the Board of Trusties for the International Centre for Research in Agroforestry (ICRAF) at Nairobi, Kenya and as the ICRAF trustee on the board of the Centre for International Forestry Research, Bogor, Indonesia. From 2009 until 2015 he served as Dean of the Faculty of Agriculture and Environment at the University of Sydney.

Preface

Nitrogen (N) and phosphorus (P) nutrition of trees has been studied for many decades, but has largely been focused on inorganic nutrient uptake and leaf level nutrient contents. In recent years it became obvious that N and P cycling at the ecosystem level is of vital importance for tree nutrition and that organic N uptake by trees is an essential part of ecosystem N cycling; in particular on N and/or P poor soils, and in cooler climates. The significance of organic P uptake by trees is still a matter of debate, especially under field conditions. The overlay of climate change on ecosystem N and P cycling has become an important issue of forest research. This overlay raises questions around competition for N and P among structural elements (overstorey *vs.* undestorey), as well as among dominant species. Many nutritionally related aspects of changing climates, such as effects on rhizosphere and phyllosphere, remain seriously under-studied. The central aim of this Special Issue is to provide new insights into some of these topics at the tree, and the ecosystem level.

Heinz Rennenberg and Mark A. Adams
Guest Editors

Nitrogen Nutrition of Trees in Temperate Forests—The Significance of Nitrogen Availability in the Pedosphere and Atmosphere

Heinz Rennenberg and Michael Dannenmann

Abstract: Nitrogen (N) is an essential nutrient that is highly abundant as N_2 in the atmosphere and also as various mineral and organic forms in soils. However, soil N bioavailability often limits the net primary productivity of unperturbed temperate forests with low atmospheric N input. This is because most soil N is part of polymeric organic matter, which requires microbial depolymerization and mineralization to render bioavailable N forms such as monomeric organic or mineral N. Despite this N limitation, many unfertilized forest ecosystems on marginal soil show relatively high productivity and N uptake comparable to agricultural systems. The present review article addresses the question of how this high N demand is met in temperate forest ecosystems. For this purpose, current knowledge on the distribution and fluxes of N in marginal forest soil and the regulation of N acquisition and distribution in trees are summarized. The related processes and fluxes under N limitation are compared with those of forests exposed to high N loads, where chronic atmospheric N deposition has relieved N limitation and caused N saturation. We conclude that soil microbial biomass is of decisive importance for nutrient retention and provision to trees both in high and low N ecosystems.

Reprinted from *Forests*. Cite as: Rennenberg, H.; Dannenmann, M. Nitrogen Nutrition of Trees in Temperate Forests—The Significance of Nitrogen Availability in the Pedosphere and Atmosphere. *Forests* **2015**, *6*, 2820-2835.

1. Introduction

Following water availability [1–3], access to nitrogen (N) sources is considered to be the main factor limiting the growth and development of plants and, hence, food and biomass production as well as land carbon (C) storage at a global scale [4–7]. This is also evident from the extensive distribution of low N soils across the globe [8]. In such soils, global climate change will further reduce N availability and accelerate N limitation due to reduced precipitation during the vegetation period [9,10] and dilution of bioavailable soil N by the increased growth in response to elevated carbon dioxide (CO_2) [6], provided N deposition from the atmosphere is low [11–15]. In agriculture, N limitation is overcome by the application of inorganic and/or organic fertilizers, amounting to a global inorganic fertilizer use of 108 Mt in 2012 [16]. At a global scale, the production of fertilizer N in conjunction with increased cultivation of N-fixing plants and industrial activities has more than doubled the annual release of reactive N forms to the biosphere [17]. Due to the cascading of mobile reactive N forms across the boundaries of agricultural ecosystems, this anthropogenic perturbation of the global N cycle has regionally increased atmospheric N deposition in forests, with N loads in Europe ranging from 5 to 60 kg N $ha^{-1} \cdot year^{-1}$ [18]. However, in forests, N fertilizer application or large atmospheric N deposition is usually not required for high biomass production and is minute compared with agricultural N use [19], despite a high distribution of forests on marginal soils with low N availability. Since perennial plants such as forest trees are assumed to have evolved on marginal soils, a perennial lifestyle has been proposed to constitute an adaptation to low nutrient availability, including nitrogen [20,21].

Most of the nitrogen in forest soils is fixed in organic compounds such as proteins, lignin, or chitin. These N forms cannot be directly used by plants, but require depolymerization by specialized microorganisms to be converted to largely bioavailable monomeric organic or mineral N forms [22,23]. According to the traditional view, the liberation of nutrients from soil organic matter (SOM) depends on the chemical recalcitrance of SOM compounds, the genetic microbial depolymerization capacity, and the climate-driven enzyme activity. However, recent evidence highlighted that SOM turnover and persistence is also governed by the accessibility of organic matter to decomposer organisms or exo-enzymes. Hence, soil aggregation, sorption/desorption processes and the formation of organo-mineral associations [24,25] are important regulators of SOM decomposition and nutrient liberation. Thus, N availability in soils varies considerably across ecosystem and soil types and is not necessarily related to total soil N content.

The high productivity of forest trees on marginal soils with low N content is not a consequence of a low N demand of forest trees. For example, mature temperate forest stands have an annual nitrogen requirement of about 100 kg N ha^{-1} year^{-1} (Figure 1), which is similar to many agricultural systems [26,27]. Such high amounts of N may become available in forests transiently as a result of accelerated microbial liberation of nutrients when gaps develop due to dying trees or windbreaks, or due to selective cutting or clear cutting activities in forests [28]. Thus, the major questions are: (i) how is this high N demand met in mature forests on marginal soil with low atmospheric N input? and (ii) which N source(s) become available in the annual growth cycle to meet this high N demand of mature forest trees? These questions are of particular significance when the competition for N between forest trees and microbial N transformation processes and microbial N use in growth and development are considered [12]. Therefore, the present review article summarizes the current knowledge on the processes and fluxes of N distribution between forest trees and microbes with particular emphasis on temperate forest ecosystems on marginal soil with low N content and low atmospheric N input. These processes and fluxes under N limitation are compared with those of forests exposed to high loads of N as observed in patchy landscapes in close proximity to intensively used agricultural and forested areas [29].

2. Distribution and Fluxes of Nitrogen in Marginal Soil

Meeting the high N demand of mature trees on marginal soil is particularly challenging. Forest ecosystems on marginal soil are characterized by a largely closed ecosystem N cycle, *i.e.*, by tightly-coupled microbial mineralization-immobilization turnover, which facilitates effective microbial N retention [12]. Under these conditions, leaching of N, particularly strongly mobile nitrate, into the hydrosphere is negligible. Furthermore, the release of volatile N compounds such as NO, N$_2$O or N$_2$, produced during microbial transformation of inorganic and organic N compounds, into the atmosphere is minute (Figure 1). The influx of N into the ecosystem from atmospheric deposition and non-symbiotic biological N fixation by heterotrophic bacteria [30,31], as well as N acquisition from the hydrosphere and its liberation from the geosphere [32] are small, but still can exceed N leaching and volatilization. Hence, the accumulation of N from these N sources can occur, but is slow. Thus, ecosystem N fluxes and N redistribution almost entirely depend on ecosystem internal sources, in particular the decomposition of leaf and root litter as well as decaying microbial biomass and older soil organic matter [12,33]. Since the mean residence time of microbial biomass with a range between days and months [26,34,35] is significantly shorter than the lifespan of leaf and root litter [36–38], microbial biomass turnover is a major driver of N redistribution in N-limited ecosystems. However, our understanding of internal gross nitrogen turnover in N-limited forests is still fragmentary because available studies on the extremely dynamic gross nitrogen mineralization-immobilization turnover have been restricted to single or a few measurement dates. Only for one N-limited forest site, *i.e.*, the Tuttlingen experimental beech forest in southern Germany, were the gross N turnover rates (ammonification, nitrification, microbial inorganic N immobilization, denitrification) determined with sufficient temporal resolution (13 sampling dates between 2002 and 2009) to constrain annual N turnover budgets [35,37,39–44] (Figure 1).

Figure 1 provides a synthesis on annual N turnover in this forest stand based on a compilation of previously published data on gross N turnover rates [35,37,39–43]. This ecosystem shows—despite very high N mineralization and significant nitrification rates—a closed N cycle characterized by competitive partitioning of N between beech trees and microbial N retention pathways so that N loss remains small. This is specifically due to the almost complete partitioning of nitrate to microbial or tree uptake (Figure 1, Tuttlingen: Beech forest, Rendzic Leptosol). Microbial mineralization-immobilization turnover is *ca.* fivefold larger than tree N uptake and plant-mediated internal N turnover, producing an annual microbial detritus >500 kg N ha^{-1}. This means that microbial biomass is processed on average several times a year, recycling and conserving a huge nutrient stock. These N cycle patterns sustain economically and ecologically valuable forests on marginal soils, which are not suitable for traditional agricultural use with herbaceous crops.

Tree uptake rates of amino acids remain difficult to estimate because on the one hand, studies on N uptake capacity using capsules with amino acid solutions attached to soil-free washed root tips show that there is significant uptake capacity [42,43]. On the other hand, recent work based on the injection of double $^{13}C/^{15}N$-labelled amino acids into intact beech-soil systems showed that only ^{15}N and not ^{13}C was recovered in beech [45]. This suggests that either the uptake of intact amino acids was not significant in the presence of microbial competition, or that amino-acid derived C had already been subjected to respiration in the mycorrhizal mantle [46].

A further feature of this N-limited beech stand was the effective closure of the N cycle in the denitrification process, *i.e.*, the removal of reactive N from the biosphere as harmless inert dinitrogen (N_2) rather than nitrous oxide (N_2O) (Figure 1), a potent greenhouse gas and dominant ozone-depleting substance in the stratosphere.

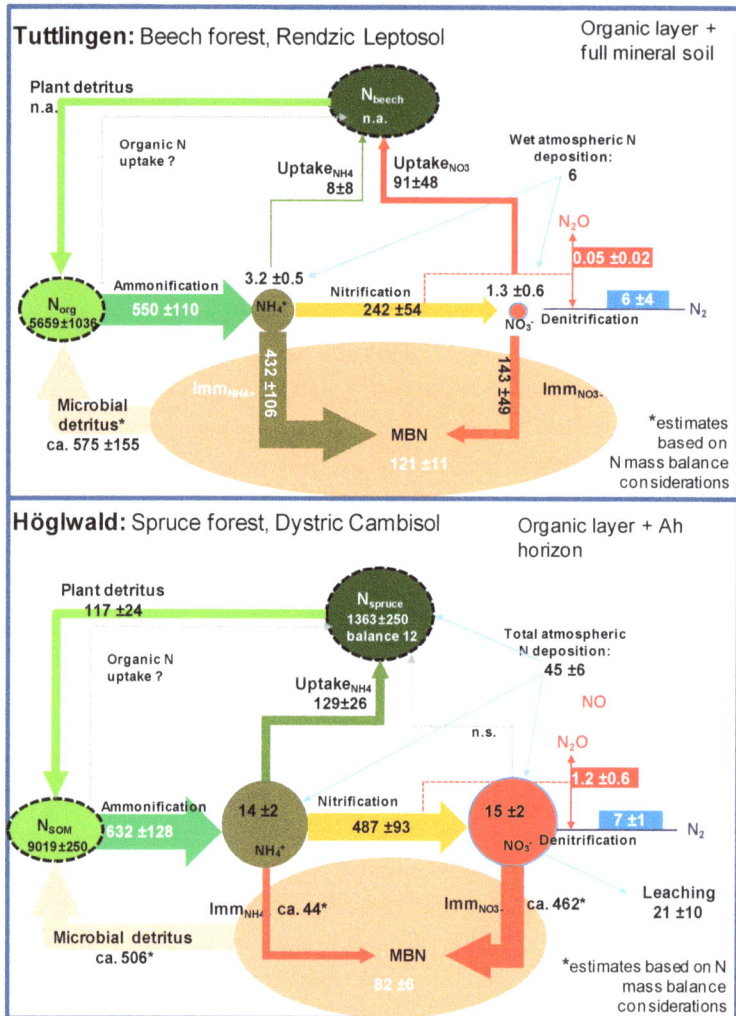

Figure 1. Gross nitrogen fluxes (kg N ha^{-1} year^{-1}) and N pools (kg N ha^{-1}) for two forest ecosystems in southern Germany: The Tuttlingen beech forest on marginal shallow Rendzic Leptosol soil, with low atmospheric N input (upper panel), and the Höglwald forest, a nitrogen-saturated spruce forest on a Dystric Cambisol soil and affected by chronic atmospheric N deposition (lower panel). Note that the thickness of the arrows relates to the size of the process rate, and the pool sizes of ammonium, nitrate and microbial biomass are reflected by the size of the pool signatures. It was not possible to provide size-scaled pool signatures for the N pools in soil organic matter (SOM) and plant biomass (both indicated with black dashed lines), because these pools exceed the labile soil N pools by several orders of magnitude. SOM does not include microbial biomass in this graph. For further information, see text. Ammonium and nitrate uptake was estimated for both forest ecosystems by multiplying the uptake capacities by the fine root biomass [26].

In such forest ecosystems, soil microbial activity provides a huge potential for the recycling and liberation of bioavailable N as well as for microbial competition with plants for N. Redistribution of N between trees and understory plants will largely depend on the competitive strength of these N consumers, with a significant advantage for generalists compared with specialists. As a consequence, the N preferences of tree roots can change from ammonium and amino acids at high N availability to no preference upon N limitation [47], thereby providing a competitive advantage at low N availability compared with more specialized players in the root-microbe system.

Using stable isotope approaches, Guo *et al.* [37,38] showed that leaf litter contributes less to the N nutrition of beech natural regeneration compared with root litter, and that microbial biomass is a much more important sink of N liberated from leaf and root litter compared with beech. Nitrogen from recent leaf litter contributed only a minor amount to the N requirement of beech, indicating that liberation of N from recalcitrant pools of old litter in soil organic matter over years and decades constitutes the dominant N source of beech in N-limited ecosystems. However, the contribution of nutrients recycled from root litter to forest nutrition remains unknown. In this context it has to be considered that microbial nitrate acquisition beyond the actual N demand of the trees for growth, and development and storage constitute an effective means to prevent nitrate leaching and gaseous N losses (Figure 1) from the ecosystem. Recent studies using tree girdling as a tool to unravel the role of C rhizodeposition in N partitioning between beech and free-living microbes showed that microbial-bound N then seems to be used as a transient storage of nutrients [42,47]. Under unfavorable conditions such as drought or reduced carbohydrate exudation of roots and thus reduced C supply to heterotrophic microorganisms, the transient storage can be abandoned [42,47]. Decaying microbial biomass will then become a new source of nutrients for the beech trees [42,48]. In a similar way "carbon expensive" mycorrhizal fungi with extensive development of rhizomorphal hyphae may be exchanged with "carbon inexpensive" mycorrhizal fungi with reduced development of rhizomorphal hyphae under a C shortage [49]. Apparently, beech trees control ecosystem N cycling and N bioavailability under N limitation, thereby operating under the general principle of "to live and let die".

3. Distribution and Fluxes of Nitrogen in Forests Exposed to High Nitrogen Loads

In forests exposed to high N loads, there are more potential sources for N acquisition compared with low N systems. The demand of plants for N can be met both by root uptake of N originating from pedospheric/hydrospheric sources and by stomatal uptake of reactive N compounds from the atmosphere [50]. Uptake of reactive compounds from atmospheric deposition is largely controlled by stomatal conductance and the concentration gradient of reactive N compounds between the atmosphere and the substomatal cavity. This concentration gradient is often determined by the removal of N from the substomatal cavity into the aqueous solution of the apoplastic space surrounding the substomatal cavity and further on into the symplasm of leaf cells [50–52]. Stomatal closure may be considered a means to down-regulate N influx. However, this down-regulation takes place at the expense of reduced carbon dioxide (CO_2) influx into the leaves, resulting in reduced photosynthetic carbon fixation and hence growth [53]. When the stomatal opening is maintained in the presence of biologically available atmospheric N, this will result in uncontrolled, compulsory N nutrition via the leaves. In the majority of tree species, reduction of N taken up by roots and its assimilation into amino acids takes place exclusively in the roots, and leaves are supplied with amino N by xylem transport [54]. Therefore, influx of atmospheric N via the stomata and its reduction and assimilation in the leaves leads to a complete change in the distribution of N metabolism between leaves and roots and in root-to-shoot allocation of N assimilation products and its metabolites. As a consequence, N deposition can change the distribution of biomass between shoot and roots in favor of the shoot, thereby enhancing drought susceptibility due to enhanced transpiration of the increased shoot biomass [55,56]. Depending on the rate of N influx via the roots plus the shoot and the N demand of the plant, N over-nutrition may be prevented by shoot-to-root interactions [54]. In contrast to the influx of atmospheric N, N uptake from pedospheric/hydrospheric sources can at least be partially controlled (see below) and can be down-regulated to the extent that atmospheric N contributes to N nutrition [54,57,58]. If high rates of N deposition result in N over-nutrition [29], this is often indicated by extremely high amounts of the N-rich amino acid arginine in phloem and xylem sap [54,59].

At the ecosystem level, long-term N deposition will turn forests from N-limited into N-saturated or even over-saturated systems [60,61]. The term "nitrogen saturation" is defined as a state where either the availability of mineral N exceeds the combined nutritional demands of plants and microbes [62], or where ecosystem N losses approximate or exceed the inputs of N [63]. Hence, nitrogen saturation is connected with a change from closed to open ecosystem N cycling. However, the extent and stage of N saturation needs to be accounted for. At an initial stage, N deposition will result in a more narrow soil C:N ratio, will remove microbial N limitation and will enrich organic matter in N, which will stimulate N mineralization [64,65]. The associated changes in N mineralization and the C:N ratio also alter the balance of ammonium consuming processes at the expense of microbial immobilization and in favor of nitrification, thus impairing microbial N retention and promoting N losses along hydrological and gaseous pathways [64,66]. In the long-term, chronic N deposition impairs soil microbial activity due to soil acidification and low C availability, thus decelerating soil organic matter decomposition, N mineralization and subsequent N turnover processes, including immobilization, nitrification and denitrification [64,67].

The consequences of chronic atmospheric N deposition for ecosystem N cycling have been analyzed in great detail for a Norway spruce forest in southern Germany surrounded by maize agriculture [26,29,44] (Figure 1). The data in Figure 1 were based on the compilation of Kreutzer et al., 2009 [26], complemented by the gross N turnover dataset of Rosenkranz et al., 2010 [44] such that this new compilation includes much better constrained gross N turnover rates compared with those of Kreutzer et al., 2010 [26].

This forest (the "Höglwald" experimental spruce forest) is characterized by long-term N input from gaseous deposition (10 and 2 kg N ha^{-1} year^{-1} for ammonia and NO$_2$, respectively) and throughfall (18 and 10 kg N ha^{-1} year^{-1} for ammonium and nitrate, respectively) amounting to a total N deposition from the atmosphere to the forest floor of ca. 40 kg N ha^{-1} year^{-1} excluding about 4 kg N ha^{-1} year^{-1} taken up by the canopy. About half of the N deposited (21 kg ha^{-1} year^{-1}) is leached into the hydrosphere, almost exclusively in the form of nitrate. In addition, ca. 16 kg N ha^{-1} year^{-1} are released into the atmosphere, mostly as nitric oxide (NO) and N$_2$, but to a minor extent also in the form of the potent greenhouse gas N_2O [26] (Figure 1). Thus, almost all the N deposited in the spruce forest ecosystem from atmospheric sources is released by leaching and volatilization into the hydrosphere and atmosphere, respectively. Despite these releases, high ecosystem internal N fluxes were observed. Figure 1 provides a synthesis of published annual N fluxes of this ecosystem [26,44]. Spruce trees were estimated to acquire 129 kg N ha^{-1} year^{-1} from pedospheric/hydrospheric NH_4^+ and 4 kg N ha^{-1} year^{-1} from atmospheric sources, but most of the N retrieved is lost annually in the form of leaf and root litter (117 kg N ha^{-1} year^{-1}). However, uptake from pedospheric/hydrospheric N sources may constitute an underestimation, since only inorganic N uptake was considered and organic N uptake that is thought to be a significant N source of trees [44,46] was neglected. Gross microbial mineralization-immobilization turnover in the Höglwald forest was about five times higher than tree N uptake and plant-mediated internal N turnover. Hence, microbial detritus (ca. 500 kg N ha^{-1} year^{-1}) constituted a much stronger N source of polymeric soil organic matter than spruce litter. Nitrogen bound in microbial biomass was significantly lower (82 kg N ha^{-1}) than microbial N immobilization or microbial N detritus formation, indicating a high turnover of this N pool with a mean residence time of about two months. Despite the relatively high N losses of this forest compared with low N natural ecosystems, it needs to be noted that annual gross nitrification rates (almost 500 kg N ha^{-1} year^{-1}) as well as microbial nitrate immobilization rates (ca. 460 kg N ha^{-1} year^{-1}) exceed annual N losses by about an order of magnitude. In view of this dominant role of microbial nitrate immobilization in N loss pathways, we conclude that microbial biomass turnover is the most important process mediating redistribution and retention of N at the ecosystem level in N over-saturated forest ecosystems [26,44,65].

4. Regulation of N Acquisition and Distribution in Trees

Nitrogen acquisition by trees and the significance of root N uptake in the biogeochemical cycle is ultimately regulated at the whole plant level. Nitrogen taken up from atmospheric sources by leaves and not used inside the leaves, as well as N taken up from pedospheric/hydrospheric sources by roots and not used inside the roots are both loaded into an N pool cycling inside the whole plant by phloem and xylem transport. This cycling of N includes bidirectional exchange not only between the xylem and phloem in the leaves and the roots, but also along the long-distance transport path [54,68]. Glutamine, the most abundant long-distance transport form of N in many tree species, also seems to be the most prominent N compound subjected to this bidirectional exchange. However, the exchange of N may also include metabolic interconversion of amino compounds [54]. The cycling pool of plant N serves whole-plant distribution of N stored in the bark and wood of stems and roots (deciduous trees) or the leaves and roots (evergreen trees) upon mobilization for the spring growth of leaves and for fructification. Furthermore, the cycling N pool also constitutes the transport path used for filling up these storage pools of N during annual growth [69–71]. In addition, it provides N for growth and development of all meristematic tissues at the advantage of immediate N availability in close proximity to these tissues. Thus, the cycling plant N pool mediates swift N supply of any part of the plant throughout the year. The size of this cycling plant N pool may change depending on N nutrition, and this pool may even be used for N storage, particularly in the form of arginine, an amino acid with high N content [59]. Mobilization of N from this storage pool may provide N faster than the mobilization from storage proteins in bark and wood [70,71]; therefore, it may be of particular significance to meet N requirements in response to environmental changes. The processes regulating the loading and unloading of the cycling N pool have currently only been studied for seasonal storage and mobilization in poplar [70,71]. Therefore, the mechanisms responsible for the up- and down-regulation of the size of this pool in response to environmental changes, including nutrient availability, constitute an exciting area of future research.

When N acquisition from pedospheric/hydrospheric and atmospheric sources as a whole exceeds the N demand of trees for growth, development, and defense, the cycling pool of N will be expanded initially by the accumulation of glutamine. This accumulation seems to result in a repression of nitrate uptake at the transcriptional or post-transcriptional level [72–74]. However, the mechanism(s) of how trees sense whole plant N status and transmit this information into changes in root uptake remain poorly understood. Whole plant N status may be integrated by cycling amino acids such as glutamine [75], but nitrate may also be involved indirectly [76,77]. Provided cycling glutamine constitutes a systemic signal, the down-regulation of nitrate uptake would require homeostasis between the cycling pool of glutamine and the glutamine pool at the site of nitrate uptake as a means to monitor the whole plant N status by the roots. Recently, a cellular mechanism of glutamine-sensing has been identified that seems to be widespread in the plant kingdom [78]. The plastid-localized PII signaling proteins can sense and integrate metabolic signals by conformational changes, thereby exerting control at all levels of metabolic regulation, including transport activity, metabolic reactions and gene expression. This protein was shown to control the key-enzyme of the ornithine synthesis pathway, N-acetyl-L-glutamate kinase, leading to arginine and polyamine synthesis [78]. It may therefore be assumed that PII signaling proteins can sense high glutamine levels and initiate arginine synthesis under these conditions. However, it remains unclear how glutamine sensing by these plastid-localized proteins reduces nitrate uptake via the root plasmalemma at a post-transcriptional or transcriptional level.

Repression of the expression of the high affinity nitrate transporter gene NRT2.1 by amino acids such as glutamine was found to be mimicked by the expression of transcription factors LBD37/38/39 [79]. In addition, an evolutionary conserved component of the RNA polymerase II complex, HN19-At1WS1, was shown to repress the transcription of root NRT2,1 in *Arabidopsis* [80]. Evidence for the action of similar mechanisms in trees remains to be seen. The release of repression of the transcription of nitrate transporters also requires a systemic signal communicating that whole plant N demand for actual growth and development exceeds N acquisition by the roots. It has been suggested that a cycling pool of cytokinins mediates such a feed-forward control in trees [73,74] and that the concerted action of feedback control by cycling glutamine and feed-forward control by cycling cytokinins regulate nitrate acquisition at the whole plant level [54,81]. However, this assumption requires further research using trees with different N nutritional states. Down-regulation of expression and activity has also been observed for ammonium transporters under a high plant N status [82], but the molecular mechanisms involved have not been reported. The significance of N nutrition for the activity and expression of root amino acid transporters in trees has not been elucidated.

Under N limitation, N availability for a particular tree will be determined by competition at the forest ecosystem level. The competitors include old growth trees, natural regeneration, other herbaceous and perennial understory species, as well as free-living microorganisms [12]. Tree-microbe competition for N occurs at several stages of the soil N cycle: (1) When plants take up monomeric organic compounds such as amino acids, they compete with the diverse microbes taking up dissolved organic nitrogen and performing ammonification and heterotrophic nitrification using organic N sources; (2) When plants take up ammonium, they reduce substrate availability for autotrophic nitrification and heterotrophic microbial ammonium immobilization; (3) Taking up nitrate, plants compete with microbial denitrification, microbial nitrate immobilization, and dissimilatory nitrate reduction to ammonium [12,23,27,83].

Little is known about root-to-root interactions in N acquisition between competitors for N in forests [43,84], and information on microbial-root interactions has only recently been established (e.g., [42,84,85]). In this context, microbial NO produced in significant amounts during nitrification and denitrification seems to be of particular significance [86]. Rhizospheric NO was shown to modulate the uptake of N compounds by tree roots [85,87,88]. Apparently, NO of microbial origin in the soil is sensed by the roots and functions as a signal that determines the competitive strength of N acquisition by the roots relative to microbial use. The mechanism of interaction of microbial NO with N uptake processes of the roots does not seem to include changes in the transcription of N transporters; activation of transporter proteins by phosphorylation/dephosphorylation has been suggested for NO stimulated N uptake by roots, as previously found for the interaction of NO produced inside plant cells with various cellular processes [89]. However, it remains unclear whether NO produced inside root cells can also interact with N uptake and if the action of extracellular and intracellular NO is mediated by similar processes [86].

5. Conclusions

The high N demand of temperate forest ecosystems on marginal soil relies almost exclusively on ecosystem internal sources. In this context, free-living soil microorganisms are responsible for both nutrient liberation and competition against trees and their associated mycorrhizal fungal symbionts. Internal ecosystem microbial N mineralization-immobilization turnover exceeds plant-mediated internal ecosystem N-cycling loops by approximately one order of magnitude. Thus, the soil microbial biomass represents a potential plant nutrient reservoir of considerable importance, and there is increasing evidence that temperate forest trees exert a direct influence on the availability of microbial-bound N for plant uptake. Chronic atmospheric N deposition relieves N limitation and microbial competition for N, resulting in mineral N accumulation in soil and ecosystem N losses along hydrological and gaseous pathways in the range of the N input rates. However, microbial N retention largely dominates over N loss and remains an important pathway of nutrient retention in N-saturated forest ecosystems. Reduced microbial competition for ammonium can result in trees taking up ammonium rather than nitrate.

While our understanding of the regulation of N uptake and distribution in trees has improved, a thorough biogeochemical quantification and understanding of soil microbial liberation, turnover and uptake of bioavailable N has thus far only been achieved for very few forest ecosystems and therefore deserves further attention in order to obtain a better understanding of the N cycling patterns in the plant-soil-microbe system of temperate forests under a changing climate.

Acknowledgments

The authors thank Javier Tejedor for providing data on soil N stocks. We are indebted to the Bundesministerium für Bildung, Wissenschaft, Forschung und Technologie (BMBF, contract numbers KFA-BEO 0339614, 0339615 and KFA-BEO 339175), the Bayerisches Staatsministerium für Ernährung, Landwirtschaft und Forsten, and the German Science foundation (DFG, contract numbers FOR 788/1, RE 515/27-1 and DA 1217/2-1) for funding this work.

Author Contributions

Heinz Rennenberg and Michael Dannenmann jointly analyzed the literature and data and wrote the manuscript.

Conflicts of Interest

The authors declare no conflict of interest.

References

1. Rosegrant, M.W.; Ringler, C.; Zhu, T. Water for agriculture: Maintaining food security under growing scarcity. *Annu. Rev. Environ. Resour.* **2009**, *34*, 205–222.
2. Strzepek, K.; Boehlert, B. Competition for water for the food system. *Philos. Trans. R. Soc. B* **2010**, *365*, 2927–2940.
3. Mekonnen, M.M.; Hoekstra, A.Y. Water footprint benchmarks for crop production: A first global assessment. *Ecol. Indic.* **2014**, *46*, 214–223.
4. De Vries, W.; Kros, J.; Kroeze, C.; Seitzinger, S.P. Assessing planetary and regional nitrogen boundaries related to food security and adverse environmental impacts. *Curr. Opin. Environ. Sustain.* **2013**, *5*, 392–402.
5. Wang, Y.P.; Houlton, B.Z. Nitrogen constraints on terrestrial carbon uptake: Implications for the global carbon-climate feedback. *Geophys. Res. Lett.* **2009**, *36*, L24403, doi:10.1029/2009gl041009.
6. Zaehle, S. Terrestrial nitrogen-carbon cycle interactions at the global scale. *Philos. Trans. R. Soc. B* **2013**, *368*, 20130125.
7. Fernández-Martinez, M.; Vicca, S.; Janssens, I.A.; Sardans, J.; Luyssaert, S.; Cmpioli, M.; Chapin, F.S., III.; Ciais, P.; Malhi, Y.; Obersteiner, M.; *et al.* Nutrient availability as the key regulator of global forest carbon balance. *Nat. Clim. Chang.* **2014**, *4*, 471–476.
8. Batjes, N.H. Total carbon and nitrogen in the soils of the world. *Eur. J. Soil Sci.* **1996**, *47*, 151–163.

9. IPCC. Climate change 2007: The physical science basis. In *Contribution of Working Group I to the Fourth Assessment Report of the Intergovernmental Panel on Climate Change*; Solomon, S., Qin, D., Manning, M., Chen, Z., Marquis, M., Averyt, K.B., Tignor, M.M.H.L., Eds.; Cambridge University Press: Cambridge, United Kingdom, 2007.

10. IPCC. Climate Change 2013: The Physical Science Basis. In *Contribution of Working Group I to the Fifth Assessment Report of the Intergovernmental Panel on Climate Change*; Stocker, T.F., Qin, D., Plattner, G.K., Tignor, M., Allen, S.K., Boschung, J., Nauels, A., Xia, Y., Bex, V., Midgley, P.M., Eds.; Cambridge University Press: Cambridge, United Kingdom and New York, NY, USA, 2013; pp. 1535.

11. Rennenberg, H.; Seiler, W.; Matyssek, R.; Gessler, A.; Kreuzwieser, J. Die Buche (*Fagus sylvatica* L.)—Ein Waldbaum ohne Zukunft im südlichen Mitteleuropa? *Allg. Forst- und Jagdzeitschrift* **2004**, *175*, 210–224.

12. Rennenberg, H.; Dannenmann, M.; Gessler, A.; Kreuzwieser, J.; Simon, J.; Papen, H. Nitrogen balance in forests: Nutritional limitation of plants under climate change stresses. *Plant Biol.* **2009**, *11*, 4–23.

13. Gessler, A.; Keitel, C.; Kreuzwieser, L.; Matyssek, R.; Seiler, W.; Rennenberg H. Potential risks for European beech (*Fagus sylvatica* L.) in a changing climate. *Trees* **2007**, *21*, 1–11.

14. Vitousek, P.M.; Howarth, R.W. Nitrogen limitation on landand in the sea: How can it occur? *Biogeochemistry* **1991**, *13*, 87–115.

15. Oren, R.; Ellsworth, D.S.; Johnsen, K.H.; Philips, N.; Ewers, B.E.; Maier, C.; Schäfer, K.V.; McCarthy, H.; Hendrey, G.; McNulty, S.G.; *et al.* Soil fertility limits carbon sequestration by forest ecosystems in a CO_2-enriched atmosphere. *Nature* **2001**, *411*, 469–472.

16. Heffer, P.; Prud'homme, M. *Fertilizer Outlook 2013–2017*; International Fertilizer Industry Association (IFA): Paris, France, 2013. Available online: http://www.fertilizer.org (accessed on 1 May 2015.).

17. Galloway, J.N.; Aber, J.D.; Erisman, J.W.; Seitzinger, S.P.; Howarth, R.H.; Cowling, E.B.; Cosby, B.J. The nitrogen cascade. *Bioscience* **2003**, *53*, 341–356.

18. Dise, N.B.; Rothwell, J.J.; Gauci, V.; van der Salm, C.; de Vries, W. Predicting dissolved inorganic nitrogen leaching in European forests using two independent databases. *Sci. Total Environ.* **2009**, *407*, 1798–1808.

19. Maynard, D.G.; Pare, D.; Thiffault, E.; Lafleur, B.; Hogg, K.E.; Kishchuk, B. How do natural disturbances and human activities affect soils and tree nutrition and growth in the Canadian boreal forest? *Environ. Rev.* **2014**, *22*, 161–178.

20. Raven, J.A.; Andrews M. Evolution of tree nutrition. *Tree Physiol.* **2010**, *30*, 1050–1071.

21. Rennenberg, H.; Schmidt, S. Perennial lifestyle—An adaptation to nutrient limitation? *Tree Physiol.* **2010**, *30*, 1047–1049.

22. Ollivier, J.; Töwe, S.; Bannert, A.; Hai, B.; Kastl, E.-M.; Meyer, A.; Su, M.X.; Kleineidam, K.; Schloter, M. Nitrogen turnover in soil and global change. *FEMS Microbiol. Ecol.* **2011**, *78*, 3–16.

23. Schimel, J.P.; Bennett, J. Nitrogen mineralization: Challenges of a changing paradigm. *Ecology* **2004**, *85*, 591–602.

24. Dungait, J.A.J.; Hopkins, D.W.; Gregory, A.S.; Whitmore, A.P. Soil organic matter turnover is governed by accessibility not recalcitrance. *Glob. Chang. Biol.* **2012**, *18*, 1781–1796.

25. Schmidt, M.W.I.; Torn, M.S.; Abiven, S.; Dittmar, T.; Guggenberger, G.; Janssens, I.A.; Kleber, M.; Kögel-Knabner, I.; Lehmann. J.; Manning, D.A.C.; *et al.* Persistence of soil organic matter as an ecosystem property. *Nature* **2011**, *478*, 49–56.

26. Kreutzer, K.; Butterbach-Bahl, K.; Rennenberg, H.; Papen H. The complete nitrogen cycle in a N-saturated spruce forest ecosystem. *Plant Biol.* **2009**, *11*, 643–649.

27. Butterbach-Bahl, K.; Gundersen, P.; Ambus, P.; Augustin, J.; Beier, C.; Boeckx, P.; Dannenmann, M.; Gimeno, B.S.; Kiese, R.; Kitzler, B.; *et al.* Nitrogen turnover processes and effects in terrestrial ecosystems. In *The European Nitrogen Assessment*, 1st ed.; Sutton, M.A., Howard, C.M., Erisman, J.W., *et al.*, Eds; Cambridge University Press: London, UK, 2011.

28. Kreutzweiser, D.P.; Hazlett, P.W.; Gunn, J.M. Logging impacts on the biogeochemistry of boreal forest soils and nutrient export to aquatic systems: A review. *Environ. Rev.* **2008**, *16*, 157–179.

29. Rennenberg, H.; Kreutzer, K.; Papen, H.; Weber, P. Consequences of high loads of nitrogen for spruce (*Picea abies* L.) and beech (*Fagus sylvatica* L.) forests. *New Phytol.* **1998**, *139*, 71–86.

30. Cleveland, C.C.; Townsend, A.R.; Schimel, D.S.; Fisherm, H.; Howarth, R.W.; Hedin, L.O.; Perakis, S.S.; Latty, E.F.; von Fischer, J.C.; Elseroad, A.; *et al.* Global patterns of terrestrial biological nitrogen (N_2) fixation in natural ecosystems. *Glob. Biogeochem. Cycles* **1999**. *13*, 623–645.

31. Cole, D.W. Soil nutrient supply in natural and managed forests. *Plant Soil* **1995**, *168*, 43–53.

32. Morford, S.L.; Houlton, B.J.; Dahlgren, R.A. Increased forest ecosystem carbon and nitrogen storage from nitrogen rich bedrock. *Nature* **2011**, *477*, 78–81.

33. Cleveland, C.C.; Houlton, B.Z.; Kolby Smith, W.; Marklein A.R.; Reed, S.C; Parton, W.; del Grosso, S.J.; Running, S.W. Patterns of new *versus* recycled primary production in the terrestrial biosphere. *Proc. Natl. Acad. Sci. U.S.A.* **2013** *110*, 12733–12737.

34. Hackl, E.; Bachmann, G.; Zechmeister-Boltenstern, S. Microbial nitrogen turnover in soils under different types of natural forest. *For. Ecol. Manag.* **2004** *188*, 101–112.

35. Dannenmann, M.; Gasche, R.; Ledebuhr, A.; Papen, H. Effects of forest management on soil N cycling in beech forests stocking on calcareous soils. *Plant Soil* **2006**, *287*, 279–300.

36. West, J.B.; Espeleta, J.F.; Donovan, L.A. Fine root production and turnover across a complex edaphic gradient of a *Pinus palustris-Aristida stricta* savanna ecosystem. *For. Ecol. Manag.* **2004**, *189*, 397–406.

37. Guo, C.J.; Simon, J.; Gasche, R.; Naumann, P.S.; Bimüller, C.; Pena, R.; Polle, A.; Kögel-Knabner, I.; Zeller, B.; Rennenberg, H.; *et al.* Minor contribution of leaf litter to N nutrition of beech (*Fagus sylvatica*) seedlings in a mountainous beech forest of Southern Germany. *Plant Soil* **2013**, *369*, 657–668.

38. Guo, C.J; Dannenmann, M.; Gasche, R.; Zeller, B.; Papen, H.; Polle, A.; Rennenberg, H.; Simon, J. Preferential use of root litter compared to leaf litter by beech seedlings and soil microorganisms. *Plant Soil* **2013**, *368*, 519–534.

39. Dannenmann, M.; Butterbach-Bahl, K.; Gasche, R.; Willibald, S.; Papen, H. Dinitrogen emissions 40 and the $N_2:N_2O$ emission ratio of a Rendzic Leptosol as influenced by pH and forest thinning. *Soil Biol. Biochem.* **2008**, *40*, 2317–2323.

40. Dannenmann, M.; Gasche, R.; Ledebuhr, A.; Holst, T.; Mayer, H.; Papen, H. The effect of forest management on trace gas exchange at the pedosphere-atmosphere interface in beech (*Fagus sylvatica* L.) forests stocking on calcareous soils. *Eur. J. For. Res.* **2007**, *126*, 331–346.

41. Dannenmann, M.; Gasche, R.; Papen, H. Nitrogen turnover and N_2O production in the forest floor of beech stands as influenced by forest management. *J. Plant Nutr. Soil Sci.* **2007**, *170*, 134–144.

42. Dannenmann, M.; Simon, J.; Gasche, R.; Holst, J.; Pena, R.; Naumann, P.S.; Kögel-Knabner, I.; Knicker, H.; Mayer, H.; Schloter, M.; *et al.* Tree girdling provides insight on the role of labile carbon in nitrogen partitioning between soil microorganisms and adult European beech. *Soil Biol. Biochem.* **2009**, *41*, 1622–1631.

43. Simon, J.; Dannenmann, M.; Gasche, R.; Holst, J.; Mayer, H.; Papen, H.; Rennenberg H. Competition for nitrogen between adult European beech and its offspring is reduced by avoidance strategy. *For. Ecol. Manag.* **2011**, *262*, 105–114.

44. Rosenkranz, P.; Dannenmann, M.; Brüggemann, N.; Papen, H.; Berger, U.; Zumbusch, E.; Butterbach-Bahl, K. Gross rates of ammonification and nitrification at a nitrogen-saturated spruce (*Picea abies* (L.) Karst.) stand in Southern Germany. *Eur. J. Soil Sci.* **2010**, *61*, 745–758.

45. Dannenmann, M.; Bimüller, C.; Gschwendtner, S.; Leberecht, M.; Tejedor, J.; Bilela, S.; Gasche, R.; Hanewinkel, M.; Baltensweiler, A.; Kögel-Knabner, I.; *et al.* Climate change impairs nitrogen cycling in European beech forests. *PLoS ONE* **2015**, Submitted for publication.

46. Näsholm, T.; Kielland, K.; Ganeteg, U. Uptake of organic nitrogen by plants. *New Phytol.* **2009**, *182*, 31–48.

47. Stoelken, G.; Simon, L.; Ehlting, B.; Rennenberg, H. The presence of amino acids affects inorganic N uptake in non-mycorrhizal seedlings of European beech (*Fagus sylvatica* L.). *Tree Physiol.* **2010**, *30*, 1118–1128.

48. Kaiser, C.; Fuchslueger, L.; Koranda, M.; Gorfer, M.; Stange, C.F.; Kitzler, B.; Rasche, F.; Strauss, J.; Sessitsch, A.; Zechmeister-Boltenstern, S.; *et al.* Plants control the seasonal dynamics of microbial N cycling in a beech forest soil by belowground C allocation. *Ecology* **2011**, *92*, 1036–1051.

49. Pena, R.; Offermann, C.; Simon, J.; Naumann, P.; Gessler, A.; Holst, J.; Dannenmann, M.; Mayer, H.; Kögel-Knabner, I.; Rennenberg, H.; *et al.* Girdling affects ectomycorrhizal diversity and reveals functional differences of EM community composition in a beech forest. *Appl. Environ. Microbiol.* **2010**, *76*, 1831–1841.

50. Rennenberg, H.; Gessler, A. Consequences of N deposition to forest ecosystems—Recent results and future research needs. *Water Air Soil Pollut.* **1999**, *116*, 47–64.

51. Gessler A.; Rienks, M.; Rennenberg, H. NH_3 and NO_2 fluxes between beech trees and the atmosphere—Correlation with climatic and physiological parameters. *New Phytol.* **2000**, *147*, 539–560.

52. Gessler, A.; Rienks, M.; Rennenberg, H. NH$_3$ and NO$_x$ exchange between spruce (*Picea abies*) trees and the atmosphere. *New Phytol.* **2002**, *156*, 179–194.

53. Grulke, N.E.; Dobrowolski, W.; Mingus, P.; Fenn, M.E. California black oak response to nitrogen amendment at a high nitrogen-saturated site. *Environ. Pollut.* **2005**, *137*, 536–545.

54. Herschbach, C.; Gessler, A.; Rennenberg, H. Long-distance transport and plant internal cycling of N- and S-compounds. *Prog. Bot.* **2012**, *73*, 161–188.

55. Ding, W.J.; Wang, R.Q.; Yuan, Y.F.; Liang, X.Q.; Liu, J. Effects of nitrogen deposition on growth and relationship of *Robinia pseudoacacia* and *Quercus acutissima* seedlings. *Dendrobiology* **2012**, *67*, 3–13.

56. Ochoa-Huesco, R.; Manrique, E. Impacts of altered precipitation, nitrogen deposition and plant competition on a Mediterranian seed bank. *J. Veg. Sci.* **2014**, *25*, 1289–1298.

57. Muller, B.; Touraine, B.; Rennenberg, H. Interaction between atmospheric and pedospheric nitrogen nutrition in spruce seedlings. *Plant Cell Environ.* **1996**, *19*, 345–355.

58. Rennenberg, H.; Herschbach, C.; Polle, A. Consequences of air pollution on shoot-root interactions. *J. Plant Physiol.* **1996**, *148*, 296–301.

59. Gessler, A.; Schneider, S.; Weber, P.; Hanemann, U.; Rennenberg, H. Soluble N compounds in trees exposed to high loads of N: A comparison between the roots of Norway spruce (*Picea abies* (L.) Karst) and beech (*Fagus sylvatica*) trees grown under field conditions. *New Phytol.* **1998**, *138*, 385–399.

60. Aber, J.D.; Goodale, C.L.; Ollinger, S.V.; Smith, M.L.; Magill, A.H.; Martin, M.E.; Hallett, R.A.; Stoddard, J.L. Is nitrogen deposition altering the nitrogen status of northeastern forests? *BioScience* **2003**, *53*, 375–389.

61. Aber, J.D.; McDowell, W.; Nadelhoffer, K.; Magill, A.; Berntson, G.; Kamakea, M.; McNulty, S.; Currie, W.; Rustad, L.; Fernandez, I. Nitrogen saturation in temperate forest ecosystems: Hypothesis revisited. *BioScience* **1998**, *48*, 921–934.

62. Aber, J.D.; Nadelhoffer, K.; Steudler, P.; Melillo, J.M. Nitrogen saturation in northern forest ecosystems. *BioScience* **1989**, *39*, 378–386.

63. Agren, G.I.; Bosatta, E. Nitrogen saturation of terrestrial ecosystems. *Environ. Pollut.* **1988**, *54*, 185–198.

64. Gao, W.; Yang, H.; Kou, L.; Li, S. Effects of nitrogen deposition and fertilization on N transformations in forest soils: A review. *J. Soils Sediments* **2015**, *15*, 863–879.

65. Corre, M.D.; Brumme, R.; Veldkamp, E.; Beese, F.O. Changes in nitrogen cycling and retention processes in soils under spruce forests along a nitrogen enrichment gradient in Germany. *Glob. Chang. Biol.* **2007**, *13*, 1509–1527.

66. Butterbach-Bahl, K.; Dannenmann, M. Soil carbon and nitrogen interactions and biosphere-atmosphere exchange of methane and nitrous oxide. In *Recarbonization of the Biosphere–Ecosystems and the Global Carbon Cycle*, 1st ed.; Lal, R., Lorenz, K., Hüttl, R.F., Schneider, B.U., von Braun, J., Eds.; Springer: Dordrecht, the Netherlands; Heidelberg, Germany; NewYork, USA; London, UK, 2012; pp. 429–443.

67. Frey, S.D.; Ollinger, S.; Nadelhoffer, K.; Bowden, R.; Brzostek, E.; Burton, A.; Caldwell, B.A.; Crow, S.; Goodale, C.L.; Grandy, A.S.; *et al.* Chronic nitrogen additions suppress decomposition and sequester soil carbon in temperate forests. *Biogeochemistry* **2014**, *121*, 305–316.

68. Gessler, A.; Weber, P.; Schneider, S.; Rennenberg H. Bidirectional exchange of amino compounds between phloem and xylem during long distance transport in Norway spruce trees (*Picea abies* [L.] Karst). *J. Exp. Bot.* **2003**, *54*, 1389–1397.

69. Rennenberg, H.; Wildhagen, H.; Ehlting, B. Nitrogen nutrition of poplar trees. *Plant Biol.* **2010**, *12*, 275–291.

70. Wildhagen, H.; Bilela, S.; Rennenberg, H. Low temperatures counteract short-day induced nitrogen storage, but not accumulation of bark storage protein transcripts in bark of Gray poplar (*Populus × canescens*). *Plant Biol.* **2013**, *15* (Suppl. 1), 44–56.

71. Wildhagen, H.; Dürr, J.; Ehlting, B.; Rennenberg, H. Seasonal nitrogen cycling in the bark is correlated with gene expression and meteorological factors in grey poplar plants. *Tree Physiol.* **2010**, *30*, 1096–1110.

72. Gessler, A.; Schulte, M.; Schrempp, S.; Rennenberg, H. Interaction of phloem-translocated amino compounds with nitrate net uptake by the roots of beech (*Fagus sylvatica*) seedlings. *J. Exp. Bot.* **1998**, *4*, 1529–1537.

73. Collier, M.; Fotelli, M.; Nahm, M.; Kopriva, S.; Rennenberg, H.; Hanke, D.; Gessler, A. Regulation of nitrogen uptake by *Fagus sylvatica* on a whole plant level—Interactions between cytokinins and soluble N compounds. *Plant Cell Environ.* **2003**, *26*, 1549–1560.

74. Dluzniewska, P.; Gessler, A.; Kopriva, S.; Strnad, M.; Novak, O.; Dietrich, H.; Rennenberg, H. Exogenous supply of glutamine and active cytokinin to the roots reduces NO3—Uptake rates in poplar. *Plant Cell Environ.* **2006**, *29*, 1284–1297.

75. Ruffel, S.; Gojon, A.; Lejay, L. Signal interactions in the regulation of root nitrate uptake. *J. Exp. Bot.* **2014**, *65*, 5509–5517.

76. Ruffel, S.; Krouk, G.; Ristova, D.; Shasha, D.; Birnbaum, K.D.; Coruzzi, G.M. Nitrogen economics of root foraging: Transitive closure of the nitrate-cytokinin relay and distinct systemic signalling for N supply *vs.* demand. *Proc. Natl. Acad. Sci. USA* **2011**, *108*, 18524–18529.

77. Forde, B.G. Local and long-range signaling pathways regulating plant responses to nitrate. *Annu. Rev. Plant Biol.* **2002**, *53*, 203–224.

78. Chellamuthu, V.-R.; Ermilova, E.; Lapina, T.; Lüddecke, J.; Minaeva, E.; Herrmann, C.; Hartmann, M,-D.; Forchhammer, K. A widespread glutamine sensing mechanism in the plant kingdom. *Cell* **2014**, *159*, 1188–1199.

79. Rubin, G.; Tohge, T.; Matsuda, F.; Saito, K.; Scheible, W.R. Members of the LBD family of transcription factors repress anthocyanin synthesis and affect additional nitrogen responses in Arabidopsis. *Plant Cell* **2009**, *21*, 3567–3584.

80. Widiez, T.; El Kafafi, E.S.; Girin, T.; Berr, A.; Ruffel, S.; Krouk, G., Vayssieres, A.; Shen, W.-H.; Coruzzi, G.M.; Gojon, A. High nitrogen insensitive 9(NHI9)-mediated systemic repression of root NO3- uptake is associated with changes in histone methylation. *Proc. Natl. Acad. Sci. USA* **2011**, *108*, 13329–13334.

81. Gessler, A.; Kopriva, S.; Rennenberg, H. Regulation of nitrate uptake of trees at the whole plant level: Interaction between nitrogen compounds, cytokinins and carbon metabolism. *Tree Physiol.* **2004**, *24*, 1313–1321.

82. Camanes, G.; Cerezo, M.; Primo-Millo, E.; Gojon, A.; Garcia-Agustin, P. Ammonium transport and CitAMT1 expression are regulated by N in Citrus plants. *Planta* **2009**, *229*, 331–342.

83. Rütting, T.; Boeckx, P.; Müller, C.; Klemedtsson, L. Assessment of the importance of dissimilatory nitrate reduction to ammonium for the terrestrial nitrogen cycle. *Biogeosciences* **2011**, *8*, 1779–1791.

84. Simon, J.; Waldhecker, P.; Brüggemann, N.; Rennenberg, H. Competition for nitrogen sources between European beech (*Fagus sylvatica*) and sycamore maple (*Acer pseudoplatanus*) seedlings. *Plant Biol.* **2010**, *12*, 453–458.

85. Simon, J.; Dong, F.; Buegger, H.; Rennenberg H. Rhizospheric NO affects N uptake and metabolism in Scots pine (*Pinus sylvestris* L.) seedlings depending on soil N availability and N sources. *Plant Cell Environ.* **2013**, *36*, 1019–1026.

86. Medinets, S.; Skiba, U.; Rennenberg, H.; Butterbach-Bahl, K. A review of soil NO transformation: Associated processes and possible physiological significance on organisms. *Soil Biol. Biochem.* **2015**, *80*, 92–117.

87. Simon, J.; Stoelken, G.; Rienks, M.; Rennenberg. Rhizospheric NO interacts with the acquisition of reduced nitrogen sources by the roots of European beech (*Fagus sylvatica*). *FEBS Lett.* **2009**, *583*, 2907–2910.

88. Dong, F.; Simon, J.; Rienks, M.; Rennenberg H. Effects of rhizospheric nitric oxide (NO) on N uptake depend on soil CO_2 concentration, soil N availability, and N source. *Tree Physiol.* **2015**, in press.

89. Leitner, M.; Vandelle, E.; Gaupells, F.; Bellin, D.; Delledonne, M. NO signals in the haze: Nitric oxide signaling in plant defense. *Curr. Opin. Plant Biol.* **2009**, *12*, 451–458.

Native and Alien Plant Species Richness Response to Soil Nitrogen and Phosphorus in Temperate Floodplain and Swamp Forests

Richard Hrivnák, Michal Slezák, Benjamín Jarčuška, Ivan Jarolímek and Judita Kochjarová

Abstract: Soil nitrogen and phosphorus are commonly limiting elements affecting plant species richness in temperate zones. Our species richness-ecological study was performed in alder-dominated forests representing temperate floodplains (streamside alder forests of *Alnion incanae* alliance) and swamp forests (alder carrs of *Alnion glutinosae* alliance) in the Western Carpathians. Species richness (*i.e.*, the number of vascular plants in a vegetation plot) was analyzed separately for native and alien vascular plants in 240 vegetation plots across the study area covering Slovakia, northern Hungary and southern Poland. The relationship between the species richness of each plant group and total soil nitrogen content, plant-available phosphorus and carbon to nitrogen (C/N) ratio was analyzed by generalized linear mixed models (GLMM) with Poisson error distribution and log-link function. The number of recorded native and alien species was 17–84 (average 45.4) and 0–9 (average 1.5) species per plot, respectively. The GLMMs were statistically significant ($p < 0.001$) for both plant groups, but the total explained variation was higher for native (14%) than alien plants (9%). The richness of native species was negatively affected by the total soil nitrogen content and plant-available phosphorus, whereas the C/N ratio showed a positive impact. The alien richness was predicted only by the total soil nitrogen content showing a negative effect.

Reprinted from *Forests*. Cite as: Hrivnák, R.; Slezák, M.; Jarčuška, B.; Jarolímek, I.; Kochjarová, J. Native and Alien Plant Species Richness Response to Soil Nitrogen and Phosphorus in Temperate Floodplain and Swamp Forests. *Forests* **2015**, *6*, 3501-3513.

1. Introduction

Variation of plant species richness in plant communities can be explained by more than one hundred plausible ecological hypotheses and theories [1,2] with little consensus regarding the nature of causal processes [3]. Experimental and observational biodiversity research recognized habitat productivity as one of the major determinants controlling local species richness of vascular plants (e.g., [4,5]). This relationship originally showed a typical hump-shaped pattern, *i.e.*, the highest species richness is at the intermediate productivity level and gradually declined towards both marginal parts of the productivity gradient [6]. However, this response appeared to be vegetation- and scale-dependent [2,7] and richness has been recently seen, not simply as a function of productivity, but as feedback to influence productivity [8].

While the relationship between habitat productivity and species richness attracted much of the research effort aimed to understand species richness patterns [3], the relationship between species richness and other measured environmental factors is far less well studied [4]. Soil nitrogen and phosphorus play an important role in the regulation of plant growth and the functioning of several ecosystem processes [9]. These nutrients are commonly absorbed from soluble inorganic sources in soil solutions. However, plant uptake strategies and mechanisms related to nutrient acquisition promote resource partitioning and may lead to local species coexistence [9,10]. Changes in soil nitrogen and phosphorus availability are reflected either in their concentrations in plant tissues, ratios with other nutrients [11] or in species richness patterns. They are considered to be a limiting soil element affecting species richness of vegetation in temperate zone [10,12,13]. The increasing soil nitrogen and phosphorus content usually reduces the richness of herb layer vascular plants (e.g., [14,15]), but unimodal, positive or non-significant effects were also found in empirical studies (e.g., [4,16,17]). These responses are most likely a consequence of diverse nitrogen and phosphorus pools among study sites. While the total soil nitrogen content corresponds especially to its supply in mineral soil and/or thickness of organic layers, the level of phosphorus is driven mainly by abiotic resources and soil age [10]. The effect of soil phosphorus on species richness is much more ambiguous because of interactions with other covariant factors modifying the supply, chemical form and availability of nutrients (e.g., soil reaction, chemistry of geological substrates). The carbon to nitrogen (C/N) ratio of soil organic matter is a useful indicator of site quality with effects on both forest species composition and richness [18,19]. However, plant species richness in floodplain forests is also driven by several other environmental factors such as soil moisture and reaction, landscape configuration and dynamic of water regime [20–22]. These factors can account for a major part of explained variation in the species richness-environmental relationship [20].

Black alder (*Alnus glutinosa* (L.) Gaertn.) and grey alder (*Alnus incana* (L.) Moench) represent indigenous tree species to Central Europe, which dominate in floodplains (streamside alder forests, phytosociological alliance *Alnion incanae* Pawłowski *et al.* 1928) and swamp forests (alder carrs, *Alnion glutinosae* Malcuit 1929) [23] (hereinafter alder forests). While *Alnus glutinosa* can dominate in both vegetation types, *Alnus incana* almost exclusively prefers streamside alder forests [23]. They grow in habitats with relatively broad ecological gradients including mineral and organic soils [24–26]. Both alder species have constant effect on understory vegetation, as they improve the soil nitrogen pool and consequently modify the C/N ratio of the soil organic matter by symbiotic N_2-fixation [27]. The alder forests are, thus, commonly referred to as non-nitrogen-limited ecosystems, although a mineralization of organic matter is locally disabled by permanently flooded soils [28]. Alder species also positively affect diversity and activity of soil microbial communities leading to a larger increase in soil phosphorus availability [29]. Therefore, the alder forests represent a proper base for the study of relationships among species richness and soil characteristics.

Invasions of alien plant species, especially neophytes, are considered one of the major threats to the diversity of natural ecosystems including floodplain forests [30–32]. Among natural and semi-natural vegetation types, floodplain forests belong to the most invaded habitats [33]. They usually contain a higher number of alien species compared with all other forest vegetation types [32,34–36]. Plant invasions may, thus, enrich the overall species richness and modify the structure of species composition. The streamside alder forests are particularly highly prone to invasions of alien species, as rivers frequently act as propagule dispersal corridors [37,38]. Moreover, there is a strong human impact on these ecosystems (e.g., forestry, agricultural, both human utilization and activities in forests and surrounding landscape), which enhances the expansion rate of alien species into disturbed habitats.

Alder forests commonly create fragmented and small-scale forest stands due to the changes and transformation of the hydrological regime and fragmentation. These fragments are obviously influenced by adjacent landscape-scale factors [39]. Although these medium to species-rich forests [28,40] belong to threatened habitats at European or national levels (European Habitat Directive 92/43/EEC) [41,42], they represent a relatively frequent habitat in the Western Carpathians [23].

The aim of present study is to assess the relationship between soil nitrogen, plant-available phosphorus, C/N ratio and plant species richness in alder forests. Our main questions were: (i) what is the role of soil nitrogen, plant-available phosphorus and C/N ratio in the variation of native and alien plant species richness? and (ii) are there differences in responses of both plant groups to these factors? As native and invasive species show differences in their competitive abilities for nutrients [43], we expect different responses of the native and invasive species richness to the explanatory variables.

2. Material and Methods

2.1. Study Area

The species richness-edaphic relationship in floodplains (streamside alder forests; *Alnion incanae*) and swamp forests (alder carrs; *Alnion glutinosae*) was studied across the central part of the Western Carpathians, including Slovakia, northern Hungary and southern Poland (Figure 1). Vegetation plots of this territory cover a wide latitudinal (47.829°–49.685°) and altitudinal (131–917 m) range, and include two ecologically and floristically distinctive European biogeographical areas—the Pannonian and Western Carpathian regions. The Pannonian climate shows drier and warmer vegetation seasons and affects the southern part of the study area, whereas the Carpathian climate is more humid and affects the central and northern part. The streamside alder forests prefer riparian zones and spring-fed areas from lowland to mountain regions, with the understory formed mainly by meso to hygrophilous species. In contrast, the crucial plant components of the understory of alder carr forests, frequent especially within lowland catchments, are mainly sedges and perennial grasses [23].

2.2. Field Sampling and Laboratory Analyses

Vegetation and soil variables were sampled in the centre of physiognomically homogeneous forest stands with a dominance of species *Alnus glutinosa* or *A. incana* in the tree layer (*i.e.*, canopy cover more than 50%). Sampling was carried out in plots with a uniform size of 400 m^2. The intensity of sampling in each alder-dominated site corresponded to the environmental heterogeneity and presence of different vegetation types [23]. In more detail, the distance between the two nearest plots of the same vegetation type was usually at least 4 km. In case the habitat conditions of the alder-dominated site were considerably variable, which was obvious in the species composition pattern, we decided to sample more than one plot. Two hundred and forty plots were sampled between June and August from 2010–2014. They were originally collected for the purpose of phytosociological classification using the standard European phytosociological approach [44]; results were published recently in Slovakia [23]. Each plot contained a list of all present vascular plant species. Species nomenclature and taxonomy were unified according to the checklist of Marhold and Hindák [45]. The plant species were divided into two groups, native and alien species based on a checklist of non-native plants in Slovakia [46]. Species richness (*i.e.*, the number of species per sampling plot) was determined for both plant groups. This dataset was stored in the Institute of Botany SAS as an xls-file.

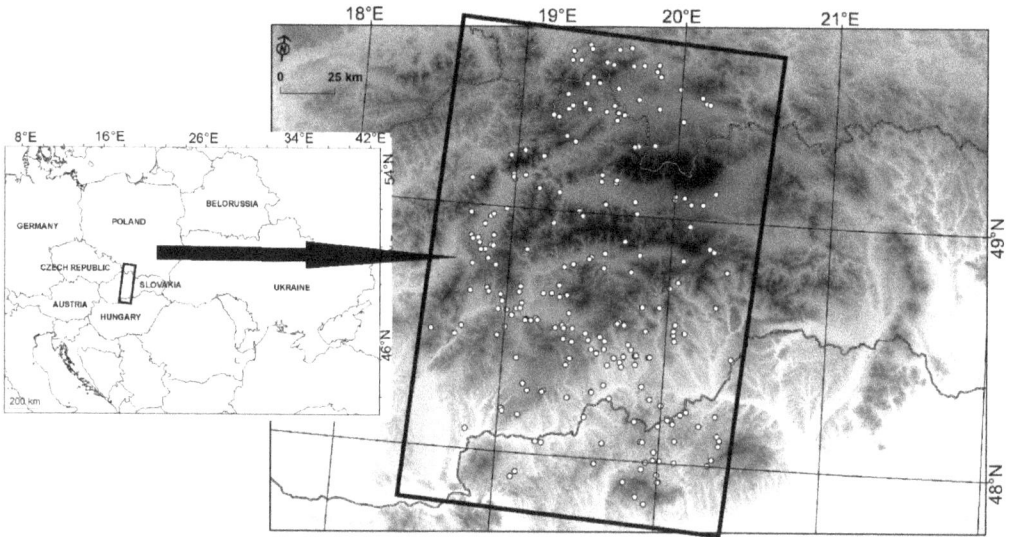

Figure 1. Study area and distribution of vegetation plots.

Soil samples were randomly taken in three places in various parts of each plot from the uppermost mineral horizon (0–10 cm depth, litter removed) and mixed to form a single sample (weight approximately 1 kg) per plot in order to reduce the soil heterogeneity. They were dried at a laboratory temperature, crushed and passed through a 2 mm sieve. Soil analyses were performed on air-dried samples following standard protocols. Total carbon (C) and total nitrogen (N) contents were determined using an NCS-FLASH 1112 analyzer (CE Instruments, UK) and subsequently, the C/N-ratio as an appropriate surrogate for the mineralization rate [47] was calculated. Plant-available phosphorus (P) was extracted using the Mehlich II solution [48] and measured by spectrophotometry (AES-ICP).

2.3. Data Analysis

The variation in species richness of native and alien vascular plants explained by the soil variables was analyzed using the generalized linear models (GLM) and generalized linear mixed models (GLMM). Poisson error distribution and log-link function were initially applied in model specification. Since over dispersion and presence of some spatial structure in the GLMs' residuals were observed (for more details see [49,50]), spatial GLMM with penalized quasi-likelihood (glmmPQL; library "MASS") [51] was used finally. This modeling procedure allows the use of spatially structured random effects. Gaussian correlation structure was used in the GLMMs. However, differences between these models—accounting for spatial autocorrelation—and GLM-models were slight.

Quadratic terms for all explanatory variables (after centering) were also meant to be included in the models. However, only the squared C/N-ratio entered the models due to collinearity ($R > 0.60$) of quadratic terms. Two models (i) containing only main effects and (ii) with second order interactions between explanatory variables were created for both native and alien species richness.

The graphic distribution of model's residuals was assessed by plotting of the standardized residuals against the values predicted by the model, as well as against each explanatory variable. The spatial autocorrelation of model's residuals was checked by the global Moran's I test (library "ape") [52]. The spatial autocorrelation in response to distance was assessed by a Moran's I correlogram with an increment of 2 (library "ncf") [53]. The proportion of explained variation in glmmPQL (pseudo-R2 value; hereinafter, R2) was calculated as the squared Pearson correlation coefficient between observed and predicted values [54]. All statistical analyses were performed in the R version 3.0.0 [55] through the integrated development environment RStudio [56].

3. Results

Altogether, 490 vascular plant species were recorded in 240 vegetation plots of alder forests. The total number of species ranged from 17 to 92 per plot. Native species created a substantial part of the total species richness, whereas the number of alien species was considerably lower, but had obvious variance among particular plots. The values of studied soil characteristics were relatively variable in plots (Table 1).

Table 1. Descriptive statistics of species richness characteristics and ecological variables ($n = 240$).

	Average ± S.D.	Min.	Max.	C.V.
Vegetation Characteristics				
Number of native species	45.4 ± 11.8	17	84	25.9
Number of alien species	1.5 ± 1.8	0	9	121.8
Soil Variables				
Total N (%)	0.599 ± 0.468	0.080	2.549	79.6
P (mg/kg)	17.702 ± 9.546	0.500	64.290	53.9
Total C (%)	7.651 ± 6.153	1.000	31.727	82.0
C/N ratio	12.882 ± 3.170	4.929	25.189	24.6

S.D.: standard deviation, CV: coefficient of variation.

The main effects models and the models with second order interactions between soil variables accounted for a comparable proportion of data variability (14% and 16%, respectively) (Tables 2 and 3). Total explained variability in species richness of vascular plants was higher for native (14%) than alien (9%) species. While the species richness of native plants was significantly related to all analyzed variables, the richness pattern of aliens was affected by only one predictor (Table 2). Higher concentrations of plant-available phosphorus and total nitrogen reduced the species richness of native plants, whereas the C/N ratio showed a positive effect. The alien species richness was negatively related to the total nitrogen content (Figure 2). Second order interactions between the soil variables did not explain the variability of species richness either in native or alien plant species (Table 3).

Table 2. Main effects models with the number of native and alien plants as a response to ecological variables, respectively.

	Native Species				Alien Species			
	Estimate	S.E.	t	P	Estimate	S.E.	t	P
(Intercept)	3.8243	0.0182	210.02	<0.001	0.4036	0.0850	4.75	<0.001
Total N	−0.0790	0.0340	−2.32	**0.021**	−0.6019	0.1506	−4.00	**<0.001**
P	−0.0068	0.0018	−3.87	**<0.001**	0.0088	0.0071	1.25	0.213
C/N	0.0217	0.0061	3.53	**<0.001**	−0.0188	0.0232	−0.81	0.419
C/N^2	−0.0016	0.0009	−1.77	0.08	−0.0043	0.0039	−1.11	0.269
R^2	0.14	$p < 0.001$			0.09	$p < 0.001$		

S.E.: standard error; df = 235; for R^2 calculation see Data analysis.

Table 3. Second order interactions models with the number of native and alien plants as response variables, respectively.

	Native Species				Alien Species			
	Estimate	S.E.	t	P	Estimate	S.E.	t	P
(Intercept)	3.8247	0.0189	202.81	<0.001	0.4176	0.0872	4.79	<0.001
Total N (1)	−0.1057	0.0354	−2.99	**0.003**	−0.4653	0.1494	−3.11	**0.002**
P (2)	−0.0060	0.0020	−2.97	**0.003**	0.0029	0.0082	0.36	0.722
C/N (3)	0.0074	0.0086	0.86	0.392	−0.0620	0.0376	−1.65	0.100
C/N^2 (4)	−0.0021	0.0015	−1.40	0.164	−0.0125	0.0068	−1.85	0.066
1 × 2	−0.0051	0.0041	−1.26	0.207	−0.0087	0.0176	−0.50	0.620

Table 3. Cont.

	Native Species				Alien Species			
	Estimate	S.E.	t	P	Estimate	S.E.	t	P
1 × 3	−0.0297	0.0205	−1.44	0.149	0.0170	0.0930	0.18	0.855
1 × 4	0.0019	0.0041	0.49	0.626	−0.0107	0.0192	−0.56	0.577
2 × 3	−0.00002	0.0008	−0.02	0.985	−0.0040	0.0030	−1.34	0.183
2 × 4	−0.0002	0.0001	−1.20	0.230	0.00004	0.0005	0.09	0.932
3 × 4	−0.0002	0.0002	0.96	0.336	0.0012	0.0007	1.56	0.120
R^2	0.16	$p < 0.001$			0.08	$p < 0.001$		

S.E.: standard error; df = 229; for R^2 calculation see Data analysis.

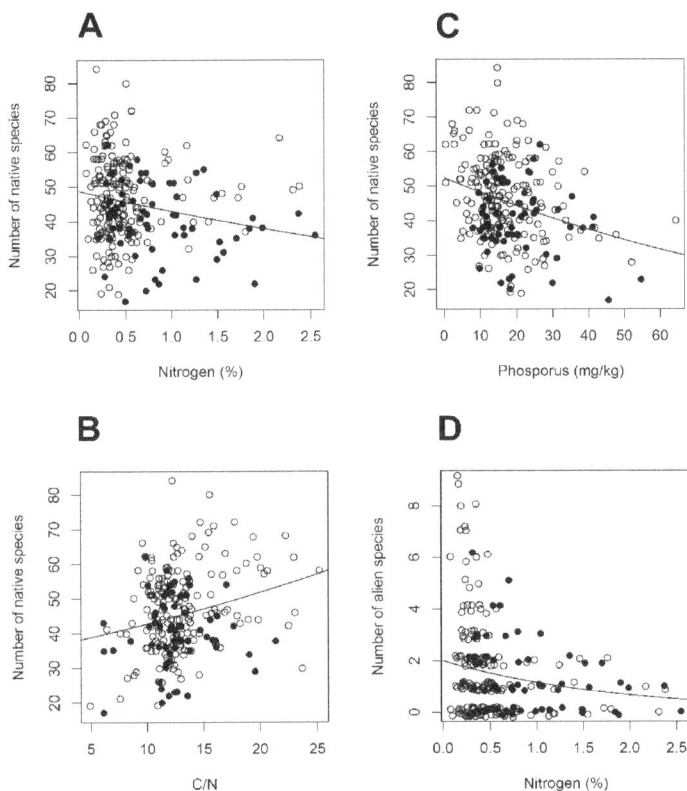

Figure 2. Relationship between species richness of native (**A–C**) and alien (**D**) vascular plants and ecological variables, which were statistically significant in the used model (empty circles—streamside alder forests, full circles—alder carrs).

4. Discussion

The present study showed a negative effect of total nitrogen on species richness for both native and alien plant species. Although some previous studies focusing on diversity research in forest understory vegetation found the unimodal response (e.g., [6]), this negative trend (*i.e.*, a higher amount of soil nitrogen led to a lower number of species) was most often identified in temperate deciduous forests [14,16,17]. The differences in the species richness trends along the nitrogen gradient can be partially addressed to habitat-related biases (e.g., restricted length of analyzed nitrogen gradient). Thus, the final response of plant species richness along the productivity gradient generally appears to be dependent on the inclusion of extreme habitats on the gradient [6,14]. We do not suppose this pattern to be relevant in our study, as the analyzed data covered the whole vegetation and nitrogen gradient of the alder forests in the Western Carpathians.

The negative effect of soil nitrogen content on species richness can be explained by a potential positive effect of nitrogen on certain highly competitive plants [16,21]. These species may have suppressed competitively weaker species. In our data, the nitrogen-rich sites are usually covered by dominant species such as *Caltha palustris, Carex acutiformis, Petasites hybridus* or *Phragmites australis*. We expected differences in total nitrogen-species richness pattern for the native and alien species, as they differ in their strategies for growth and survival in habitats [43]. However, the negative response of species richness to soil nitrogen was consistent between groups. Similarly, Jones and Chapman [17] found a negative decreasing trend in both cases of native and alien species in Pennsylvania oak forests. In addition, the highest variability of alien species richness was documented at lower parts of the nitrogen gradient in our study, which could affect the number of alien species-nitrogen relationships. Similarly to alder forests, the highest variability of species richness at lower part of nitrogen gradient has also been observed in other forests [17].

The C/N ratio had a positive effect on the species richness of native plants. Our findings differed from those observed in German floodplain forests [19,57], where no relationship between C/N ratio and plant species richness was found. Similarly, positive trends and/or slightly convex curves were also identified in deciduous forests by Schuster and Diekmann [4]. The C/N ratio of soil organic matter was related to a nitrogen immobilization and mineralization during decomposition by microorganisms [18,58]. Alder forests of European temperate zone create two main vegetation groups, which vary in their environmental conditions, including their affinity for soil substrates. Species-richer streamside alder forests show better soil mineralization [28] and higher C/N ratio than species-poorer alder carrs. In our data, the ten most species-rich sites corresponded to the streamside alder forests (average C/N ratio = 14.15), whereas ten most species-poor sites were mostly swamp forests with a C/N ratio of 11.12 (see also Figure 2). Increasing species richness along the C/N ratio gradient in the total data set of alder forests is therefore logical and expected, although it does not agree with findings reported from mesic deciduous forests. In addition, a positive correlation between C/N ratio and species richness most likely corresponds to a negative effect of nitrogen on species richness.

Plant-available phosphorus was the most variable soil characteristic in our study (Table 1). Similar soil phosphorus concentrations were also documented in temperate floodplain and swamp alder forests of the south-western Poland [59]. Diverse forms of soil phosphorus support niche differentiation and facilitate plant species coexistence [60], but plant species richness commonly declines along an increasing soil phosphorus gradient [15,61]. The negative species richness-phosphorus relationship was also found in the present study for native vascular plants. High soil phosphorus content promotes vigorous growth, especially in fast-colonizing, competitive species, and reduces the species richness of vascular plants in broadleaved deciduous forests [16,62]. Phosphorus-rich sites in studied alder forests were dominated mainly by highly productive clonal species, most often by *Urtica dioica*. The high availability of soil phosphorus favors species *Urtica dioica* during successful development and strong competition [63]. Indeed, clonal species with lateral spreading are able to reallocate resources among individual ramets growing in patches with diverse nutrient-availability [22]. Thus, their competitive advantage at high phosphorus levels can suppress the colonization and successful establishment of other plant species and, consequently, can lead to the species richness decline. Although these species occurred at phosphorus-poorer sites as well, they reached only an additional cover value within the herb layer vegetation. Changes in dominant species at phosphorus-rich grassland sites are usually accompanied by changes in overall species composition and vegetation [61], but we did not observe this pattern in alder forests.

It is important to note that the soil variables (total nitrogen, plant-available phosphorus and C/N ratio) used as explanatory variables in our study explained together only between 8 and 16% of the variability in alien and native species richness in alder forests' understories, respectively. For a complex interpretation of variation in species richness or composition of vascular plants in streamside and alder carr forests, other habitat quality and landscape configuration-related factors have to be taken into account as well [20,21,39]. In more detail, habitat quality parameters, such as hydrological characteristics (e.g., water table, flooding), soil nutrient/acidity complex, temperature and light availability represent significant variables controlling vegetation structure of alder forests [20,22]. Moreover, the portion of surrounding country, distance to settlements or distance from the stream source plays the important role, as well [20,22,39]. These drivers could explain the major part of the variation in species richness-environmental relationships within temperate floodplain and swamp forests.

Acknowledgments

We would like to thank Dušan Senko for the preparation of the map. The research was supported by the Science Grant Agency of the Ministry of Education of the Slovak Republic and the Slovak Academy of Sciences (VEGA 2/0019/14 and 2/0051/15) and COST project "European Information System for Alien Species".

Author Contributions

We declare the following concerning the authors' contributions to the research: fieldwork: Richard Hrivnák, Michal Slezák, Judita Kochjarová, Ivan Jarolímek; manuscript preparation: Richard Hrivnák, Michal Slezák, Benjamín Jarčuška, Ivan Jarolímek; data analyses: Benjamín Jarčuška.

Conflicts of Interest

The authors declare no conflict of interest.

References

1. Palmer, M.W. Variation in species richness: Towards a unification of hypotheses. *Folia Geobot. Phytotaxon.* **1994**, *29*, 511–530.
2. Grace, J.B. The factors controlling species density in herbaceous plant communities: An assessment. *Perspect. Plant Ecol. Evol. Syst.* **1999**, *2*, 1–28.
3. Gillman, L.N.; Wright, S.D. The influence of productivity on the species richness of plants: A critical assessment. *Ecology* **2006**, *87*, 1234–1243.
4. Schuster, B.; Diekmann, M. Species richness and environmental correlates in deciduous forests of Northwest Germany. *For. Ecol. Manag.* **2005**, *206*, 197–205.

5. Axmanová, I.; Chytrý, M.; Zelený, D.; Li, C.-F.; Vymazalová, M.; Danihelka, J.; Horsák, M.; Kočí, M.; Kubešová, S.; Lososová, Z.; *et al.* The species richness-productivity relationship in the herb layer of European deciduous forests. *Glob. Ecol. Biogeogr.* **2012**, *21*, 657–667.

6. Pausas, J.G.; Austin, M.P. Patterns of plant species richness in relation to different environments: An appraisal. *J. Veg. Sci.* **2001**, *12*, 153–166.

7. Waide, R.B.; Willing, M.R.; Steiner, C.F.; Mittelbach, G.; Gough, L.; Dodson, S.I.; Juday, G.P.; Parmenter, R. The relationship between productivity and species richness. *Annu. Rev. Ecol. Syst.* **1999**, *30*, 257–300.

8. Adler, P.B.; Seabloom, E.W.; Borer, E.T.; Hillebrand, H.; Hautier, Y.; Hector, A.; Harpole, W.S.; O'Halloran, L.R.; Grace, J.B.; Anderson, T.M.; *et al.* Productivity is a poor predictor of plant species richness. *Science* **2011**, *333*, 1750–1753.

9. Lambers, H.; Brundrett, M.C.; Raven, J.A.; Hopper, S.D. Plant mineral nutrition in ancient landscape: High plant species diversity on infertile soils is linked to functional diversity for nutritional strategies. *Plant Soil* **2010**, *334*, 11–31.

10. Lambers, H.; Raven, J.A.; Shaver, G.R.; Smith, S.E. Plant nutrient-acquisition strategies change with soil age. *Trends Ecol. Evol.* **2008**, *23*, 95–103.

11. Güsewell, S. N:P ratios in terrestrial plants: variation and functional significance. *New Phytol.* **2004**, *164*, 243–266.

12. Attiwill, P.M.; Adams, M.A. Nutrient cycling in forests. *New Phytol.* **1993**, *124*, 561–582.

13. Vitousek, P.M.; Porder, S.; Houlton, B.Z.; Chadwick, O.A. Terrestrial phosphorus limitation: Mechanisms, implications, and nitrogen-phosphorus interactions. *Ecol. Appl.* **2010**, *20*, 5–15.

14. Dupré, C.; Wessberg, C.; Diekmann, M. Species richness in deciduous forests: Effects of species pools and environmental variables. *J. Veg. Sci.* **2002**, *13*, 505–516.

15. Dumortier, M.; Butaye, J.; Jacquemyn, H.; van Camp, N.; Lust, N.; Hermy, M. Predicting vascular plant species richness of fragmented forests in agricultural landscapes in central Belgium. *For. Ecol. Manag.* **2002**, *158*, 85–102.

16. Hofmeister, J.; Hošek, J.; Modrý, M.; Roleček, J. The influence of light and nutrient availability on herb layer species richness in oak-dominated forests in central Bohemia. *Plant Ecol.* **2009**, *205*, 57–75.

17. Jones, R.O.; Chapman, S.K. The roles of biotic resistance and nitrogen deposition in regulating non-native understory plant diversity. *Plant Soil* **2011**, *345*, 257–269.

18. Yakamura, T.; Sahunalu, P. Soil carbon/nitrogen ratio as a site quality index for some South-east Asian forests. *J. Trop. Ecol.* **1990**, *6*, 371–378.

19. Härdtle, W.; von Oheimb, G.; Westphal, C. The effects of light and soil conditions on the species richness of the ground vegetation of deciduous forests in northern Germany (Schleswig-Holstein). *For. Ecol. Manag.* **2003**, *182*, 327–338.

20. Douda, J. The role of landscape configuration in plant composition of floodplain forests across different physiographic areas. *J. Veg. Sci.* **2010**, *21*, 1110–1124.

21. Mölder, A.; Schneider, E. On the beautiful diverse Danube? Danubian floodplain forest vegetation and flora under the influence of river eutrophication. *River Res. Appl.* **2011**, *27*, 881–894.

22. Douda, J.; Doudová-Kochánková, J.; Boublík, K.; Drašnarová, A. Plant species coexistence at local scale in temperate swamp forest: Test of habitat heterogeneity hypothesis. *Oecologia* **2012**, *169*, 523–534.

23. Slezák, M.; Hrivnák, R.; Petrášová, A. Numerical classification of alder carr and riparian alder forests in Slovakia. *Phytocoenologia* **2014**, *44*, 283–308.

24. McVean, D.N. Biological flora of the British Isles. *Alnus glutinosa* (L.) Gaertn. *J. Ecol.* **1953**, *41*, 447–466.

25. Schwabe, A. Monographie *Alnus incana*-reicher Waldgesellschaften in Europa Variabilität und Ähnlichkeit einer azonal verbreiteten Gesellschaftsgruppe. *Phytocoenologia* **1985**, *13*, 197–302.

26. Claessens, H.; Oosterbaan, A.; Savill, P.; Rondeux, J. A review of the characteristics of black alder (*Alnus glutinosa* (L.) Gaertn.) and their implications for silvicultural practises. *Forestry* **2010**, *83*, 163–175.

27. Eickenscheidt, T.; Heinichen, J.; Augustin, J.; Freibauer, A.; Drösler, M. Nitrogen mineralization and gaseous nitrogen losses from waterlogged and drained organic soils in a black alder (*Alnus glutinosa* (L.) Gaertn.) forest. *Biogeosciences* **2014**, *11*, 2961–2976.

28. Ellenberg, H. *Vegetation Ecology of Central Europe*; Cambridge University Press: New York, NY, USA, 2009; pp. 1–731.

29. Lõhmus, K.; Truu, M.; Truu, J.; Ostonen, I.; Kaar, E.; Vares, A.; Uri, V.; Alama, S.; Kanal, A. Functional diversity of culturable bacterial communities in the rhizosphere in relation to fine-root and soil parameters in alder stands on forests, abandoned agricultural, and oil-shale mining areas. *Plant Soil* **2006**, *283*, 1–10.

30. Kowarik, I. *Biologische Invasionen: Neophyten und Neozoen in Mitteleuropa*; Ulmer: Stuttgart, Germany, 2003; pp. 1–492.

31. Weber, E. *Invasive Plant Species of the World: A Reference Guide to Environmental Weeds*; CAB International Publ.: Wallingford, UK, 2003; pp. 1–548.

32. Pyšek, P.; Bacher, S.; Chytrý, M.; Jarošík, V.; Wild, J.; Celesti-Grapow, L.; Gassó, N.; Kenis, M.; Lambdon, P.W.; Nentwig, W.; *et al.* Contrasting patterns in the invasions of European terrestrial and freshwater habitats by alien plants, insects and vertebrates. *Glob. Ecol. Biogeogr.* **2010**, *19*, 317–331.

33. Richardson, D.M.; Holmes, P.M.; Esler, K.J.; Galatowitsch, S.M.; Stromberg, J.G.; Kirkman, S.P.; Pyšek, P.; Hobbs, R.J. Riparian vegetation: Degradation, alien plant invasions, and restoration prospects. *Divers. Distrib.* **2007**, *13*, 126–139.

34. Medvecká, J.; Jarolímek, I.; Senko, D.; Svitok, M. Fifty years of plant invasion dynamics in Slovakia along a 2500 m altitudinal gradient. *Biol. Invasions* **2014**, *16*, 1627–1638.

35. Walter, J.; Essl, F.; Englisch, T.; Kiehn, M. Neophytes in Austria: Habitat preferences and ecological effects. *Neobiota* **2005**, *6*, 13–25.

36. Vilá, M.; Pino, J.; Font, X. Regional assesment of plant invasions across different habitat types. *J. Veg. Sci.* **2007**, *18*, 35–42.

37. Burkart, M. River corridor plants (Stromtalpflanzen) in Central European lowland: A review of a poorly understood plant distribution pattern. *Glob. Ecol. Biogeogr.* **2001**, *10*, 449–468.

38. Zając, A.; Tokarska-Guzik, B.; Zając, M. The role of rivers and streams in the migration of alien plants into the Polish Carpathians. *Biodivers. Res. Conserv.* **2011**, *23*, 43–56.

39. Pielech, R.; Anioł-Kwiatkowska, J.; Szczęśniak, E. Landscape-scale factors driving plant species composition in mountain streamside and spring riparian forests. *For. Ecol. Manag.* **2015**, *347*, 217–227.

40. Chytrý, M.; Dražil, T.; Hájek, M.; Kalníková, V.; Preislerová, Z.; Šibík, J.; Ujházy, K.; Axmanová, I.; Bernátová, D.; Blanár, D.; *et al.* The most species-rich plant communities of the Czech Republic and Slovakia (with new world records). *Preslia* **2015**, *87*, 217–278.

41. Stanová, V.; Valachovič, M. *Katalóg Biotopov Slovenska*; DAPHNE—Inštitút aplikovanej ekológie: Bratislava, Slovakia, 2002; pp. 1–225.

42. Bölöni, J.; Molnár, Z.; Kun, A. *Magyarország Élőhelyei Vegetációtípusok Leírása és Határozója. Ánér 2011*; MTA Ökológiai és Botanikai Kutatóintézete: Vácrátót, Hungary, 2011; pp. 1–441.

43. Davis, M.A.; Grime, J.P.; Thomson, K. Fluctuating resources in plant communities: A general theory of invisibility. *J. Ecol.* **2000**, *88*, 528–534.

44. Dengler, J.; Chytrý, M.; Ewald, J. Phytosociology. In *Encyclopedia of Ecology*; Jørgensen, S.E., Fath, B.D., Eds.; Elsevier: Oxford, UK, 2008; Volume 4, pp. 2767–2779.

45. Marhold, K.; Hindák, F. *Zoznam nižších a vyšších rastlín Slovenska*; Veda: Bratislava, Slovakia, 1998; pp. 1–688.

46. Medvecká, J.; Kliment, J.; Májeková, J.; Halada, Ľ.; Zaliberová, M.; Gojdičová, E.; Feráková, V.; Jarolímek, I. Inventory of alien species of Slovakia. *Preslia* **2012**, *84*, 257–309.

47. Finzi, A.C.; van Breemen, N.; Canham, C.D. Canopy tree-soil interactions within temperate forests: species effects on soil carbon and nitrogen. *Ecol. Appl.* **1998**, *8*, 440–446.

48. Mehlich, A. New extractant for soil test evaluation of phosphorus, potassium, magnesium, calcium, sodium, manganese, and zinc. *Commun. Soil Sci. Plant Anal.* **1978**, *9*, 477–492.

49. Dormann, C.F.; McPherson, J.M.; Araújo, M.B.; Bivand, R.; Bolliger, J.; Carl, G.; Davies, R.G.; Hirzel, A.; Jetz, W.; Kissling, W.D.; *et al.* Methods to account for spatial autocorrelation in the analysis of species distributional data: a review. *Ecography* **2007**, *30*, 609–628.

50. Beale, C.M.; Lennon, J.J.; Yearsley, J.M.; Brewer, M.J.; Elston, D.A. Regression analysis of spatial data. *Ecol. Lett.* **2010**, *13*, 246–264.

51. Venables, W.N.; Ripley, B.D. *Modern Applied Statistics with S*; Springer: New York, NY, USA, 2002; pp. 1–495.

52. Paradis, E.; Claude, J.; Strimmer, K. APE: Analyses of phylogenetics and evolution in R language. *Bioinformatics* **2004**, *20*, 289–290.

53. Bjornstad, O.N. *Ncf: Spatial Nonparametric Covariance Functions*; R package version 1.1-5, 2013. Available online: http://onb.ent.psu.edu/onb1/Rs (accessed on 2 April 2015).

54. Kissling, W.D.; Carl, G. Spatial autocorrelation and the selection of simultaneous autoregressive models. *Glob. Ecol. Biogeogr.* **2008**, *17*, 59–71.

55. R Core Team. *R: A Language and Environment for Statistical Computing*; R Foundation for Statistical Computing: Vienna, Austria, 2013. Available online: http://www.R-project.org/ (accessed on 13 June 2014).

56. RStudio. *RStudio: Integrated Development Environment for R (version 0.98.1091)*; Computer Software: Boston, MA, USA, 2014. Available online: http://www.rstudio.org/ (accessed on 31 March 2015).

57. Härdtle, W.; von Oheimb, G.; Meyer, H.; Westphal, C. Patterns of species composition and species richness in moist (ash-alder) forests of northern Germany (Schleswig-Holstein). *Feddes Repert.* **2003**, *114*, 574–586.

58. Swift, M.J.; Heal, O.W.; Anderson, J.M. *Decomposition in Terrestrial Ecosystems*; Blackwell Scientific Publications: Oxford, UK, 1979; pp. 1–372.

59. Orczewska, A. The impact of former agriculture on habitat conditions and distribution patterns of ancient woodland plant species in recent black alder (*Alnus glutinosa* (L.) Gaertn.) woods in south-western Poland. *For. Ecol. Manag.* **2009**, *258*, 794–803.

60. Turner, B.L. Resource partitioning for soil phosphorus: A hypothesis. *J. Ecol.* **2008**, *96*, 698–702.

61. Merunková, K.; Chytrý, M. Environmental control of species richness and composition in upland grasslands of the southern Czech Republic. *Plant Ecol.* **2012**, *213*, 591–602.

62. Honnay, O.; Hermy, M.; Coppin, P. Impact of habitat quality on forest plant species colonization. *For. Ecol. Manag.* **1999**, *115*, 157–170.

63. Pigott, C.D. Analysis of the response of *Urtica dioica* to phosphate. *New Phytol.* **1971**, *70*, 953–966.

Soil Nitrogen Transformations and Availability in Upland Pine and Bottomland Alder Forests

Tae Kyung Yoon, Nam Jin Noh, Haegeun Chung, A-Ram Yang and Yowhan Son

Abstract: Soil nitrogen (N) processes and inorganic N availability are closely coupled with ecosystem productivity and various ecological processes. Spatio-temporal variations and environmental effects on net N transformation rates and inorganic N concentrations in bulk soil and ion exchange resin were examined in an upland pine forest (UPF) and a bottomland alder forest (BAF), which were expected to have distinguishing N properties. The annual net N mineralization rate and nitrification rate (kg $N \cdot ha^{-1} \cdot year^{-1}$) were within the ranges of 66.05–84.01 and 56.26–77.61 in the UPF and -17.22–72.24 and 23.98–98.74 in the BAF, respectively. In the BAF, which were assumed as N-rich conditions, the net N mineralization rate was suppressed under NH_4^+ accumulated soils and was independent from soil temperature. On the other hand, in the UPF, which represent moderately fertile N conditions, net N transformation rates and N availability were dependent to the generally known regulation by soil temperature and soil water content. Stand density might indirectly affect the N transformations, N availability, and ecosystem productivity through different soil moisture conditions. The differing patterns of different inorganic N indices provide useful insight into the N availability in each forest and potential applicability of ion exchange resin assay.

Reprinted from *Forests*. Cite as: Yoon, T.K.; Noh, N.J.; Chung, H.; Yang, A-R.; Son, Y. Soil Nitrogen Transformations and Availability in Upland Pine and Bottomland Alder Forests. *Forests* **2015**, *6*, 2941-2958.

1. Introduction

Soil nitrogen (N) has become a focus of terrestrial ecosystem studies for two main reasons: (1) N availability is a limiting factor for terrestrial ecosystem productivity [1–3] and (2) N losses from terrestrial ecosystems impact the atmosphere especially by greenhouse gas emission, aquatic ecosystems by eutrophication, and drinking water by nitrate contamination [4,5]. N mineralization, the microbial conversion of organic N to inorganic N, has been intensively studied because it is believed to be the principal control of N availability to plants in terrestrial ecosystems [6]. Recent progress in the knowledge of and techniques related to organic N and the priming of rhizospheres for N availability has shaken the classical assumption that N mineralization dominates as the primary source of available N, [7–10]; nevertheless, in field N studies, N mineralization and other inorganic N indices remain meaningful standard measures which provide useful insight.

Environmental controls of N mineralization and availability, particularly temperature and soil moisture, have been investigated through a theoretical model [11], literature review [9,12], laboratory experiments [13–15], and field manipulations [16–19]. For example, soil moisture varies significantly in space and time, and affects various biogeochemical processes in different ways [20–22]. Therefore, understanding soil moisture and its effect on N and nutrient processes is essential in soil ecology. Various studies have analyzed the effect of the soil moisture gradient on N mineralization and developed empirical models that describe the response of N mineralization to soil moisture based on laboratory incubation experiments. However, these laboratory studies may insufficiently reflect the effects of soil moisture on N mineralization in nature, where differing soil moisture ranges have indirect effects on N mineralization through other variables, such as the vegetation community, soil chemistry, and substrate availability [23–25]. To understand the complex interactions among such environmental factors and N availability, field studies incorporating the repeated measurement on various environmental conditions are necessary since they can better reflect the temporal and spatial variation of soil moisture and substrate gradients.

Numerous N indices techniques, each with respective advantages and limitations, have been used to trace the real availability of N under various conditions [6,26]. These techniques include ascertaining net N mineralization using a buried bag or resin core, net N mineralization potential using laboratory incubation, gross N mineralization and immobilization using a stable N isotope, and the total or inorganic N concentration of bulk soil. Although ion exchange resins (IERs) that efficiently adsorb "available" inorganic N from soil would provide an alternative useful measure of N availability by imitating the N uptake processes of plant roots [6,27,28], they have been rarely applied to soil N studies, and the knowledge and insight so far gained from this measure of N availability has been limited. Furthermore, because no single N index can give complete insight into N availability [6], a measurement of multiple N indices increases the understanding of N availability and its relationship with environmental variables. Therefore, this study uses multiple N indices, including IER assays.

The aim of the current study was to investigate net N transformation and N availability in an upland pine forest and a bottomland alder forest, which are expected to have distinguishing N properties due to the different species functional type and topography. Broad-leaved deciduous and coniferous forests are generally assumed to be developed on fertile and infertile soils, respectively [29–31], even though several recent syntheses failed to find the general pattern and harmonized empirical evidences supporting this assumption [32,33]. Consequently, distinctive litter N properties between broad-leaved deciduous and coniferous trees respectively affect soil N availability [34]. Topography also discriminates various soil processes coupled with hydrology. Therefore, the upland pine forest and the bottomland alder forest could be assumed to represent N-moderate and N-rich environments, respectively, and the forests would exhibit distinctive N cycle processes. Specifically, our objectives were: (1) to determine the spatio-temporal patterns of net N transformation rates and N availability and their relationship with environmental factors under different stand densities or soil moisture conditions; (2) to analyze the similarities and differences of multiple N indices, including IER assays, to better understand N availability; and (3) to characterize the soil N transformation, availability, and uptake in each forest.

2. Materials and Methods

2.1. Study Sites

This study was carried out in an upland pine forest (UPF) and a bottomland alder forest (BAF) in central Korea (Table 1). The UPF is located in a rural mountainous area, 30 km north of the city

Seoul at Gwangneung Experimental Forest (37°47'01" N, 127°10'37" E, 410–440 m a.s.l.). The canopy is dominated by Japanese red pine (*Pinus densiflora* Sieb. et Zucc.) with deciduous middle and understory species (e.g., *Acer palmatum, Carpinus laxiflora, C. cordata, Fraxinus rhynchophylla, Quercus mongolica, Q. variabilis*, and *Styrax obassia*). The UPF area is typical of the upland pine forest that is the most common forest type in South Korea and represents 23.1% of the total national forest by area. The BAF is located at Heonilleung Ecosystem Landscape Conservation Area (37°27'52" N, 127°04'53" E, 40 m a.s.l.) in a southern suburb of Seoul. The canopy is dominated by Japanese alder (*Alnus japonica* Steud.) with several middle and understory species (e.g., *Euonymus oxyphyllus, F. rhynchophylla, Ginkgo biloba, Ligustrum obtusifolium, Q. aliena, Q. mongolica*, and *Rosa multiflora*). Japanese alder is a typical species tolerant of hydric conditions in East Asian bottomlands [35,36], although its cover is currently small due to land use change. Interestingly, both forests have been well preserved from a harvest for hundreds of years due to the preservation of the royal tombs established within them during the 15th century [37,38]. The forests were sheltered from disturbances during the first half of the 20th century, in which the rule of Japanese Imperialism (1910–1945) and the Korean War (1950–1953) severely degraded nation-wide forests. The forests could provide reference status of the upland and bottomland forests in Korea, accompanied with abundant previous studies [25,39–42].

The sites are only 35 km apart and at the sites, precipitation was similar (Table 1). Temperature, however, differs and the UPF, due to its higher altitude, is slightly colder than the BAF. Their soils are classified into silt loam or sandy loam and are acidic (pH: 4.45–5.14) at both sites. The UPF had less soil water content (SWC) than the BAF during the all study period (Figure S1f), reflecting the topographical characteristics of each site. Moreover, the UPF and the BAF would be determined as moderately fertile (medium) and N-rich (high) conditions, respectively, based on the needle or leaf litter and soil C:N ratio, and total N (sum of organic and inorganic N) storage in vegetation and soil, using the nutrient availability criteria by Vicca *et al.* [43].

UPF stands were classified by stand density and BAF stands were classified by soil moisture, in order to characterize the spatial variability of stand structure and ecological processes in each type of forest (for UPF see [39,41] and for BAF see [25,42]). Stand density and soil hydrology are assumed to be the main environmental driver of upland forest and bottomland ecosystems, respectively; this assumption is generally applied to ecosystem management practices in each ecosystem. In the UPF, low density (LD; 983 trees·ha^{-1}) and high density (HD; 1517 trees·ha^{-1}) stands were selected and in the BAF, low moisture (LM), middle moisture (MM), and high moisture (HM) stands were selected with SWC of <0.50, 0.50–1.00, and 1.00–2.00 kg·kg^{-1}, respectively. Note that the MM and HM stands, which were underlain with poorly drained soils, can be classified as forested wetland. Three replicate plots, each covering 400 m^2 in the UPF and 100 m^2 in the BAF, were established in each stand.

Table 1. Characteristics of the upland pine forest (UPF) and bottomland alder forest (BAF) study sites. Numbers in parentheses indicate the standard error of the means. Climate data was obtained from the nearest meteorological stations and site description data was taken from previous studies of the UPF [39,41] and BAF [25,42].

	Upland Pine Forest (UPF)		Bottomland Alder Forest (BAF)		
	Low Density (LD)	High Density (HD)	Low Moisture (LM)	Middle Moisture (MM)	High Moisture (HM)
Site (1980–2010 average for climate data)					
Location	37°47'01" N, 127°10'37" E		37°27'52" N, 127°04'53" E		
	Rural area 30 km north of Seoul		Southern suburb of Seoul		
Altitude (m)	410–440		40		
Aspect, slope (°)	West, 13–22		South, <1		
Temperature (°C)	11.4 (monthly range: −5.0–26.5)		13.5 (monthly range: −1.9–27.2)		
Annual precipitation (mm)	1471		1473		
Vegetation (in 2007 for the UPF and in 2009 for the BAF)					
Canopy species	*Pinus densiflora*		*Alnus japonica*		
Density (trees·ha^{-1})	983	1517	900	575	550
Diameter at breast height (cm)	20.5	17.6	32.6	27.1	24.2
Aboveground biomass (Mg C·ha^{-1})	88.1	86.7	187.9	78.9	61.8
Needle or leaf litter C:N ratio	72.7 (2.0)	73.2 (2.2)	21.8 (0.3)	20.8 (0.2)	18.6 (0.2)
Total N storage (Mg N·ha^{-1})	0.78	0.78	1.65	0.74	0.60
Soil (0–15 cm; during the study period)					
Soil texture	Silt loam	Silt loam	Sandy loam	Sandy loam	Silt loam
Bulk density (g·cm^{-3})	1.05 (0.04)	1.02 (0.03)	0.86 (0.06)	0.81 (0.03)	0.50 (0.14)
Soil water content (kg·kg^{-1})	0.25 (0.02)	0.20 (0.03)	0.35 (0.06)	0.72 (0.17)	1.23 (0.10)
pH	4.54 (0.03)	4.56 (0.01)	4.45 (0.07)	5.04 (0.10)	5.14 (0.04)
Total C concentration (g·kg^{-1})	33.4 (2.3)	29.7 (2.1)	20.8 (2.0)	31.2 (1.8)	61.7 (2.3)
C:N ratio	15.8 (0.4)	17.8 (0.4)	9.6 (0.3)	12.4 (0.2)	12.3 (0.1)
Total N storage (0–30 cm; Mg N·ha^{-1})	3.62 (0.24)	2.98 (0.12)	4.64 (0.52)	5.06 (0.70)	7.14 (0.20)

2.2. Field Experiment and Laboratory Analysis

Three inorganic N indices of top mineral soils, net N transformation rates, inorganic N concentrations in bulk soil and IER assays, were investigated at five points, every 2 m along a single transect across each plot from March 2008 to April 2009. Net N transformation rates including net N mineralization and net nitrification were determined using a field incubation method [44]. Two soil cores from the 0–15 cm mineral layer were sampled from the five points in each plot. The soil sample for determining soil status before field incubation was taken immediately to the laboratory at a cool temperature (<4 °C) and analyzed for inorganic N concentrations in bulk soil. The other soil sample, the field incubation sample, was put into a gas permeable polyethylene bag and returned to its original location. After 45 days, the incubated soil core was collected and also taken to the laboratory for analysis. This soil sampling and incubation procedure was repeated every 45 days, except for a 90 day winter interval due to the difficulty of sampling and incubating frozen soil.

Two IER bags, which were made of stocking nylon and contained 14 mL of either anion or cation resins (Sybron IONAC ASB-1P OH⁻ and C-251 H⁺, Sybron International, Birmingham, NJ, USA), were placed 10 cm below the surface of the mineral soil in each plot, near the five sampling points of the initial and incubated soils. The IER bags were retrieved and replaced every 90 days.

In the laboratory, the moist initial and incubated soil samples were subjected to an extraction process with 2 M KCl solution for 24 h, immediately upon arrival from the sites. The IERs were air-dried first, and then the inorganic N ions were extracted. The extracts were filtered through Whatman No. 42 filter paper and the NH_4^+ and NO_3^- concentrations were determined colorimetrically using a flow-injection auto analyzer (QuikChem AE, Lachat Instruments, Loveland, CO, USA). The NH_4^+ and NO_3^- concentrations extracted from moist initial and incubated soil samples were corrected to a dry soil basis using the gravimetric SWC ((moist soil weight – dry soil weight)/dry soil weight) of each soil.

The net N mineralization and net nitrification rates were calculated as the difference between the NH_4^+ and NO_3^- concentrations of the incubated and initial soil samples. Annual net N mineralization and annual net nitrification per unit area in the upper 15 cm mineral soil layer was calculated by summing the net N mineralization and net nitrification values of incubation periods spanning a year [6,45].

Annual N uptake was estimated from aboveground net primary productivity (ANPP) multiplied by the concentrations of N in woody tissues and leaves, determined using an elemental analyzer (Vario Macro CN analyzer, Elementar Analysensysteme GmbH, Hanau, Germany). ANPP was calculated as the sum of total aboveground biomass increment and litter production (See [41] for details in the UPF.). Aboveground biomass in the BAF was calculated using a DBH-based allometric equation [46] and C concentrations in each compartment determined by the elemental analyzer.

2.3. Statistical Analysis

The values of the five soil samples from each plot were averaged. Thus, plots, of which there were three in each stand, were considered as true replicates. Correlation and regression analyses were conducted on the data derived from individual plots to take into account the spatial variability within each stand. Significant differences in net N transformation rates, inorganic N concentrations in bulk soil and IER, and SWC across stands and seasons were tested using the two-way analysis of variance (ANOVA). Duncan's multiple range test was performed to determine the differences between the means of each stand, when the ANOVA model was significant ($p < 0.05$). Correlation and regression analyses were conducted among the N indices and soil environment variables. All statistical analyses were conducted using R [47].

3. Results

The mean NH_4^+, NO_3^-, and total inorganic N concentrations in bulk soils varied significantly between the stands (Figure 1, $p < 0.01$). NH_4^+ was abundant at the HM and the MM BAF stands (Figure 1a); whereas NO_3^- was rich at the LM BAF stand (Figure 1b). Total inorganic N concentration was lowest at the UPF stands (Figure 1c). As a proportion of total inorganic N, the N in NH_4^+ form represented 61% in the LD UPF stand, 63% in the HD UPF stand, and 36%, 69%, and 78% in the LM, MM, and HM BAF stands, respectively. Conversely, as a proportion of total inorganic N, the N in NO_3^- form represented 39% in the LD UPF stand and 37% in the HD UPF stand and 64%, 31%, and 22% in the LM, MM, and HM BAF stands, respectively. There were also differences in the seasonal patterns of NH_4^+ and NO_3^- levels and of total inorganic N concentrations in bulk soils between the sites. Whereas in the UPF, inorganic N concentrations increased during the growing season (May to August), in the BAF, they were lower (Figure 1a–c, Figure S1).

The total IER-inorganic N and IER-NO_3^- concentrations did not vary significantly (Figure 1e,f). However, IER-NH_4^+ concentration (mg $N \cdot bag^{-1}$) did differ between the stands; it was highest at the HD UPF stand (5.17), followed by the LD UPF stand (4.29), and then the LM (3.01), HM (1.75), and MM (0.66) BAF stands in turn (Figure 1d). IER-NO_3^- accounted for a major portion of IER-inorganic N (61%–93%) and, conversely, IER-NH_4^+ accounted for a very small portion of IER-inorganic N, especially at the MM (7%) and HM (15%) BAF stands. In MM and HM, the majority proportion of IER inorganic NO_3^- strongly contrasted that of NH_4^+ in bulk soil.

The net N transformation rates were higher under the drier stands (LD and HD UPF and LM BAF) than the wetter ones (MM and HM BAF) (Figure 1g,h). Negative net N mineralization rate was observed in the MM and HM BAF stands (Figure 1g). The seasonal pattern of net N transformation rates at the UPF stands and the LM BAF stand closely followed that of soil temperature (Figure S1g,h), in that the rates increased in the growing season and decreased in the non-growing season. On the other hand, the seasonal pattern was not clear at MM and HM BAF stands; negative net N mineralization rates were observed during most seasons (Figure S1g).

Table 2 shows the correlations between inorganic N indices and environmental variables. Generally, inorganic N indices in the UPF were considerably dependent to soil temperature and SWC; meanwhile, those in the BAF were less dependent to the environmental variables. IER-NO_3^-, IER-inorganic N, net N mineralization, and net nitrification were positively correlated with soil temperature in both the UPF and BAF. Correlation coefficients were higher in the UPF ($r = 0.57$–0.71) than the BAF ($r = 0.24$–0.47). NH_4^+ and inorganic N only in the UPF and IER-NH_4^+ only in the BAF were correlated with soil temperature. NH_4^+, inorganic N, net N mineralization and net nitrification in the UPF ($r = 0.46$–0.60) and NH_4^+ in the BAF ($r = 0.57$) were positively correlated with SWC. On the other hand, NO_3^- ($r = -0.49$), net N mineralization ($r = -0.37$), and net nitrification ($r = -0.29$) in the BAF were negatively correlated with SWC.

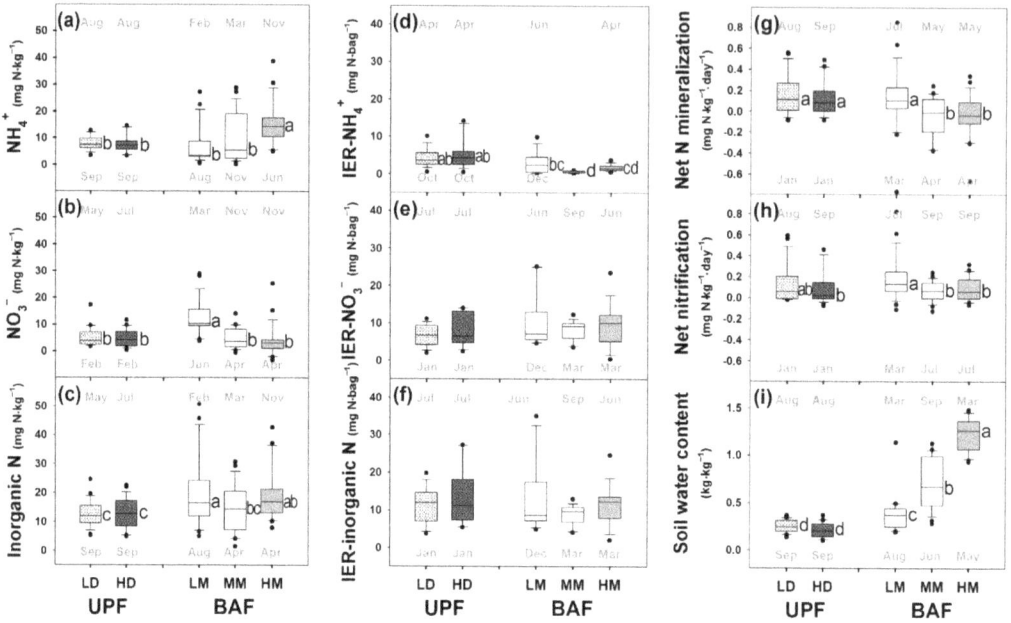

Figure 1. Inorganic N indices in an upland pine forest (UPF) with different stand densities and a bottomland alder forest (BAF) with different soil moisture content. (a–c) Inorganic N concentrations in bulk soil; (d–f) Inorganic N concentrations attached to ion exchange resin (IER); (g,h) Net N transformation rates; (i) Soil water content. Months written at the top and the bottom of each panel indicate the months corresponding to the highest and lowest values. Letters beside boxes indicate the significant difference, if there is one, among the means determined by Duncan's multiple range tests.

Table 2. Environmental dependence of inorganic N indices in an upland pine forest (UPF) and a bottomland alder forest (BAF). A correlation coefficient (r) that satisfies the statistical significance is presented ($p < 0.05$). A dash indicates insignificant.

	Soil Temperature		Soil Water Content	
	UPF	BAF	UPF	BAF
NH_4^+	0.31	–	0.54	0.57
NO_3^-	–	–	–	−0.48
Inorganic N	0.34	–	0.43	–
IER-NH_4^+	–	0.37	–	–
IER-NO_3^-	0.71	0.46	–	–
IER-inorganic N	0.57	0.47	–	–
Net N mineralization	0.58	0.24	0.56	−0.37
Net nitrification	0.67	0.32	0.60	−0.29

Spatio-temporal variations of net N mineralization and NH_4^+ were related, especially in the BAF (Figure 2). In the BAF, a significant negative relationship between net N mineralization and NH_4^+

was observed (Figure 2b, $p < 0.001$, $R^2 = 0.45$), whereas, in the UPF, the positive and weak relationship was only close to marginally significant (Figure 2a, $p = 0.10$, $R^2 = 0.06$). Seasonal patterns of the net N mineralization rate and of NH_4^+ concentration were coupled in the UPF (Figure 2c,d), but inversely coupled in the BAF (Figure 2e–g).

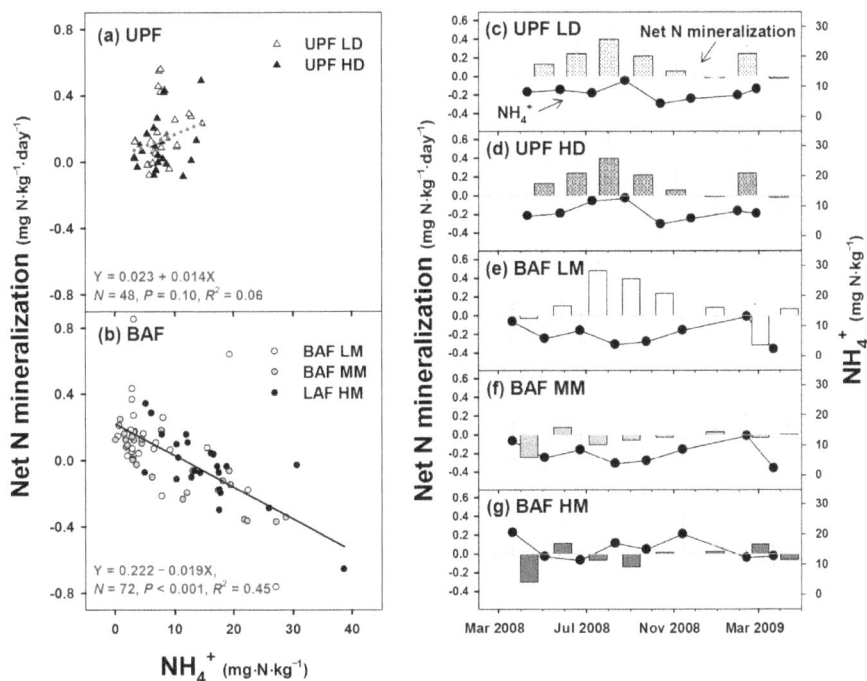

Figure 2. (**a,b**) Relationship between NH_4^+ concentrations and net N mineralization rates in an upland pine forest (UPF) and a bottomland alder forest (BAF); (**c–g**) Seasonal patterns of net N mineralization rates and NH_4^+ concentrations in the UPF and the BAF.

The annual net N mineralization rate (kg $N \cdot ha^{-1} \cdot year^{-1}$) was higher at the LD ($84.01 \pm 18.18$) and HD ($66.05 \pm 11.25$) UPF stands and the LM BAF stand (72.24 ± 11.13) than at the MM (-17.22 ± 43.10) and HM (-10.26 ± 17.60) BAF stands (Table 3, $p < 0.01$). The annual net N nitrification rate (kg $N \cdot ha^{-1} \cdot year^{-1}$), from highest to lowest, was LM (98.74 ± 16.01) $\geq$ LD (77.61 ± 18.26) = HD (56.26 ± 14.56) $\geq$ MM (32.69 ± 12.67) = HM (23.98 ± 9.80) (Table 3). Figure 3 describes the relationship between SWC and annual net N transformation rates. In the BAF, annual N transformation rates showed a linear inverse correlation with SWC. In contrast, in the UPF the correlation was close to being positive. Annual N uptake (kg $N \cdot ha^{-1} \cdot year^{-1}$) was estimated to be 73.19, 60.18, 191.77, 141.21, and 128.76 for LD, HD, LM, MM, and HM stands, respectively (Table 3).

Table 3. Summary of net N transformation rates, N availability, N uptake, and ANPP.

	Annual Net N Mineralization Rate [†]	Annual Net Nitrification Rate [†]	IER-NH₄⁺ [‡]	IER-NO₃⁻ [‡]	IER-Total Inorganic N [‡]	Annual N Uptake [†]	ANPP [§]
UPF							
LD	84.01 ± 18.18	77.61 ± 18.26	4.29 ± 0.49	6.81 ± 0.58	11.10 ± 0.97	73.19	6.17
HD	66.05 ± 11.25	56.26 ± 14.56	5.17 ± 0.79	8.17 ± 0.88	13.34 ± 1.53	60.18	5.03
BAF							
LM	72.24 ± 11.13	98.74 ± 16.01	3.03 ± 0.59	10.73 ± 1.43	13.76 ± 1.98	191.77	5.01
MM	−17.22 ± 43.10	32.69 ± 12.67	0.66 ± 0.05	8.14 ± 0.52	8.80 ± 0.54	141.21	6.06
HM	−10.26 ± 17.60	23.98 ± 9.80	1.75 ± 0.17	9.62 ± 1.13	11.37 ± 1.07	128.76	4.51

IER; ion exchange resin, ANPP; aboveground net primary productivity; Unit: [†] kg N·ha⁻¹·year⁻¹; [‡] mg N·bag⁻¹; [§] Mg C·ha⁻¹·year⁻¹.

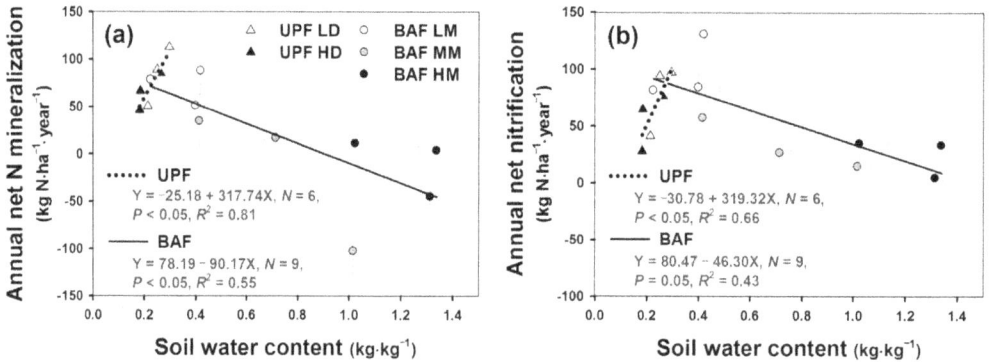

Figure 3. Relationship between gravimetric soil water content and (**a**) annual net N mineralization or (**b**) nitrification rates.

4. Discussion

The results in the UPF corresponded to the general established patterns of N availability in forest soils [12–16,18]. For example, net N transformation rates increased with soil temperature and SWC (Table 2, Figure 2). Net N transformation rates and inorganic N availability were higher during the growing season and lower during the non-growing season (Figure 1). Spatio-temporal patterns of three inorganic N indices, N transformation rates, IER-inorganic N, and bulk soil inorganic N concentrations, corresponded with each other (Figure 1). Results for the LM BAF stand were also similar to general established patterns.

However, some abnormal results, mainly from MM and HM BAF stands, led to some unanswered questions. Why was net N mineralization negatively related to bulk NH₄⁺ concentration (Figure 2)? Why was IER-NH₄⁺ low at the MM and HM BAF stands where bulk soil NH₄⁺ concentration was high (Figure 1)? Moreover, could negative net N mineralization occur in ecosystems that require a substantial amount of N to sustain plant uptake (Figure 1, Table 3)? In addition, for the UPF stands, what was the effect of stand density on soil N transformation rates and N availability?

4.1. Restricted Net N Mineralization under NH_4^+ Rich Condition

Net N transformation rates were independent of soil temperature in the MM and HM BAF stands. Net N transformation rates in the wetter stands (MM and HM) were restricted by the high SWC as compared to the drier stand (LM) (Figures 1 and 3, Table 2). SWC could limitedly explain the spatio-temporal variations in soil N transformation rates and inorganic N indices (Table 2), unlike the soil carbon processes and soil properties which were strongly regulated by SWC in the previous studies [25,42]. Correlation between SWC and net N mineralization and net nitrification rates were much weaker in the BAF ($r = -0.29$ and -0.37) than in the UPF ($r = 0.56$ and 0.60) (Table 2). While soil moisture strongly regulated soil properties and soil C processes in the BAF [25,42], it only weakly affected N transformations processes.

Instead, the net N mineralization rate in the BAF was best explained by the negative relationship with NH_4^+ concentration (Figure 2b,f,g). Ideally, NH_4^+ will increase as a result of a net N mineralization process even though a proportion of mineralized NH_4^+ can be lost by plant uptake or nitrification. However, net N mineralization rate was low for soils with a high level of NH_4^+ in the BAF. This unexpected result can be speculated that a high level of NH_4^+ in bulk soil may have reduced the microbial activities involved in the mineralization process. Theoretically, the enzyme activities involved in mineralization could be negatively affected by nutrient availability because of the cost-efficiency strategy of microbes, known as the "microbial nitrogen mining" hypothesis [48,49]. When N is less available, microbes actively mine recalcitrant organic matter to acquire it; whereas, in N rich environments, microbes can assimilate inorganic N without costly mineralization activity. Although the relationship between inorganic N availability and microbial activities have been reported to be both positive [50,51] and negative [52–54] in empirical studies mainly based on fertilization experiments in forests, wetlands, or agricultural land, a meta-analysis of Treseder [55] suggested that N addition reduced microbial biomass by 15% and, consequently, also microbial activities such as soil respiration. Those articles could provide a theoretical proof and empirical evidence supporting the decrease of net N mineralization under high NH_4^+ concentrations in the BAF. Consequently, the intensive restricting effect of NH_4^+ on net N mineralization probably overspread the effect of soil temperature or SWC.

Meanwhile, it could be questioned how the high soil NH_4^+ concentration could be supported with a deficient inorganic N supply from extremely low net N mineralization. In the wet soils of the MM and HM stands, a major source of inorganic N was not N mineralization which was nearly zero year-round due to the inhibiting effect of high soil moisture on N mineralization. Instead, inorganic N could be supplied through atmospheric N deposition [56,57], accumulation of external N through run-off from upland forests [4,58], and symbiotic and non-symbiotic N fixation [59,60]. For instance, Shin et al. [57] reported notable atmospheric N deposition in Seoul (16.5 kg $N \cdot ha^{-1} \cdot year^{-1}$ for NH_4^+ and NH_3 deposits and 18.1 kg $N \cdot ha^{-1} \cdot year^{-1}$ for NO_3^-, NO_2 and HNO_3 deposits); the BAF is located within the range of the urban effect on atmospheric N deposition [61]. N fixation of pure alder forests was known to be 37–150 kg $N \cdot ha^{-1} \cdot year^{-1}$ [62]. Even though lack of atmospheric N deposition and N fixation data in our study limits the certainty and depth of this implication, evidences from other studies suggest that there may be sufficient NH_4^+ sources without N mineralization. In addition, at MM and HM, the anaerobic conditions could have restricted the nitrification process, especially since the mean net nitrification rate at MM and HM was two to three times lower than that at LM (Figure 1), and this would tend to conserve NH_4^+ as the major form of inorganic N. The dominance of NH_4^+ in the inorganic N pool agreed with previous studies from alder forests [18,63] and wetlands [17,64].

4.2. Discrepancy among Inorganic N in Bulk Soil and IER, and Net N Transformation Rates: Which N Indices Could Describe the Real Status of N Availability?

The net N mineralization rate and IER-NH_4^+ were extremely low whereas soil NH_4^+ concentration was high at MM and HM BAF stands. Why was there a discrepancy between bulk soil and IER-NH_4^+ concentration? Understanding this discrepancy may elucidate the real status of N availability from the balance among microbial mineralization and immobilization, soil adsorption and desorption, and plant root uptake in the field. The low value of IER-NH_4^+ at MM and HM suggests that the rich NH_4^+ in bulk soil were not actually assimilated by plants. Due to the strong affinity of NH_4^+ for negatively charged soil organic matter, NH_4^+ may be strongly associated with soil organic matter and unavailable to plants [18] at the MM and HM stands where soil organic matter concentration was high (Table 1). This hypothesis is supported by the positive relationship between soil NH_4^+ and total carbon concentration ($p < 0.001$, $R^2 = 0.53$). On the other hand, the pattern of NO_3^- concentration in soil and IER showed an opposite pattern to that of NH_4^+; soil NO_3^- concentration was low and IER-NO_3^- was high. The low soil NO_3^- concentration is likely due to its high uptake by plants, as suggested by the high IER-NO_3^-. Abundant moisture at MM and HM could increase the availability of NO_3^- since it is mobilized by soil moisture.

This result implies that IER or ion exchange membranes could be comprehensive and practical indices for N availability in reality [27,65], even though the lack of standardized application remains the challenge [66]. The assumption that plant available N is mainly driven through N mineralization, was not applicable in the MM and HM BAF stands; net N transformation rates failed to meet the interest in N availability. On the other hand, IER assay data would be the best indirect measure of N availability unless the direct measure of plant uptake using stable isotope. An additional advantage of IER or ion exchange membrane assays is to determine availabilities of other nutrients such as phosphorus and cations simultaneously with inorganic N [66–68].

4.3. Stand Density Effects on Soil N Transformation and Availability

The annual net N mineralization rate of the UPF (84.01 and 66.05 kg N·ha^{-1}·year^{-1} for LD and HD stands, respectively) was within reasonable range of the generally accepted rate for temperate pine forests (43.6–87.0 kg N·ha^{-1}·year^{-1} [30,45,69,70]). The differences in net N transformation rates and N availability between LD and HD stands were insignificant according to the statistical tests (Figure 1). Nevertheless, the difference may not be negligible; annual net N mineralization rate, annual net nitrification rate, and IER-inorganic N concentration were 40%, 38%, and 20% higher, respectively, at LD than at HD (Table 3). The higher net N transformation rates and N availability at LD than at HD seem to correspond to the difference in SWC (*t*-test; $p < 0.05$). The higher SWC at LD, which was probably due to lower evaporation under the lower stand density and a less steep topography (Table 1), would promote soil N cycling. In addition, the difference of N processes matched that of ANPP in the both stands [41]. ANPP was 23% higher at the LD stands than the HD stands (Table 3). This result supports the relationship between N availability and primary production in forests [1–3]. It may provide an implication for stand density control (e.g., forest tending and thinning) with respect to nutrient availability. Stand density control as a management practice can be applied to improve nutrient use efficiency for tree production associated with C sequestration in unmanaged pine forests with high stand density in Korea [71,72]. In addition, this application may contribute to maintenance of stability in stand dynamics by reducing density-dependent tree mortality [73].

5. Conclusions

The UPF and BAF are presented, distinguishing net N transformation rates, availabilities, and environmental controls on those N processes. First, the UPF, which were assumed as moderately fertile environments, confirmed the generally accepted knowledge of net N transformation rates and N availability. Second, soil N transformation rates and N availability at the LM BAF stand represent active N cycling under an N-rich environment. The annual plant N uptake was the highest and the annual net N mineralization rate accounted for 37% of that uptake. The lower net N mineralization (72.24 kg N·ha^{-1}·year^{-1}) than net nitrification (98.74 kg N·ha^{-1}·year^{-1}) at the LM stand indicates that most of the mineralized NH_4^+ was converted to NO_3^- during the field incubation. Thus, the relatively low moisture condition at LM supported NO_3^- accumulation due to the high net nitrification rate and low NO_3^- loss via leaching and denitrification [19,74,75]. Third, the N-rich and wet MM and HM BAF stands exhibited restricted net N transformation rates and N availability unlike high inorganic N concentration in bulk soils and low C:N ratio in leaf litters and soils.

In the UPF, annual net N mineralization slightly exceeded the annual N uptake (Table 3). As taking account into the potential N input from atmospheric N deposition, plant N uptake did not consume the all N input. The surplus of inorganic N would not be retained in the stands; instead, it might leach to N runoff by a ground water and stream. The N saturation in the UPF agreed with other coniferous stands in this region [76]. Meanwhile, in the BAF annual net N mineralization failed to satisfy the annual N uptake (Table 3); the external N input might be retained in the stands. It suggests the importance of bottomland or riparian forests for N retention function [4,77].

The significant findings are summarized as follows. First, net N transformation rates and N availability were not regulated by SWC and soil temperature, instead regulated by initial soil NH_4^+ in the MM and HM BAF stands. The net N mineralization rate was inhibited under NH_4^+-rich soils. The NH_4^+ accumulation and plant uptake-N might be supplied from external N sources. Second, net N transformation rates, inorganic N concentrations in bulk soil, and those on IER corresponded in the UPF, but did not in the BAF. Discrepancies between the patterns of N availability indices in the BAF may enhance our knowledge of the real status of N availability. Third, differences in soil moisture conditions between high and low density stands might affect N transformation and N availability, and, consequently, ecosystem productivity.

Acknowledgments

We appreciate the cooperation of Forest Practice Research Center, Korea Forest Research Institute and Heolleung Office, Culture Heritage Administration of Korea for allowing us access to the UPF and the BAF sites, respectively. We also appreciate Ah Reum Lee, Su Jin Heo, Kyung Won Seo, Sue Kyoung Lee, and Koong Yi for their assistance both in the laboratory and the field. This work was supported by research grants from the Korea Environmental Industry & Technology Institute (C314-00131-0408, G214-00181-0402).

Author Contributions

Tae Kyung Yoon and Nam Jin Noh equally led the field and laboratory work, data analyses, and manuscript writing. Tae Kyung Yoon and Nam Jin Noh have the main authorship of the results from the BAF and UPF, respectively. Haegeun Chung collaborated on discussion of the data and the manuscript writing, especially from the perspective of microbiology. A-Ram Yang contributed to the field and laboratory work and manuscript writing. Yowhan Son supervised the whole process of the study including the experimental design, discussion of the data, and manuscript writing.

Conflicts of Interest

The authors declare no conflict of interest.

References

1. Binkley, D.; Vitousek, P. Soil nutrient availability. In *Plant Physiological Ecology*; Pearcy, R., Ehleringer, J., Mooney, H., Rundel, P., Eds.; Chapman and Hall Ltd.: New York, NY, USA, 1989; pp. 75–96.
2. Vitousek, P.M.; Howarth, R.W. Nitrogen limitation on land and in the sea: How can it occur? *Biogeochemistry* **1991**, *13*, 87–115.
3. LeBauer, D.S.; Treseder, K.K. Nitrogen limitation of net primary productivity in terrestrial ecosystems is globally distributed. *Ecology* **2008**, *89*, 371–379.
4. Saunders, D.L.; Kalff, J. Nitrogen retention in wetlands, lakes and rivers. *Hydrobiologia* **2001**, *443*, 205–212.
5. Cameron, K.C.; Di, H.J.; Moir, J.L. Nitrogen losses from the soil/plant system: A review. *Ann. Appl. Biol.* **2013**, *162*, 145–173.
6. Binkley, D.; Hart, S.C. The components of nitrogen availability assessments in forest soils. In *Advances in Soil Science*; Stewart, B.A., Ed.; Springer: New York, NY, USA, 1989; Volume 10, pp. 57–112.
7. Schimel, J.P.; Bennett, J. Nitrogen mineralization: Challenges of a changing paradigm. *Ecology* **2004**, *85*, 591–602.
8. Frank, D.A.; Groffman, P.M. Plant rhizospheric N processes: What we don't know and why we should care. *Ecology* **2009**, *90*, 1512–1519.

9. Ros, G.H.; Temminghoff, E.J.M.; Hoffland, E. Nitrogen mineralization: A review and meta-analysis of the predictive value of soil tests. *Eur. J. Soil Sci.* **2011**, *62*, 162–173.

10. Zhu, B.; Gutknecht, J.L.M.; Herman, D.J.; Keck, D.C.; Firestone, M.K.; Cheng, W. Rhizosphere priming effects on soil carbon and nitrogen mineralization. *Soil Biol. Biochem.* **2014**, *76*, 183–192.

11. Manzoni, S.; Porporato, A. Soil carbon and nitrogen mineralization: Theory and models across scales. *Soil Biol. Biochem.* **2009**, *41*, 1355–1379.

12. Paul, K.I.; Polglase, P.J.; O'Connell, A.M.; Carlyle, J.C.; Smethurst, P.J.; Khanna, P.K. Defining the relation between soil water content and net nitrogen mineralization. *Eur. J. Soil Sci.* **2003**, *54*, 39–48.

13. Sierra, J. Temperature and soil moisture dependence of n mineralization in intact soil cores. *Soil Biol. Biochem.* **1997**, *29*, 1557–1563.

14. Guntiñas, M.E.; Leirós, M.C.; Trasar-Cepeda, C.; Gil-Sotres, F. Effects of moisture and temperature on net soil nitrogen mineralization: A laboratory study. *Eur. J. Soil Biol.* **2012**, *48*, 73–80.

15. Sleutel, S.; Moeskops, B.; Huybrechts, W.; Vandenbossche, A.; Salomez, J.; De Bolle, S.; Buchan, D.; de Neve, S. Modeling soil moisture effects on net nitrogen mineralization in loamy wetland soils. *Wetlands* **2008**, *28*, 724–734.

16. Zak, D.R.; Grigal, D.F. Nitrogen mineralization, nitrification and denitrification in upland and wetland ecosystems. *Oecologia* **1991**, *88*, 189–196.

17. Cartaxana, P.; Caçador, I.; Vale, C.; Falcão, M.; Catarino, F. Seasonal variation of inorganic nitrogen and net mineralization in a salt marsh ecosystem. *Mangroves Salt Marshes* **1999**, *3*, 127–134.

18. Uri, V.; Lõhmus, K.; Tullus, H. Annual net nitrogen mineralization in a grey alder (*Alnus incana* (L.) Moench) plantation on abandoned agricultural land. *For. Ecol. Manag.* **2003**, *184*, 167–176.

19. Hefting, M.; Clément, J.C.; Dowrick, D.; Cosandey, A.C.; Bernal, S.; Cimpian, C.; Tatur, A.; Burt, T.P.; Pinay, G. Water table elevation controls on soil nitrogen cycling in riparian wetlands along a European climatic gradient. *Biogeochemistry* **2004**, *67*, 113–134.

20. Western, A.W.; Grayson, R.B.; Bloschl, G. Scaling of soil moisture: A hydrologic perspective. *Annu. Rev. Earth Pl. Sc.* **2002**, *30*, 149–180.

21. Porporato, A.; D'Odorico, P.; Laio, F.; Rodriguez-Iturbe, I. Hydrologic controls on soil carbon and nitrogen cycles. I. Modeling scheme. *Adv. Water Resour.* **2003**, *26*, 45–58.

22. Robinson, D.A.; Campbell, C.S.; Hopmans, J.W.; Hornbuckle, B.K.; Jones, S.B.; Knight, R.; Ogden, F.; Selker, J.; Wendroth, O. Soil moisture measurement for ecological and hydrological watershed-scale observatories: A review. *Vadose Zone J.* **2008**, *7*, 358–389.

23. Updegraff, K.; Pastor, J.; Bridgham, S.D.; Johnston, C.A. Environmental and substrate controls over carbon and nitrogen mineralization in northern wetlands. *Ecol. Appl.* **1995**, *5*, 151–163.

24. Ehrenfeld, J.G.; Han, X.; Parsons, W.F.J.; Zhu, W. On the nature of environmental gradients: Temporal and spatial variability of soils and vegetation in the New Jersey pinelands. *J. Ecol.* **1997**, *85*, 785–798.

25. Yoon, T.K.; Noh, N.J.; Han, S.; Kwak, H.; Lee, W.-K.; Son, Y. Small-scale spatial variability of soil properties in a Korean swamp. *Landsc. Ecol. Eng.* **2015**, *11*, 723–734.

26. Khanna, P.K.; Raison, R.J. *In situ* core methods for estimating soil mineral-N fluxes: Re-evaluation based on 25 years of application and experience. *Soil Biol. Biochem.* **2013**, *64*, 203–210.

27. Friedel, J.K.; Herrmann, A.; Kleber, M. Ion exchange resin–soil mixtures as a tool in net nitrogen mineralisation studies. *Soil Biol. Biochem.* **2000**, *32*, 1529–1536.

28. Johnson, D.W.; Verburg, P.S.J.; Arnone, J.A. Soil extraction, ion exchange resin, and ion exchange membrane measures of soil mineral nitrogen during incubation of a tallgrass prairie soil. *Soil Sci. Soc. Am. J.* **2005**, *69*, 260–265.

29. Vitousek, P. Nutrient cycling and nutrient use efficiency. *Am. Nat.* **1982**, *119*, 553–572.

30. Gower, S.T.; Son, Y. Differences in soil and leaf litterfall nitrogen dynamics for five forest plantations. *Soil Sci. Soc. Am. J.* **1992**, *56*, 1959–1966.

31. Aerts, R. The advantages of being evergreen. *Trends Ecol. Evol.* **1995**, *10*, 402–407.

32. Reich, P.B.; Grigal, D.F.; Aber, J.D.; Gower, S.T. Nitrogen mineralization and productivity in 50 hardwood and conifer stands on diverse soils. *Ecology* **1997**, *78*, 335–347.

33. Mueller, K.E.; Hobbie, S.E.; Oleksyn, J.; Reich, P.B.; Eissenstat, D.M. Do evergreen and deciduous trees have different effects on net N mineralization in soil? *Ecology* **2012**, *93*, 1463–1472.

34. Binkley, D.; Giardina, C. Why do tree species affect soils? The warp and woof of tree-soil interactions. *Biogeochemistry* **1998**, *42*, 89–106.

35. Fujita, H.; Fujimura, Y. Distribution pattern and regeneration of swamp forest species with respect to site conditions. In *Ecology of Riparian Forests in Japan*; Sakio, H., Tamura, T., Eds.; Springer: Tokyo, Japan, 2008; pp. 225–236.

36. Eom, H.J.; Chang, K.S.; Kim, H.; Chang, C.-S. Notes on a new overlooked taxon of *Alnus* (*Betulaceae*) in Korea. *For. Sci. Technol.* **2011**, *7*, 42–46.

37. Whang, B.-C.; Lee, M.-W. Landscape ecology planning principles in korean Feng-Shui, Bi-Bo woodlands and ponds. *Landsc. Ecol. Eng.* **2006**, *2*, 147–162.

38. Yoon, T.K.; Noh, N.J.; Kim, R.-H.; Seo, K.W.; Lee, S.K.; Yi, K.; Lee, I.K.; Lim, J.-H.; Son, Y. Mass dynamics of coarse woody debris in an old-growth deciduous forest of Gwangneung, Korea. *For. Sci. Technol.* **2011**, *7*, 145–150.

39. Noh, N.J.; Son, Y.; Lee, S.K.; Yoon, T.K.; Seo, K.W.; Kim, C.; Lee, W.-K.; Bae, S.W.; Hwang, J. Influence of stand density on soil CO_2 efflux for a *Pinus densiflora* forest in Korea. *J. Plant Res.* **2010**, *123*, 411–419.

40. Noh, N.J.; Son, Y.; Jo, W.; Yi, K.; Park, C.W.; Han, S. Preliminary study on estimating fine root growth in a natural *Pinus densiflora* forest using a minirhizotron technique. *For. Sci. Technol.* **2012**, *8*, 47–50.

41. Noh, N.J.; Kim, C.; Bae, S.W.; Lee, W.K.; Yoon, T.K.; Muraoka, H.; Son, Y. Carbon and nitrogen dynamics in a *Pinus densiflora* forest with low and high stand densities. *J. Plant Ecol.* **2013**, *6*, 368–379.

42. Yoon, T.K.; Noh, N.J.; Han, S.; Lee, J.; Son, Y. Soil moisture effects on leaf litter decomposition and soil carbon dioxide efflux in wetland and upland forests. *Soil Sci. Soc. Am. J.* **2014**, *78*, 1804–1816.

43. Vicca, S.; Luyssaert, S.; Peñuelas, J.; Campioli, M.; Chapin, F.S.; Ciais, P.; Heinemeyer, A.; Högberg, P.; Kutsch, W.L.; Law, B.E.; *et al.* Fertile forests produce biomass more efficiently. *Ecol. Lett.* **2012**, *15*, 520–526.

44. Eno, C.F. Nitrate production in the field by incubating the soil in polyethylene bags. *Soil Sci. Soc. Am. J.* **1960**, *24*, 277–279.

45. Son, Y.; Lee, I.K. Soil nitrogen mineralization in adjacent stands of larch, pine and oak in central Korea. *Ann. For. Sci.* **1997**, *54*, 1–8.

46. Johansson, T. Biomass equations for determining fractions of common and grey alders growing on abandoned farmland and some practical implications. *Biomass Bioenerg.* **2000**, *18*, 147–159.

47. R Development Core Team. *R: A Language and Environment for Statistical Computing*; R Foundation for Statistical Computing: Vienna, Austria, 2011. Available online: http://www.r-project.org (accessed on 15 June 2015).

48. Craine, J.M.; Morrow, C.; Fierer, N. Microbial nitrogen limitation increases decomposition. *Ecology* **2007**, *88*, 2105–2113.

49. Shi, W. Agricultural and ecological significance of soil enzymes: Soil carbon sequestration and nutrient cycling. In *Soil Enzymology*; Shukla, G.; Varma, A., Eds.; Springer: Berlin-Heidelberg, Germany, 2011; Volume 22, pp. 43–60.

50. Regina, K.; Nykänen, H.; Maljanen, M.; Silvola, J.; Martikainen, P.J. Emissions of N_2O and NO and net nitrogen mineralization in a boreal forested peatland treated with different nitrogen compounds. *Can. J. For. Res.* **1998**, *28*, 132–140.

51. Höfferle, Š.; Nicol, G.W.; Pal, L.; Hacin, J.; Prosser, J.I.; Mandić-Mulec, I. Ammonium supply rate influences archaeal and bacterial ammonia oxidizers in a wetland soil vertical profile. *FEMS Microbiol. Ecol.* **2010**, *74*, 302–315.

52. Carpenter-Boggs, L.; Pikul, J.L.; Vigil, M.F.; Riedell, W.E. Soil nitrogen mineralization influenced by crop rotation and nitrogen fertilization. *Soil Sci. Soc. Am. J.* **2000**, *64*, 2038–2045.

53. Fisk, M.; Fahey, T. Microbial biomass and nitrogen cycling responses to fertilization and litter removal in young northern hardwood forests. *Biogeochemistry* **2001**, *53*, 201–223.

54. Bowden, R.D.; Davidson, E.; Savage, K.; Arabia, C.; Steudler, P. Chronic nitrogen additions reduce total soil respiration and microbial respiration in temperate forest soils at the Harvard forest. *For. Ecol. Manag.* **2004**, *196*, 43–56.

55. Treseder, K.K. Nitrogen additions and microbial biomass: A meta-analysis of ecosystem studies. *Ecol. Lett.* **2008**, *11*, 1111–1120.

56. Matson, P.; Lohse, K.A.; Hall, S.J. The globalization of nitrogen deposition: Consequences for terrestrial ecosystems. *AMBIO* **2002**, *31*, 113–119.

57. Shin, A.Y.; Sung, M.Y.; Choi, J.S.; On, J.S.; Roh, S.A.; Ahn, J.Y.; Han, J.S.; Lee, G.W. A study on the acid deposition of inorganic ions and flux in Seoul, Gangwha, Icheon, 2008. *J. Korean Soc. Urban Environ.* **2012**, *12*, 14–22. (In Korean)

58. Jacks, G.; Joelsson, A.; Fleischer, S. Nitrogen retention in forest wetlands. *AMBIO* **1994**, *23*, 358–362.

59. Son, Y. Non-symbiotic nitrogen fixation in forest ecosystems. *Ecol. Res.* **2001**, *16*, 183–196.

60. Vitousek, P.; Cassman, K.; Cleveland, C.; Crews, T.; Field, C.; Grimm, N.; Howarth, R.; Marino, R.; Martinelli, L.; Rastetter, E., *et al.* Towards an ecological understanding of biological nitrogen fixation. *Biogeochemistry* **2002**, *57–58*, 1–45.

61. Han, J.S.; Hong, Y.D.; Ahn, J.Y.; Chung, I.R.; Shin, A.Y.; Lim, B.K.; Noh, S.A.; Son, J.S.; Park, J.H. *Acid Deposition Monitoring and Impact Assessment (V): Centered on Forest Ecosystem*; National Institute of Environmental Research: Incheon, Korea, 2008.

62. Lee, Y.; Son, Y. Diurnal and seasonal patterns of nitrogen fixation in an *Alnus hirsuta* plantation of central Korea. *J. Plant Biol.* **2005**, *48*, 332–337.

63. Kim, D.K. Comparisons of Physico-Chemical Properties of Soil between the *Alnus japonica* Wetlands and Adjacent Slope. Master's Thesis, Kongju National University, Gongju, Korea, Febuary 2014. (In Korean)

64. Bedard-Haughn, A.; Matson, A.L.; Pennock, D.J. Land use effects on gross nitrogen mineralization, nitrification, and N_2O emissions in ephemeral wetlands. *Soil Biol. Biochem.* **2006**, *38*, 3398–3406.

65. Durán, J.; Delgado-Baquerizo, M.; Rodríguez, A.; Covelo, F.; Gallardo, A. Ionic exchange membranes (IEMs): A good indicator of soil inorganic N production. *Soil Biol. Biochem.* **2013**, *57*, 964–968.

66. Qian, P.; Schoenau, J.J. Practical applications of ion exchange resins in agricultural and environmental soil research. *Canad. J. Soil Sci.* **2002**, *82*, 9–21.

67. Van Raij, B. Bioavailable tests: Alternatives to standard soil extractions. *Commun. Soil Sci. Plant Anal.* **1998**, *29*, 1553–1570.

68. Meason, D.F.; Idol, T.W. Nutrient sorption dynamics of resin membranes and resin bags in a tropical forest. *Soil. Sci. Soc. Am. J.* **2008**, *72*, 1806–1814.

69. Nadelhoffer, K.J.; Aber, J.D.; Melillo, J.M. Leaf-litter production and soil organic matter dynamics along a nitrogen-availability gradient in southern Wisconsin (U.S.A.). *Can. J. For. Res.* **1983**, *13*, 12–21.

70. Binkley, D.; Valentine, D. Fifty-year biogeochemical effects of green ash, white pine, and Nrway spruce in a replicated experiment. *For. Ecol. Manag.* **1991**, *40*, 13–25.

71. Kim, C.; Son, Y.; Lee, W.-K.; Jeong, J.; Noh, N.-J.; Kim, S.-R.; Yang, A.R.; Ju, N.-G. Influence of forest tending (Soopkakkugi) works on litterfall and nutrient inputs in a *Pinus densiflora* stand. *For. Sci. Technol.* **2012**, *8*, 83–88.

72. Kang, J.-S.; Shibuya, M.; Shin, C.-S. The effect of forest-thinning works on tree growth and forest environment. *For. Sci. Technol.* **2014**, *10*, 33–39.

73. Franklin, J.F.; Spies, T.A.; van Pelt, R.; Carey, A.B.; Thornburgh, D.A.; Berg, D.R.; Lindenmayer, D.B.; Harmon, M.E.; Keeton, W.S.; Shaw, D.C.; *et al.* Disturbances and structural development of natural forest ecosystsems with silvicultural implications, using Douglas-fir forests as an example. *For. Ecol. Manag.* **2002**, *1–3*, 399–423.

74. Zak, D.R.; Groffman, P.M.; Pregitzer, K.S.; Christensen, S.; Tiedje, J.M. The vernal dam: Plant-microbe competition for nitrogen in northern hardwood forests. *Ecology* **1990**, *71*, 651–656.

75. Aulakh, M.S.; Singh, K.; Singh, B.; Doran, J.W. Kinetics of nitrification under upland and flooded soils of varying texture. *Commun. Soil Sci. Plant Anal.* **1996**, *27*, 2079–2089.

76. Yoo, J. Hydrological Properties and Nutrient Budgets in a Coniferous Forest Catchment in Gwangneung, Gyeonggi Province, Korea. Ph.D. Thesis, Korea University, Seoul, Korea, Febuary 2010. (In Korean)

77. Fisher, J.; Acreman, M.C. Wetland nutrient removal: A review of the evidence. *Hydrol. Earth Syst. Sci.* **2004**, *8*, 673–685.

Juvenile Southern Pine Response to Fertilization Is Influenced by Soil Drainage and Texture

Timothy J. Albaugh, Thomas R. Fox, H. Lee Allen and Rafael A. Rubilar

Abstract: We examined three hypotheses in a nutrient dose and application frequency study installed in juvenile (aged 2–6 years old) *Pinus* stands at 22 sites in the southeastern United States. At each site, eight or nine treatments were installed where nitrogen was applied at different rates (0, 67, 134, 268 kg ha^{-1}) and frequencies (0, 1, 2, 4 and 6 years) in two or four replications. Phosphorus was applied at 0.1 times the nitrogen rate and other elements were added as needed based on foliar nutrient analysis to insure that nutrient imbalances were not induced with treatment. Eight years after treatment initiation, the site responses were grouped based on texture and drainage characteristics: soil group 1 consisted of poorly drained soils with a clayey subsoil, group 2 consisted of poorly to excessively drained spodic soils or soils without a clay subsoil, and group 3 consisted of well-drained soils with a clayey subsoil. We accepted the first hypothesis that site would be a significant factor explaining growth responses. Soil group was also a significant factor explaining growth response. We accepted our second hypothesis that the volume growth-cumulative dose response function was not linear. Volume growth reached an asymptote in soil groups 1 and 3 between cumulative nitrogen doses of 300–400 kg ha^{-1}. Volume growth responses continued to increase up to 800 kg ha^{-1} of cumulatively applied nitrogen for soil group 2. We accepted our third hypothesis that application rate and frequency did not influence the growth response when the cumulative nitrogen dose was equivalent. There was no difference in the growth response for comparisons where a cumulative nitrogen dose of 568 kg ha^{-1} was applied as 134 kg ha^{-1} every two years or as 269 kg ha^{-1} every four years, or where 269 kg ha^{-1} of nitrogen was applied as four applications of 67 kg ha^{-1} every two years or as two applications of 134 kg ha^{-1} every four years. Clearly, the sites examined here were limited by nitrogen and phosphorus, and applications of these elements to young stands effectively ameliorated these limitations. However, there were differences in the response magnitude that were related to soil texture and drainage. Juvenile fertilizer applications resulted in high stocking levels early in the rotation; this condition should be considered when undertaking juvenile fertilization programs.

Reprinted from *Forests*. Cite as: Albaugh, T.J.; Fox, T.R.; Allen, H.L.; Rubilar, R.A. Juvenile Southern Pine Response to Fertilization Is Influenced by Soil Drainage and Texture. *Forests* **2015**, *6*, 2799-2819.

1. Introduction

Light interception drives the growth of forest plantations [1–3]. In the southeast United States, the primary commercial tree species are *Pinus taeda* L. and *P. elliottii* Engelm. which account for 16% of the world's timber production [4]. Theoretical estimates of light interception and productivity indicate that a maximum leaf area index of approximately 4 [5,6] is possible in southeast US pine stands, but stands often fall short of this level primarily due to nitrogen and phosphorus limitations observed at mid-rotation [7–9]. Empirical trials have identified an average response of 3.5 m^3 ha^{-1} yr^{-1} over an eight-year period to the application of 224 and 28 kg ha^{-1} of elemental nitrogen and phosphorus, respectively, in mid-rotation stands across a wide range of sites [10].

Early in the rotation, nitrogen and phosphorus are typically more readily available from mineralization of the litter layer from the previous rotation. This effect, known as the assart effect [11,12], likely dissipates and nitrogen availability returns to background levels within five years [13]. In studies where the forest floor was manipulated (doubled in some plots, removed in other plots), improved nitrogen availability was found through age 10 [14]. Similarly, silvicultural treatments such as intensive site preparation, vegetation control and fertilization applied at planting may result in large growth gains and increased leaf area through age 10 [15,16]. However, in typical stands, the assart effect and effects from early rotation or time of planting treatments would likely be diminished before a midrotation fertilizer treatment would be applied (8–10 years of age) [17]. Once leaf area levels are reduced through a lack of resources, it may take up to three years to build the crown back up to a high level of leaf area [18]. Given these conditions, it is likely that the productive potential of many sites may not be achieved without the application of additional resources (nitrogen and phosphorus, in this case) prior to midrotation.

The midrotation application of 224 and 28 kg ha^{-1} of elemental nitrogen and phosphorus, respectively, has resulted in a positive growth response in 85% of research trials with a range in response from 0.7 to 7.0 m^3 ha^{-1} yr^{-1} over an eight-year period [19]. However, the growth response reaches a peak between two and four years after treatment application [20]. Whereas the growth response was linearly related to the amount of nitrogen applied (up to 336 kg ha^{-1}), the growth response per unit of applied nitrogen was higher at lower doses [20]. At a conceptual and practical level, applying nutrients when and in the amounts needed by plants has resulted in near 100% uptake of applied nutrients in laboratory and field experiments [21–23]. In this framework there may be benefit to applying nutrients in lower doses more frequently. Consequently, there is uncertainty about the appropriate dose and frequency of application if one were to consider ameliorating potential nutrient limitations in juvenile stands.

Given these circumstances, our interest was in quantifying juvenile pine plantation response to nutrient additions. If a difference in response to treatment was observed across sites, we wanted to identify site-soil variables that would help group the responses patterns. We wanted to determine the shape of the growth response curve especially at nutrient applications greater than 336 kg nitrogen ha^{-1} and how best to apply nutrients (low, frequent doses or larger, less frequent doses). Specifically, we tested the following hypotheses: site would influence juvenile plantation response to nitrogen and phosphorus additions (there would be a site effect), the relationship between volume growth response and applied nitrogen would not be linear (there is a maximum amount of applied nitrogen beyond which growth improvements would be small and other resources would become limiting) and the overall applied nitrogen dose and not the application frequency would determine the growth response (e.g. the growth response from 538 kg ha^{-1} nitrogen would be the same whether the nitrogen was applied in two doses of 269 or four doses of 134 kg ha^{-1} nitrogen).

2. Materials and Methods

Twenty-two sites were selected to represent a range of soil, physiography, geology and drainage conditions in the southeastern United States (Tables 1 and 2). Selected sites were well stocked, growing vigorously and had minimal woody competition. Silvicultural practices used to create these conditions varied by site and included tillage (poorly drained sites were typically bedded), chemical site preparation (typical for upland sites), herbaceous vegetation control (common on most sites) and fertilization (phosphorus was applied on phosphorus deficient sites). Vegetation control was not included in our treatments, at treatment initiation planted pines had already captured or were poised to capture the site in most cases. Stand age at treatment initiation ranged from two to six years. Installation occurred from 1998 to 2003. Selected sites were assigned to a Cooperative Research in Forest Fertilization (CRIFF) program soils group based on their drainage and subsurface soil texture [24] (Table 2). All sites were cutover natural pine stands or plantations. Site 22 was planted to *P. elliottii* and all other sites were planted to *P. taeda*. Prior to treatment initiation, stocking across all sites ranged from 1216 to 2246 trees ha^{-1}, diameter at breast height ranged from 0.9 to 10.2 cm, height ranged from 1.4 to 7.2 m, basal area ranged from 0.1 to 13.5 m^2 ha^{-1} and stem volume ranged from 5.3 to 49.6 m^3 ha^{-1} (Table 3).

Table 1. Location, age and year of study initiation for the sites examined in this study.

Site	County	State	Latitude (decimal degrees)	Longitude (decimal degrees)	Year of Initiation (year)	Age at Initiation (years)
1	Kershaw	SC	34.45	−80.50	2000	4
2	Oglethorpe	GA	33.89	−82.91	2000	6
3	Brunswick	VA	36.68	−77.99	1999	6
4	Berkeley	SC	33.19	−80.19	1999	5
5	Coosa	AL	32.91	−86.38	2002	6
6	Floyd	GA	34.15	−85.38	2001	3
7	Angelina	TX	31.13	−94.46	2003	3
8	Wilkes	GA	33.81	−82.96	2000	3
9	Nassau	FL	30.68	−81.75	1999	5
10	Sabine	LA	31.72	−93.56	1999	5
11	Vernon	LA	31.34	−93.18	2000	6
12	Marengo	AL	32.37	−87.84	2001	3
13	Brantley	GA	31.34	−81.82	1998	3
14	Brantley	GA	31.34	−81.83	1998	2
15	Marion	GA	32.17	−84.63	2000	4
16	Talbot	GA	32.68	−84.74	2003	5
17	Bradley	AR	33.49	−92.13	2000	5
18	Marengo	AL	32.25	−87.55	2000	4
19	Newton	TX	30.48	−93.78	2001	2
20	Montgomery	NC	35.28	−79.94	2003	4
21	Montgomery	MS	32.55	−89.64	2003	6
22	Dixie	FL	29.65	−83.17	2003	3

Table 2. Soils, physiography and geology for the sites examined in this study. Cooperative Research in Forest Fertilization (CRIFF) program soils groups are based on drainage and subsurface texture where groups A and B are poorly drained with clay subsurface, C and G are poorly to well drained with no clay in the subsurface and E and F are well drained with a clay subsurface. The current study soil group combines CRIFF A and B soils (current study group 1), CRIFF C and G soils (current study group 2) and CRIFF E and F soils (current study group 3).

site	CRIFF Soils Group	Current study soil group	Physiographic Province	Soil Series	Drainage	Geologic formation	Soil taxonomy
1	G	2	Sandhills	Blanton	Well	Cretaceous, Upper, Lumbee, Black Creek	Typic Quartzipsamments
2	E	3	Piedmont	Iredell	Moderately well	Diabase Ultramafic	Fine, mixed, active, thermic Oxyaquic Vertic Hapludalfs
3	E	3	Piedmont	Cecil	Well	Biotite Gneiss	Fine, Kaolinitic, Thermic Typic Kanhapludults
4	B	1	Atlantic Coastal Plain	Lynchburg	Somewhat poorly	Quaternary, Pleistocene, Penholoway	Fine-loamy, siliceous, semiactive, thermic Aeric Paleaquults
5	E	3	Piedmont	Louisa	Well	Biotite Gneiss	Loamy, Micaceous, Thermic, Shallow Typic Dystrudepts
6	E	3	Ridge and Valley	Townley	Moderately well	Conasauga Shale	Fine, Mixed, Semiactive, Thermic, Typic Hapludults
7	E	3	Upper Gulf Coastal Plain	Kurth	Moderately well	Tertiary, Eocene, Jackson, Manning	Fine-Loamy, siliceous, semiactive, thermic Oxyaquic Glossudalfs
8	E	3	Piedmont	Appling	Well	Metadacite	Fine, kaolinitic, thermic Typic Kanhapludults
9	A	1	Flatwoods	Meggett	Poorly	Quaternary, Pleistocene, Wicomico	Fine, Mixed, Active, Thermic, Typic Albaqualf
10	E	3	Upper Gulf Coastal Plain	Sacul	Moderately well	Tertiary, Paleocene, Wilcox, Undifferentiated	Fine, mixed, active, thermic aquic hapludults
11	A	1	Lower Gulf Coastal Plain	Mayhew	Somewhat poorly	Tertiary, Miocene, Fleming, Carnahan Bayou	Fine, smectitic, thermic Chromic Dystraquerts

Table 2. *Cont.*

site	CRIFF Soils Group	Current study soil group	Physiographic Province	Soil Series	Drainage	Geologic formation	Soil taxonomy
12	A	1	Upper Gulf Coastal Plain	Lenoir	Somewhat poorly	Quaternary, Holocene, High Terrace	Fine, mixed, semiactive, thermic Aeric Paleaquults
13	C	2	Flatwoods	Seagate	Somewhat poorly	Quaternary, Pleistocene, Penholoway	Sandy over loamy, Siliceous, Active, Thermic Typic Haplohumods
14	C	2	Flatwoods	Pelham	Poorly	Quaternary, Pleistocene, Penholoway	Loamy, Siliceous, Subactive, Thermic Arenic Paleaquults
15	F	3	Lower Gulf Coastal Plain	Troup	Well	Cretaceous, Upper, Navarro, Providence	Loamy, Kaolinitic, thermic Grossarenic Kandiudults
16	E	3	Piedmont	Cecil	Well	Gneiss	Fine, Kaolinitic, Thermic Typic Kanhapludults
17	B	1	Upper Gulf Coastal Plain	Stough	Somewhat poorly	Quaternary, Pleistocene, High Terrace	Coarse, Loamy, Siliceous, Thermic Fragiaquic Paleudults
18	A	1	Upper Gulf Coastal Plain	Brantley	Well	Cretaceous, Upper, Taylor, Ripley	Fine, Mixed active, Thermic Ultic Haplidults
19	B	1	Lower Gulf Coastal Plain	Evadale	Poorly	Quaternary, Pleistocene, Beaumont	Fine Smectitic, Thermic Typic glossaqualfs
20	E	3	Piedmont	Herndon	Well	Carolina Slate	Fine, kaolinitic, thermic Typic Kanhapludults
21	B	1	Upper Gulf Coastal Plain	Shabuta	Well	Tertiary, Eocene, Claiborne, Cook Mountain	Fine, mixed, semiactive, thermic Typic Paleudults
22	C	2	Flatwoods	Sapelo	Somewhat poorly	Tertiary, Eocene, Ocala Limestone	Sandy, siliceous, thermic Ultic Alaquods

Table 3. Tree and stand measurements (diameter at breast height, height, basal area, volume and stocking) and foliar nutrient concentration (nitrogen, phosphorus, potassium, calcium, and magnesium) mean and standard deviation (SE) prior to treatment initiation for the 22 sites where fertilizers were applied in varying doses and frequencies. Diameter at breast height was not measured at Site 14 prior to treatment initiation.

Current study soil group	Site	Diameter Mean (cm)	SE	Height mean (m)	SE	Basal area mean (m² ha⁻¹)	SE	Volume mean (m³ ha⁻¹)	SE	Stocking mean (trees ha⁻¹)	SE	Nitrogen mean (%)	SE	Phosphorus mean (%)	SE	Potassium mean (%)	SE	Calcium mean (%)	SE	Magnesium mean (%)	SE
1	4	10.2	0.1	7.2	0.0	13.5	0.2	49.6	1.8	1,591	18	1.19	0.01	0.14	0.00	0.40	0.01	0.18	0.00	0.10	0.00
1	9	7.7	0.1	5.1	0.1	11.2	0.3	40.6	1.1	2,246	24	1.18	0.02	0.11	0.00	0.39	0.01	0.31	0.01	0.15	0.00
1	11	7.6	0.1	5.9	0.1	8.3	0.2	33.2	1.0	1,731	54	0.95	0.01	0.08	0.00	0.38	0.01	0.17	0.01	0.08	0.00
1	12	4.4	0.1	3.8	0.0	2.4	0.1	13.0	0.2	1,412	17	0.99	0.02	0.11	0.00	0.51	0.01	0.16	0.00	0.07	0.00
1	17	5.7	0.2	3.7	0.0	3.4	0.3	13.6	0.5	1,247	32	1.08	0.01	0.11	0.00	0.51	0.01	0.14	0.00	0.07	0.00
1	18	3.2	0.1	3.0	0.0	1.3	0.0	10.7	0.1	1,416	23	1.27	0.02	0.15	0.00	0.57	0.01	0.22	0.01	0.09	0.00
1	19	1.3	0.0	1.6	0.0	0.2	0.0	6.6	0.3	1,266	29	1.09	0.02	0.11	0.00	0.37	0.01	0.30	0.01	0.09	0.00
1	21	9.4	0.1	6.0	0.1	9.5	0.3	34.3	1.2	1,331	10	1.21	0.02	0.13	0.00	0.58	0.02	0.19	0.01	0.09	0.00
2	1	0.9	0.0	1.4	0.0	0.1	0.0	5.3	0.4	1,441	23	0.96	0.01	0.11	0.00	0.35	0.01	0.21	0.01	0.06	0.00
2	13	3.9	0.1	3.2	0.0	2.2	0.1	14.0	0.3	1,699	34	1.06	0.02	0.13	0.00	0.29	0.01	0.17	0.01	0.11	0.00
2	14			2.1	0.0					1,749	20	1.05	0.02	0.13	0.00	0.29	0.02	0.31	0.01	0.13	0.00
2	22	3.8	0.1	2.9	0.0	1.9	0.1	12.2	0.2	1,576	31	0.64	0.01	0.06	0.00	0.26	0.01	0.21	0.01	0.10	0.00
3	2	5.9	0.1	3.5	0.0	4.7	0.2	17.8	0.5	1,635	36	1.14	0.01	0.11	0.00	0.44	0.01	0.23	0.01	0.12	0.00
3	3	7.3	0.1	4.8	0.0	7.3	0.2	26.4	0.5	1,678	24	1.20	0.01	0.11	0.00	0.53	0.01	0.19	0.01	0.07	0.00
3	5	8.1	0.2	5.8	0.1	7.9	0.3	29.9	0.9	1,458	19	1.22	0.02	0.11	0.00	0.43	0.01	0.20	0.01	0.09	0.00
3	6	3.5	0.0	3.0	0.0	1.7	0.0	12.6	0.1	1,655	13	1.01	0.02	0.11	0.00	0.49	0.01	0.17	0.01	0.05	0.00
3	7	4.2	0.1	3.1	0.1	1.9	0.1	10.7	0.3	1,268	29	1.15	0.03	0.12	0.01	0.57	0.05	0.26	0.02	0.09	0.01
3	8	1.5	0.0	2.1	0.0	0.4	0.0	11.1	0.2	1,817	23	1.26	0.01	0.12	0.00	0.54	0.01	0.25	0.01	0.10	0.00
3	10	7.4	0.1	4.9	0.0	9.4	0.3	34.3	0.8	2,125	42	1.30	0.02	0.10	0.00	0.59	0.01	0.14	0.00	0.09	0.00
3	15	2.7	0.0	2.7	0.0	1.2	0.0	12.3	0.1	1,725	14	1.13	0.03	0.13	0.00	0.42	0.02	0.21	0.01	0.09	0.00
3	16	8.2	0.1	5.8	0.0	7.2	0.1	27.0	0.5	1,315	21	1.21	0.01	0.11	0.00	0.42	0.01	0.29	0.00	0.12	0.00
3	20	5.3	0.1	3.3	0.0	2.9	0.1	12.0	0.3	1,216	25	1.32	0.02	0.12	0.00	0.59	0.01	0.16	0.01	0.06	0.00
Adequate nutrient concentration levels from [25]												1.20		0.12		0.35		0.15		0.08	

2.1. Experimental Design

We installed a randomized complete block design with two or four replications at each site. Sites were blocked on height, basal area and stocking prior to treatment initiation to ensure homogeneity. Plots within a block had less than 10% difference for these variables. Measurement plots varied in size from 0.025 to 0.081 ha (0.042 ha average) and were surrounded by a 12-m treated buffer such that the average treated plot size was 0.18 ha. Treatments were a combination of application frequency and nutrient (nitrogen and phosphorus) dose (Table 4). Application frequency was every 1, 2, 4, or 6 years. All nutrient applications mentioned in this document are elemental rates. Nitrogen was applied at 0, 67, 134, 202, 269 kg nitrogen ha^{-1}. Phosphorus was added with nitrogen at amounts 0.1 times the nitrogen rate. Nutrients were added as urea, diammonium phosphate, triple super phosphate, coated urea fertilizer, and nitrogen and phosphorus blends. The application frequencies and nutrient doses were selected to span a range of nutrient applications such that the same total dose at a given time would be achieved from different combinations of nutrient dose and application frequency. For example, applying 67 kg nitrogen ha^{-1} every year for eight years resulted in a cumulative dose of 538 kg nitrogen ha^{-1}. This same cumulative dose was also achieved by applying 134 kg nitrogen ha^{-1} every two years or 269 kg nitrogen ha^{-1} every four years over an eight-year period (Table 4).

Table 4. Nitrogen (N) applications completed at the 22 study sites. The 106 treatment was only applied at Sites 3 and 4. Phosphorus was applied at 0.1 times the nitrogen rate. Other elements were added when foliar nutrient analysis indicated a limitation.

Treatment code	Dose of elemental N applied each time (kg N ha^{-1})	Frequency of application (years)	Cumulative N dose 8 years after initiation (kg N ha^{-1})
0	0	0	0
106	67	1	538
206	67	2	269
212	134	2	538
218	202	2	806
412	134	4	269
418	202	4	403
424	269	4	538
624	269	6	538

Foliage samples were collected in each plot prior to treatment initiation (reported here) at all sites and every year thereafter (data not shown). Samples were composited by plot and analyzed using a CHN analyzer (CE Instruments NC2100 elemental analyzer) for nitrogen (CE Instruments, 1997), and a nitric acid digest and ICP (Varian Liberty II ICO-AES) analysis was used for phosphorus, potassium, calcium, magnesium, manganese, boron, copper, sulfur, and zinc (Huang and Schulte, 1985). These data were used to monitor the nutrient status of elements other than those normally applied (nitrogen and phosphorus) to determine if other elements might limit growth. The goal of applying additional nutrients was to insure that nutrient imbalances were not generated as a result of the nitrogen and phosphorus applications. When nutrient concentration levels were near or below the recommended ranges [25], additional nutrients were added at the next application time for nitrogen and phosphorus. No additional applications were made at sites 10 and 16. Boron was added as borate, solubor or in a blend at all other sites at 0.005 times the nitrogen rate. Potassium, as potassium chloride, was added at sites 17, 21, and 22 at 0.40 times the nitrogen rate. At site 9, manganese was applied as manganese sulfate at 0.1 times the nitrogen rate. Sulfur was added at site 15 as sulfur coated urea at 0.05 times the nitrogen rate. Potassium, sulfur, magnesium and manganese were added to sites 13 and 14 at 0.40, 0.05, 0.04 and 0.1 times the nitrogen rate, respectively, as potassium chloride, manganese sulfate and blends.

Diameter at breast height (1.3 m) and tree height were measured prior to treatment initiation and annually thereafter. Individual tree volume was calculated by converting a published volume equation for unthinned trees [26] to metric units as

$$V = ((0.21949 + (0.0012103044 \times D \times D \times H)) \times 0.02831685) \tag{1}$$

where V is individual tree volume (m^3 tree^{-1}), D is diameter at breast height (cm) and H is height (m). Individual tree volume and basal area were summed by plot and scaled to a hectare basis. In this case we are assuming that treatment did not influence taper. Treatment response at a given time period was calculated as the mean of treatment growth minus control growth for all blocks at a site. Relative treatment response was the treatment response divided by the control growth.

2.2. Statistical Analyses

PROC MIXED [27] was used to examine our first hypothesis regarding the treatment by site effects eight years after treatment initiation for the volume growth response. Treatment code (Table 4) was a fixed effect and treated as a categorical variable, whereas block by site was treated as a random effect for this analysis. A second analysis examined all studies together eight years after treatment. PROC MIXED was used with treatment code as a fixed effect categorical variable; site, block by site and site by treatment were treated as random effects, and initial basal area was used as a covariate. When examining all studies together, the following treatments were included: 0, 206, 212, 218, 412, 418, 424, and 624. The 106 treatment was excluded because it was installed at only two sites. This analysis was repeated with the addition of soil group, where treatment, soil group and treatment by soil group were fixed effects and random effects were as before. Soil groups were determined by combining sites with similar CRIFF soils groups. CRIFF groups A and B (poorly drained soils with clay subsoil), groups C, D and G (soils with spodic horizons or no clay subsoil) and groups E and F (well drained soils with clay subsoil) were assigned as soil groups 1, 2, and 3, respectively, for this analysis. Multiple treatment comparisons within each soil group were completed using the Tukey adjustment in PROC MIXED. Residuals were examined for all analyses and no biases were detected.

Our second hypothesis examined the height, diameter, basal area and volume growth increment rate response using 2, 4 and 8 years since treatment initiation data. Although it did not receive any fertilizer applications, the control treatment was assigned a cumulative dose of 1.12 kg ha^{-1} of nitrogen to avoid defining the function as zero. The soil groups were coded using indicator variables such that the model was fit with data from soil group 3 (the group with the most sites) and then adjusted for soil groups 1 and 2. Random effects on the equation coefficients were examined using PROC NLMIXED. The final response function fitted was an exponential function in the form of:

$$y = (b_0 + b_{01}S_1 + b_{02}S_2)e^{(b_1 + b_{11}S_1 + b_{12}S_2)/N} + \varepsilon \tag{2}$$

where y is height, diameter, basal area or volume growth response, N is the cumulative amount of nitrogen applied at that time, b_0 (asymptote response) and b_1 (steepness or shape of response) are coefficients, S_1 is an indicator variable that equals 1 for soil group 1 and S_2 is an indicator variable that equals 1 for soil group 2, b_{01} and b_{02} were adjustments to b_0 for soil group 1 and 2, respectively, b_{11} and b_{12} were adjustments to b_1 for soil group 1 and 2, respectively, and ε is the residual random error (iid $N(0,\sigma2)$). The SAS macro %NLINMIX [28] was used to determine the coefficients of the model. This macro uses nonlinear mixed models to account for repeated measures (the same plots contributed data in multiple years).

Our third hypothesis examined whether the frequency of application had an effect on growth response. We used the same model as that used for the second analysis of our first hypothesis where all studies were examined together eight years after treatment initiation. We included single degree of freedom contrasts, where treatments were compared with the same amount of total dose but different application frequencies used to achieve the specific dose. We compared the 212 (134 kg ha^{-1} nitrogen applied every two years) to the 424 (269 kg ha^{-1} nitrogen applied every four years) treatment and the 206 (67 kg ha^{-1} nitrogen applied every two years) to the 412 (134 kg ha^{-1} nitrogen applied every four years) treatment. We did not examine the 624 treatment in the 212 and 424 comparison because there were only two years for response to occur after the second application in year 6 for the 624 treatment. All statistical tests were evaluated with alpha equal to 0.05.

3. Results

Eight of the 22 sites had significant treatment effects when examining treatment effects at individual sites (Table 5). For the individual responsive sites, the average volume growth response over the control for the 424 treatment was 8.4 m^3 ha^{-1} yr^{-1} (87%) (Table 6). Corresponding increases in diameter, height and basal area were 0.5 cm yr^{-1} (47%), 0.2 m yr^{-1} (23%) and 1.2 m^2 ha^{-1} yr^{-1} (60%), respectively. When examining treatment effects across site, site ($p < 0.001$), site by treatment ($p = 0.002$), initial basal area ($p = 0.004$) and treatment ($p < 0.001$) were significant.

Table 5. Summary of statistical significance (*p* values) for annual volume growth eight years after study initiation for the 22 sites where nitrogen and phosphorus were added at different rates and frequencies in *Pinus* stands in the southeast United States.

Site	CRIFF Soils Group	Current study soil group	*p* value
1	G	2	0.000
2	E	3	0.355
3	E	3	0.219
4	B	1	0.095
5	E	3	0.368
6	E	3	0.261
7	E	3	0.464
8	E	3	0.286
9	A	1	0.030
10	E	3	0.939
11	A	1	0.115
12	A	1	0.631
13	C	2	0.012
14	C	2	0.010
15	F	3	0.497
16	E	3	0.002
17	B	1	0.096
18	A	1	0.361
19	B	1	0.050
20	E	3	0.148
21	B	1	0.027
22	C	2	0.007

Table 6. Control treatment diameter at breast height (diameter), height, basal area, volume, green weight and stocking growth (Growth), 424 treatment absolute response (Response = treated minus control), and percentage response (%) eight years after treatment initiation for 22 sites in the southeast United States. The control treatment received no fertilization during this time and the 424 treatment received two applications of 269 kg nitrogen ha^{-1} and 27 kg phosphorus ha^{-1}.

Current study soil group	Site	Diameter			Height			Basal area			Volume			Green weight			Stocking		
		Growth cm yr^{-1}	Response cm yr^{-1}	%	Growth m yr^{-1}	Response m yr^{-1}	%	Growth m^2 ha^{-1} yr^{-1}	Response m^2 ha^{-1} yr^{-1}	%	Growth m^3 ha^{-1} yr^{-1}	Response m^3 ha^{-1} yr^{-1}	%	Growth Mg ha^{-1} yr^{-1}	Response Mg ha^{-1} yr^{-1}	%	Growth trees ha^{-1} yr^{-1}	Response trees ha^{-1} yr^{-1}	%
1	4	1.4	0.1	9	1.3	-0.1	-4	0.1	1.3	2516	7.7	7.8	106	8.4	7.1	89	-144.7	25.2	-19
1	9	0.6	0.8	122	1.2	0.1	11	2.4	1.1	47	21.9	2.9	13	20.7	3.0	14	-12.0	-27.6	272
1	11	0.8	0.4	48	0.8	0.3	32	2.3	1.2	52	15.6	9.0	59	14.7	8.8	61	0.0	-38.3	
1	12	1.7	-0.3	-18	1.4	-0.3	-17	4.0	-0.2	-5	29.4	-2.2	-7	27.7	-2.0	-7	-7.0	-6.6	104
1	17	1.5	0.1	4	1.1	0.1	9	3.8	0.3	7	22.1	6.4	31	20.8	6.0	31	-3.9	-2.3	67
1	18	1.7	0.1	7	1.5	-0.2	-15	3.7	0.0	-1	25.0	-2.5	-10	23.6	-2.1	-9	-8.7	-23.8	280
1	19	1.8	0.2	14	1.2	0.2	15	2.8	0.7	26	14.1	6.8	55	13.1	6.7	59	-2.2	-25.2	1143
1	21	1.1	0.6	53	1.2	0.1	6	3.2	1.1	33	27.6	2.5	9	26.0	2.4	9	-7.7	-5.2	23
2	1	1.4	0.4	31	0.9	0.2	25	1.9	1.2	60	7.7	7.2	84	6.8	6.2	80	-15.0	8.6	-40
2	13	1.0	0.6	63	1.0	0.4	46	2.0	1.6	80	10.8	14.2	131	10.2	13.5	132	-8.0	-22.2	258
2	14	1.4	0.2	17	1.1	0.4	38	2.2	1.8	95	10.2	16.0	170	9.8	14.9	165	-19.2	17.5	-93
2	22	0.9	0.5	55	0.9	0.4	44	1.4	1.5	106	6.8	15.7	223	6.6	14.8	216	-32.4	0.6	-1
3	2	1.2	0.5	42	0.9	0.2	20	3.5	0.1	4	18.4	0.3	2	17.3	0.1	0	-3.5	-18.2	528
3	3	1.2	0.2	16	1.2	-0.1	-6	4.0	0.5	12	28.8	0.7	2	27.1	0.8	3	-2.5	-16.6	369
3	5	0.9	0.5	49	1.0	0.2	18	2.6	-0.7	-21	19.7	-2.5	-9	18.6	-1.8	-7	-3.0	-68.3	2389
3	6	1.6	-0.1	-7	1.1	0.0	3	4.0	0.3	7	21.0	6.6	31	19.8	6.2	31	-1.6	-3.3	1
3	7	1.7	0.1	5	1.2	-0.1	-6	3.9	0.6	15	23.5	1.5	7	22.1	1.4	7	-2.5	-4.2	164
3	8	1.7	0.0	2	1.1	0.0	3	3.9	0.1	4	19.6	2.2	14	18.4	2.2	15	-7.3	-5.3	79
3	10	1.1	0.0	5	1.0	0.2	23	3.9	0.0	1	26.7	3.5	14	25.3	3.5	14	-30.1	-24.0	84
3	15	1.4	0.4	27	1.0	0.3	32	3.0	0.6	20	14.9	6.8	42	14.1	6.5	42	-9.9	-5.6	56
3	16	1.2	0.3	22	1.3	0.0	0	3.2	0.9	29	27.3	2.1	10	25.8	1.9	10	-13.2	8.7	-66
3	20	1.7	0.3	20	1.2	0.0	4	4.1	0.0	0	25.6	-1.5	-7	24.1	-1.4	-7	-4.2	-6.8	-8

When including soil group in the treatment effects across site analysis, initial basal area ($p \leq$ 0.001), treatment ($p \leq 0.001$), and treatment by soil group were significant. The average control volume growth eight years after treatment initiation was 19.7, 10.7 and 22.6 m^3 ha^{-1} yr^{-1} for soil groups 1, 2 and 3, respectively (Figure 1). Average treated growth was 22.6, 21.8, and 25.6 m^3 ha^{-1} yr^{-1} for soil groups 1, 2, and 3, respectively. Means separation tests indicated that the control tre... 8, 41

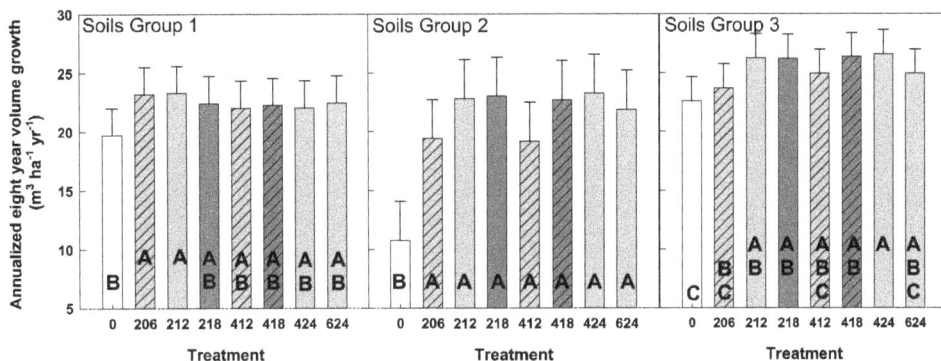

Figure 1. Eight-year average volume growth in soil groups 1, 2 and 3 for eight treatments where nitrogen was applied at different rates and frequencies across 22 sites in the southeastern United States. Different letters indicate significant differences in treatment means within a soil group. Error bars are one standard error.

The asymptote (b$_0$) for all rate response functions was about the same for soil groups 1 and 3 (Table 7 and Figure 2A–D). The asymptote for soils group 2 was generally much greater than those for soil groups 1 and 3. For example, over the range of data, volume response for soil groups 1 and 3 reached an asymptote at about 3.5 m^3 ha^{-1} yr^{-1} whereas soil group 2 sites achieved a volume response up to 12 m^3 ha^{-1} yr^{-1} (Figure 2C). In contrast, the shape parameter (b$_1$) was significantly different for soil groups 1 and 3 for volume and basal area. The volume and basal area growth response increased much more rapidly in soil group 1 than in soil group 3 for applied nitrogen levels less than 300 kg ha^{-1}. For example, at a cumulative applied dose of 100 kg ha^{-1} nitrogen, the estimated eight-year volume response reached 76% and 31% of the volume response predicted at 800 kg nitrogen ha^{-1} for soil groups 1 and 3, respectively (Figure 2C). The shape parameter was not different for soil groups 2 and 3. There was little additional growth response for any of the growth variables for soil groups 1 and 3 sites after a cumulative dose of 300 kg ha^{-1} of nitrogen had been applied. For soil group 2, the growth response continued to increase through the range of applied nitrogen (up to 800 kg ha^{-1}) for height and volume growth; however, the rate of increase in the growth response was reduced at higher nitrogen doses. The response models for basal area did not converge when including cumulative nitrogen doses greater than 400 kg ha^{-1}. Consequently, for the basal area response model presented here, data where the cumulative nitrogen dose was greater than 400 kg ha^{-1} were excluded from the analysis.

Table 7. Parameter estimates, standard error, t values, lower and upper 95% confidence intervals (CI) for the height, diameter, volume and basal area growth response to increasing nitrogen dose functions. b_0 is the asymptote for soil group 3, b_{01} and b_{02} are adjustments to b_0 for soil groups 1 and 2, respectively. b_1 is the shape parameter for soil group 3, b_{11} and b_{12} are adjustments to b_1 for soil groups 1 and 2, respectively.

Parameter	Estimate	Standard error	$Pr > t$	Lower 95% CI	Upper 95% CI
			Height		
b_0	0.066	0.042	0.120	−0.017	0.149
b_{01}	−0.014	0.054	0.799	−0.120	0.092
b_{02}	0.323	0.064	<0.001	0.197	0.450
b_1	−350.1	174.0	0.045	−692.4	−8.521
b_{11}	337.1	177.8	0.059	−12.25	686.4
b_{12}	293.2	174.3	0.093	−49.30	635.7
			Diameter		
b_0	0.259	0.041	<0.001	0.178	0.340
b_{01}	0.032	0.060	0.602	−0.087	0.150
b_{02}	0.434	0.071	<0.001	0.295	0.573
b_1	−45.01	21.63	0.038	−87.54	−2.506
b_{11}	25.75	28.01	0.358	−29.27	80.78
b_{12}	39.58	22.58	0.080	−4.798	83.95
			Volume		
b_0	4.350	1.066	<0.001	2.256	6.445
b_{01}	−0.730	1.560	0.643	−3.788	2.343
b_{02}	11.89	2.052	<0.001	7.862	15.92
b_1	−134.1	24.97	<0.001	−183.1	−85.03
b_{11}	103.6	29.70	<0.001	45.31	162.0
b_{12}	−55.90	28.88	0.053	−112.7	0.815
			Basal area		
b_0	1.033	0.196	<0.001	0.648	1.419
b_{01}	−0.183	0.273	0.502	−0.721	0.354
b_{02}	1.346	0.349	<0.001	0.661	2.032
b_1	−107.1	25.74	<0.001	−157.7	−56.54
b_{11}	80.04	31.06	0.010	18.97	141.1
b_{12}	34.99	28.89	0.227	−21.80	91.78

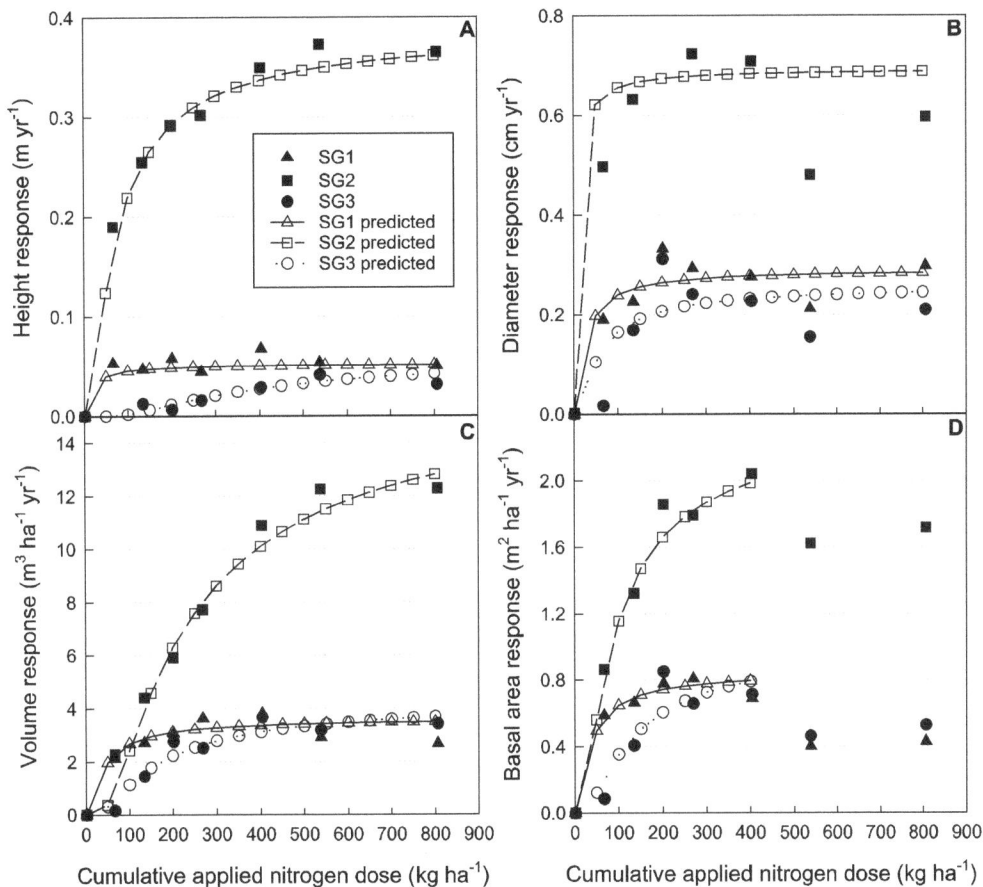

SG3) where nitrogen was applied at different rates and frequencies across 22 sites in the southeastern United States.

There were no significant differences observed in the frequency of application comparisons as long as the total applied dose was the same. The response from treatment 212, where 134 kg ha^{-1} nitrogen was applied every two years (average of 5.1 m^3 ha^{-1} yr^{-1} for eight years), was no different than the response from treatment 424, where 268 kg ha^{-1} nitrogen was applied every four years ($p = 0.687$) (average of 4.9 m^3 ha^{-1} yr^{-1} for eight years) (Figure 3A). A similar result was observed in the comparison between treatments 206 and 412 ($p = 0.555$). The average responses from treatment 206, where 67 kg ha^{-1} nitrogen was applied every two years, and treatment 412, where 134 kg ha^{-1} nitrogen was applied every four years, were 3.1 and 3.4 m^3 ha^{-1} yr^{-1} for eight years, respectively (Figure 3B).

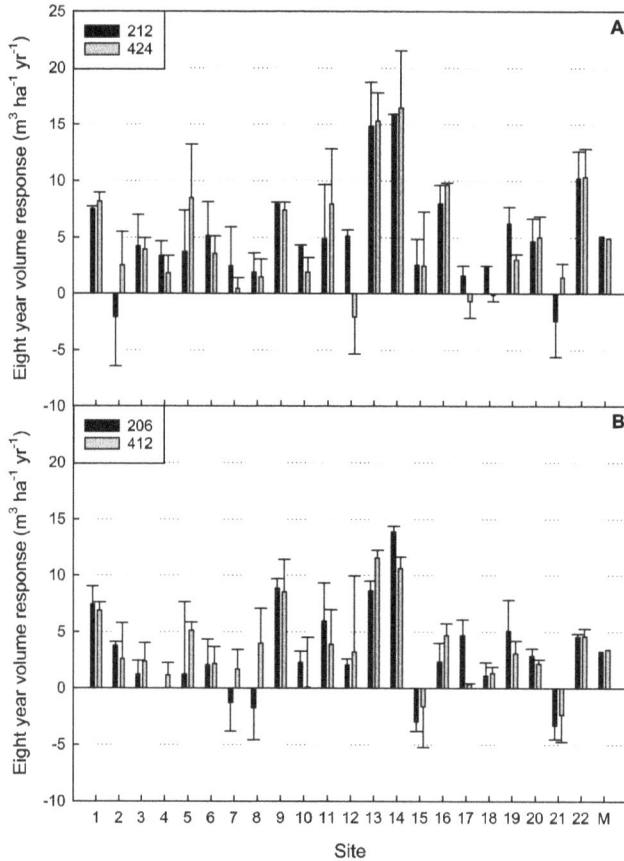

Figure 3. Comparison of treatments where the same cumulative nitrogen dose was applied at different rates and frequency of application across 22 sites in the southeastern United States and the mean (M) across all sites. Panel A shows the comparison where 538 kg ha^{-1} of nitrogen was applied as four applications of 134 kg ha^{-1} every two years or as two applications of 269 kg ha^{-1} every four years. Panel B shows the comparison where 269 kg ha^{-1} of nitrogen was applied as four applications of 67 kg ha^{-1} every two years or as two applications of 134 kg ha^{-1} every four years. Error bars are one standard error.

4. Discussion

We accepted our first hypothesis because site was a significant factor in the analyses examining treatment effects for individual sites, where eight sites had a significant treatment effect (Table 5), and in the across site analysis where site was a significant factor. Soil group was also a significant factor in the across site analysis. Grouping sites on soils is a useful tool for managers interested in determining which stands are the best candidates for adding resources [24]. Most of the studies had two replications, which reduces the ability to detect differences due to treatment at individual sites.

However, use of the two-replication studies allowed more installations across the landscape, which improved our ability to examine the second and third hypotheses at the regional scale. Regardless, the growth responses at the soil group 2 sites were large, such that all of the soil group 2 sites had significant responses to treatment when examined as individual studies, and the soil group 2 sites had a larger response to treatment than the other soil groups (Figure 2). From a management perspective, while the response to fertilization was dramatic at the soil group 2 sites, these sites required more applied nutrients to achieve the same level of absolute growth than the sites in soil groups 1 and 3 (Figures 1 and 2C). This study provides information managers can use to determine which stands will likely respond to treatment from a biological perspective while having an understanding of what it will take to achieve that biological potential from an economic perspective.

We accepted our second hypothesis that the volume growth rate response curve was not linear (Figure 2). Volume growth response reached an asymptote for soil groups 1 and 3 at a cumulative dose between 300–400 kg ha^{-1} of applied nitrogen. The response curve for soil group 2 continued to increase through the maximum cumulative applied nitrogen dose of 800 kg ha^{-1}. The rate of growth increase was greater for soil group 1 than for soil group 3 (significant shape parameter b_{11} in Table 7). Soil group 1 sites are poorly drained and may have been able to capitalize on available water to take up the newly available nutrients more readily than those in soil group 3, which were located on well-drained sites. The two-year response data used to develop the response function may underestimate the true response because the trees only had a short time to respond, although this effect would likely be experienced across all sites. However, without the two-year response data, the rate response analysis would have no low doses and the model becomes insensitive to changes in the b_1 coefficient.

The observed growth responses from soil groups 1 and 3 were in the same range as previous reports where response to nitrogen was linear through 336 kg ha^{-1} of applied nitrogen [10,19]. The large response observed from soil group 2 was in the same range as the data from the literature for an equivalent amount of nitrogen. In our study, the application of 224 kg ha^{-1} nitrogen resulted in a response of 7 m^3 ha^{-1} yr^{-1} for eight years, and Rojas [19] found studies with responses of approximately 7.5 m^3 ha^{-1} yr^{-1} over eight years from an equivalent nitrogen application. The response at the soil group 2 sites was impressive; however, the control plots at these site were growing relatively slowly at 10.7 m^3 ha^{-1} yr^{-1}. Average treated growth was (22.6, 21.8 and 25.6 m^3 ha^{-1} yr^{-1} for soil groups 1, 2, and 3, respectively) lower than modeled potential productivity of loblolly pine in the southeast United States (>30 m^3 ha^{-1} yr^{-1}) [5] and that observed in individual studies where treatments were applied with the intent to eliminate all resource limitations (up to 35 m^3 ha^{-1} yr^{-1}) (e.g., [8,29,30]). Clearly, the sites in our study were nitrogen and phosphorus limited. After these limitations were ameliorated with our treatments, other resource limitations such as water, other nutrients, light and space would influence productivity, following Liebig's Law of the Minimum.

We accepted our third hypothesis that the overall applied nitrogen dose and not application frequency determined the growth response. None of our treatments were single dose applications, however, our data are in agreement with a study where the same dose of nitrogen was applied in either one or two applications two years apart, with both treatments providing similar responses in southern pine [31]. As mentioned previously, our responses were in the same range as other studies where applications up to 336 kg nitrogen ha^{-1} were applied in single doses [10,19]. Our results are consistent with those found in *Eucalyptus grandis* where a similar range of doses (60 to 240 kg ha^{-1} of applied nitrogen) was applied at 1, 2, 3, and 6 application frequencies to achieve the same cumulative dose, and only one of eleven tests indicated a difference in growth response at the end of the study period (3 years in this case) [32]. Reductions in fertilizer applications on a regional scale are likely a result of high material prices [33]. At the same time, urea prices (the most common form of nitrogen used in forestry application) can fluctuate by large amounts in relatively short periods of time [34]. Consequently, this result provides managers more flexibility in planning for nutrient additions.

The age at study initiation ranged from two to six years old. For stands at the low end of this range, the assart effect may still have been providing nutrients, and yet these young stands were responsive to additional nutrients applied in this study. The assart effect is generally applicable to nitrogen availability. However, sites that would likely be phosphorus limited (poorly drained coastal plain sites) received phosphorus at planting and competing vegetation was relatively low at study initiation. Consequently, the rapid early response in our studies was likely from amelioration of nitrogen limitations. Even if resources continued to be available from the assart effect, the magnitude of resource availability from this effect would be relatively small compared with the application rates in this study. As noted in previous studies, loblolly pine is a very plastic species and responds well to large amounts of nutrient inputs [8,9,30,35]. This rapid growth early in the rotation (typical rotation length of 20–25 years) resulted in a situation where some stands reached stocking levels that would indicate the need for a thin. Based on Reineke's [36] maximum stand density index of 450 for loblolly, and Drew and Flewelling's [37] estimate that density dependent mortality begins at 50%–55% of maximum stand density index, sites with basal areas greater than approximately 23 m^2 ha^{-1} would begin to have intraspecific competition. All of the 424 treatment stands had basal areas greater than this amount and some sites (3, 10, and 21) had basal areas greater than 40 m^2 ha^{-1} eight years after treatment initiation. If juvenile fertilization is included in a silvicultural prescription, then it is likely that thinning may need to be considered earlier than what might be considered normal. Density-dependent mortality may have resulted in the failure of the basal area rate response model to converge for cumulative nitrogen doses greater than 400 kg ha^{-1} of nitrogen. In these cases, mortality may have reduced basal area responses below what would be expected if over-stocking had not occurred.

5. Conclusions

Juvenile southern pine stands will respond to fertilizer. Sites in this study were limited by nitrogen and phosphorus and they responded to nutrient applications at young ages. Soil texture and drainage were useful in categorizing the sites for potential response. Soils with clay subsoil reached an asymptotic response between 300 and 400 kg ha^{-1} of applied nitrogen. Cumulative nitrogen dose, and not application frequency or dose, determined the response. When undertaking juvenile fertilization, the silvicultural prescription should include consideration for rapid stand development.

Acknowledgments

We appreciate support from Forest Productivity Cooperative members for their role in the establishment and management of the trials central to this publication. We gratefully acknowledge the support provided by the Department of Forest Resources and Environmental Conservation at Virginia Polytechnic Institute and State University, the Departamento de Silvicultura, Facultad de Ciencias Forestales, Universidad de Concepción and the Department of Forestry and Environmental Resources at North Carolina State University. Funding for this work was provided in part by the Virginia Agricultural Experiment Station and the McIntire-Stennis Program of the National Institute of Food and Agriculture, U.S. Department of Agriculture. We thank all Forest Productivity Cooperative staff and students involved in the installation and monitoring of these sites and analysis of interim data. The use of trade names in this paper does not imply endorsement by the associated agencies of the products named, nor criticism of similar ones not mentioned.

Author Contributions

All authors contributed to the design, installation, and maintenance of the study. Timothy J. Albaugh, Thomas R. Fox and H. Lee Allen participated in data analysis. Timothy J. Albaugh wrote the paper.

Conflicts of Interest

The authors declare no conflict of interest.

References

1. Linder, S. Responses to water and nutrients in coniferous ecosystems. In *Potentials and limitations of Ecosystem Analysis*; Schulze, E.D., Zwolfer, H., Eds.; Springer-Verlag: Berlin, Germany, 1987; pp. 180–202.
2. Cannell, M.G.R. Physiological basis of wood production: A review. *Scand. J. For. Res.* **1989**, *4*, 459–490.
3. Landsberg, J.J.; Sands, P.J. *Physiological Ecology of Forest Production Principles, Processes and Models*; Academic Press: London, UK, 2011.
4. Wear, D.N.; Greis, J.G. *Southern Forest Resource Assessment*; General Technical Report SRS-53; U.F. Service: Asheville, NC, USA, 2002; pp. 1–635.
5. Sampson, D.A.; Allen, H.L. Regional influences of soil available water-holding capacity and climate, and leaf area index on simulated loblolly pine productivity. *For. Ecol. Manag.* **1999**, *124*, 1–12.
6. Landsberg, J.J.; Johnsen, K.H.; Albaugh, T.J.; Allen, H.L.; McKeand, S.E. Applying 3-PG, a simple process-based model designed to produce practical results, to data from loblolly pine experiments. *For. Sci.* **2001**, *47*, 43–51.
7. Vose, J.M.; Allen, H.L. Leaf-area, stemwood growth, and nutrition relationships in loblolly-pine. *For. Sci.* **1988**, *34*, 547–563.

8. Albaugh, T.J.; Allen, H.L.; Dougherty, P.M.; Kress, L.W.; King, J.S. Leaf area and above- and belowground growth responses of loblolly pine to nutrient and water additions. *For. Sci.* **1998**, *44*, 317–328.

9. Colbert, S.R.; Jokela, E.J.; Neary, D.G. Effects of annual fertilization and sustained weed-control on dry-matter partitioning, leaf-area, and growth efficiency of juvenile loblolly and slash pine. *For. Sci.* **1990**, *36*, 995–1014.

10. Fox, T.R.; Allen, H.L.; Albaugh,T.J.; Rubilar, R.A.; Carlson, C.A. Tree nutrition and forest fertilization of pine plantations in the southern United States. *South. J. Appl. For.* **2007**, *31*, 5–11.

11. Kimmins, J.P. Biogeochemistry, cycling of nutrients in ecosystems. In *Forest Ecology, a Foundation for Sustainable Management*, 2nd edition; Kimmins, J.P., Ed.; Prentice Hall: Saddle River, NJ, USA, 1997; pp. 71–129.

12. Tamm, C.O. Determination of nutrient requirements of forest stands. *Int. Rev. For. Res.* **1964**, *1*, 115–170.

13. Vitousek, P.M.; Andariese, S.W.; Matson, P.A.; Morris, L.A.; Sanford, R.L. Effects of harvest intensity, site preparation, and herbicide use on soil nitrogen transformations in a young loblolly pine plantation. *For. Ecol. Manag.* **1992**, *49*, 277–292.

14. Zerpa, J.L.; Allen, H.L.; Campbell, R.G.; Phelan, J.; Duzan, H.W., Jr. Influence of variable organic matter retention on nutrient availability in a 10-year-old loblolly pine plantation. *For. Ecol. Manag.* **2010**, *259*, 1480–1489.

15. Nilsson, U.; Allen, H.L. Short- and long-term effects of site preparation, fertilization and vegetation control on growth and stand development of planted loblolly pine. *For. Ecol. Manag.* **2003**, *175*, 367–377.

16. Leggett, Z.; Kelting, D.L. Fertilization effects on carbon pools in loblolly pine plantations on two upland sites. *Soil Sci. Soc. Am. J.* **2006**, *70*, 279–286.

17. Allen, H.L. Manipulating loblolly pine productivity with early cultural treatment. In *Sustained Productivity of Forest Soils*; Gessel, S.P., Lacate, D.S., Weetman, G.F., Powers, R.F., Eds.; University of British Columbia: Vancouver, BC, Canada, 1990; pp. 301–317.

18. Albaugh, T.J.; Allen, H.L.; Fox, T.R. Individual tree crown and stand development in Pinus taeda under different fertilization and irrigation regimes. *For. Ecol. Manag.* **2006**, *234*, 10–23.

19. Rojas, J.C. Factors influencing responses of loblolly pine stands to fertilization. Ph.D. Thesis, Department of Forestry, North Carolina State University, Raleigh, NC, USA, 2005.

20. Hynynen, J.; Burkhart, H.D.; Allen, H.L. Modeling tree growth in fertilized midrotation loblolly pine plantations. *For. Ecol. Manag.* **1998**, *107*, 213–229.

21. Ingestad, T. Towards optimum nutrition. *Ambio.* **1974**, *3*, 49–54.

22. Ingestad, T. Nitrogen and plant growth; maximum efficiency of nitrogen fertilizers. *Ambio.* **1977**, *6*, 146–151.

23. Linder, S. Foliar analysis for detecting and correcting nutrient imbalances in Norway spruce. *Ecol. Bull.* **1995**, *44*, 178–190.

24. Jokela, E.J.; Long, A.J. *Using Soils to Guide Fertilizer Recommendations for Southern Pines*; Circular 1230: Gainesville, FL, USA, 2012; p. 12.

25. Albaugh, J.M.; Blevins, L.L.; Allen, H.L.; Albaugh, T.J.; Fox, T.R.; Stape, J.L.; Rubilar, R.A. Characterization of foliar macro- and micronutrient concentrations and ratios in loblolly pine plantations in the Southeastern United States. *South. J. Appl. For.* **2010**, *34*, 53–64.

26. Tasissa, G.; Burkhart, H.D.; Amateis, R.L. Volume and taper equations for thinned and unthinned loblolly pine trees in cutover, site-prepared plantations. *South. J. Appl. For.* **1997**, *21*, 146–152.

27. SAS. *SAS Version 9.1 TS*; SAS Institute, Inc.: Cary, NC, USA, 2002.

28. SAS, %NLINMIX macro to fit nonlinear models. Available online: http://support.sas.com/kb/25/032.html (accessed on 23 June 2015).

29. Borders, B.E.; Bailey, R.L. Loblolly pine—Pushing the limits of growth. *South. J. Appl. For.* **2001**, *25*, 69–74.

30. Will, R.E.; Narahari, N.V.; Shiver, B.D.; Teskey, R.O. Effects of planting density on canopy dynamics and stem growth for intensively managed loblolly pine stands. *For. Ecol. Manag.* **2005**, *205*, 29–41.

31. Jokela, E.J.; Stearns-Smith, S.C. Fertilization of established southern pine stands: Effects of single and split nitrogen treatments. *South. J. Appl. For.* **1993**, *17*, 135–138.

32. Albaugh, T.J., Rubilar, R.A.; Fox, T.R.; Allen, H.L.; Urrego, J.B.; Zapata, M.; Stape, J.L. Response of *Eucalyptus grandis* in Colombia to mid-rotation fertilization is dependent on site and rate but not frequency of application. *For. Ecol. Manag.* **2015**, *350,* 30–39.

33. Albaugh, T.J.; Vance, E.D.; Gaudreault, C.; Fox,T.R.; Allen, H.L.; Stape, J.L.; Rubilar, R.A. Carbon emissions and sequestration from fertilization of pine in the southeastern United States. *For. Sci.* **2012**, *58*, 419–429.

34. USDA-ERS. Data sets of U.S. Fertilizer Use and Price, USDA Economic Research Service, 2013. Available online: http://www.ers.usda.gov/data-products/fertilizer-use-and-price.aspx (accessed on 30 June 2015).

35. Clark, A., III.; Saucier, J.R. Influence of planting density, intensive culture, geographic location, and species on juvenile wood formation in southern pine. *Ga. For. Res. Pap.* **1991**, *85*, 1–14.

36. Reineke, L.H. Perfecting a stand-density index for even-aged forests. *J. Agric. Res.* **1933**, *46*, 627–638.

37. Drew, T.J.; Flewelling, J.W. Stand density management: an alternative approach and its application to Douglas-fir plantations. *For. Sci.* **1979**, *25*, 518–532.

Importance of Arboreal Cyanolichen Abundance to Nitrogen Cycling in Sub-Boreal Spruce and Fir Forests of Central British Columbia, Canada

Ania Kobylinski and Arthur L. Fredeen

Abstract: The importance of N_2-fixing arboreal cyanolichens to the nitrogen (N)-balance of sub-boreal interior hybrid spruce (*Picea glauca* × *engelmannii*) and subalpine fir (*Abies lasiocarpa*) forests was examined at field sites in central BC, Canada. Host trees were accessed by a single-rope climbing technique and foliage as well as arboreal macrolichen functional groups were sampled by branch height in eight random sample trees from each of two high (High Cyano) and two low (Low Cyano) cyanolichen abundance sites for a total of 32 sample trees. Natural abundances of stable isotopes of N (^{15}N, ^{14}N) and carbon (^{13}C, ^{12}C) were determined for aggregate host tree and epiphytic lichen samples, as well as representative samples of upper organic and soil horizons (Ae and Bf) from beneath host trees. As expected, N_2-fixing cyanolichens had 2–6-fold greater N-contents than chlorolichens and a δ^{15}N close to atmospheric N_2, while foliage and chlorolichens were more depleted in ^{15}N. By contrast, soils at all trees and sites were ^{15}N-enriched (positive δ^{15}N), with declining (not significant) δ^{15}N with increased tree-level cyanolichen abundance. Lichen functional groups and tree foliage fell into three distinct groups with respect to δ^{13}C; the tripartite cyanolichen *Lobaria pulmonaria* (lightest), host-tree needles (intermediate), and bipartite cyanolichens, hair (*Alectoria* and *Bryoria* spp.) and chlorolichens (heaviest). Branch height of host trees was an effective predictor of needle δ^{13}C. Our results showed a modest positive correlation between host tree foliage N and cyanolichen abundance, supporting our initial hypothesis that higher cyanolichen abundances would elevate host tree foliar N. Further study is required to determine if high cyanolichen abundance enhances host tree and/or stand-level productivity in sub-boreal forests of central BC, Canada.

Reprinted from *Forests*. Cite as: Kobylinski, A.; Fredeen, A.L. Importance of Arboreal Cyanolichen Abundance to Nitrogen Cycling in Sub-Boreal Spruce and Fir Forests of Central British Columbia, Canada. *Forests* **2015**, *6*, 2588-2607.

1. Introduction

Nitrogen (N) is commonly limiting to conifer forests in much of the Pacific Northwest, USA [1] as well as the central interior of BC, Canada [2]. Biological N_2-fixing species have been shown to contribute and mediate important inputs of N to these forests [3]. In central British Columbia, mature spruce and fir forests, particularly in the wetter ecological zones, can have well-developed and vertically stratified communities of lichen epiphyte, with N_2-fixing cyanolichens predominating (when present) in the lower canopy branches [4,5]. This stratification and interaction between upper and lower canopy zones can be further altered by vertical gradients in lichen epiphyte guild to create significant variation in N content [4] and δ^{15}N [6] across a vertical profile of a host tree. Although the presence and guild of epiphytic lichens have been shown to enhance certain aspects of N-cycling, empirical evidence for their forest level significance has been equivocal. For example, the removal of lichens from an oak forest system had no effect on tree growth [7]. Thus, while atmospheric N contributions from cyanolichens are observed in sub-boreal spruce and fir ecosystems [8], it is difficult to predict the nature of combined cyanolichen and host conifer tree ecological interactions (*i.e.*, mutualistic to competitive) over tree to forest spatial and as well as temporal scales.

Stable isotopes have been used experimentally to infer many properties about elemental cycles of important nutrients (most notably C and N) at a variety of temporal and spatial scales and ecological contexts. The ratio of the heavier less common stable isotope to that of the lighter abundant isotope (e.g., $^{15}N{:}^{14}N$ and $^{13}C{:}^{12}C$) can be expressed as atom %, but are typically represented in delta (δ) notation, which is not the absolute isotope ratio, but the difference between the sample measurement and an internationally accepted reference standard [9] in parts per thousand or per mil (‰) [10]. In theory, natural abundances of isotopes can be used to make inferences about the contributions of N_2-fixing species to forest trees by exploiting the naturally occurring differences in $^{15}N{:}^{14}N$ ratios between plant-available mineral N sources in the soil and that of atmospheric N_2 utilized by N_2-fixing species [11]. Non-N_2-fixing plants receive their entire N supply from soil N pools and can be expected to be isotopically heavier or lighter than N_2-fixing plants and lichens in concordance with the soil $\delta^{15}N$. In either case, natural abundances of stable N isotopes have frequently been used to estimate direct or indirect contributions of N_2-fixation to plant N content [12].

Most terrestrial materials have $\delta^{15}N$ compositions ranging between −20‰ and +30‰ [13], but lichens have a more restricted range of ^{15}N compositions (−21.5‰ [14]) to (+18‰ [15]), and plants even moreso (e.g., −5‰ to +2.9‰ [16]). The natural abundances and ratios of stable isotope compositions are potentially useful to the researchers quantifying these processes in that they can shift due to isotopic fractionation.

Isotopic fractionation can occur from a variety of processes. For example, N_2-fixation in organisms such as epiphytic cyanolichens does not discriminate between the ^{15}N and ^{14}N and therefore would normally represent N isotope ratios close to the atmospheric $\delta^{15}N$ standard of 0‰. By contrast, non-N_2-fixing elements and processes of forest systems such as ectomycorrhizae (ECM), mineralization of organic nitrogen, nitrification, microbial assimilation of inorganic N, and denitrification can fractionate stable isotopes of N belowground. Nutrient uptake in temperate and boreal trees is predominantly dependent on ECM hyphae growing into the soil from the mycorrhizal root tips [17]. Högberg *et al.* [18] found that ECM roots of Norway spruce (*Picea abies*) and beech (*Fagus sylvatica*) were 2‰ enriched in ^{15}N relative to non-mycorrhizal roots. The study also found ECM fungi were enriched in ^{15}N compared to their host plants, further suggesting that ^{15}N ECM discrimination was fungal in origin. ECM fungi were also more enriched in ^{13}C relative to total soil C [19], demonstrating ^{13}C ECM discrimination. Biologically mediated reactions that control elemental dynamics in soils can result in ^{15}N and ^{13}C enrichment or depletion, making inferences on forest ecosystem N- and C-cycling based on natural abundances of stable isotopes challenging.

Bulk soil $\delta^{15}N$ values are generally higher than atmospheric N (positive, ~5–10‰), a result of faster losses of the lighter isotope in soil N during decomposition [20], usually with increasing (more positive) deltas with increasing depth of organic ('L', 'F', and 'H') and mineral horizons [21]. Thus, fractionation of isotopes during litter decomposition in forests causes surface soils to have lower $\delta^{15}N$ values than deeper soil horizons [22]. Even slight fractionations occurring over decades of transfers of N from mineral soil to forest biomass can be sufficient enough to increase $\delta^{15}N$ of soil organic matter by ~6–8‰ [23]. Surface soils located beneath trees have also been found to have lower $\delta^{15}N$ values than those in open areas as a result of litter deposition [24], making decomposing epiphytic lichens an important factor influencing soil N isotope composition.

Nitrogen-fixing plants in western North American forest systems typically have more positive $\delta^{15}N$ values than non-N_2-fixing plants, and/or are closer to the of atmospheric N_2 [25,26]. The differences between these plant $\delta^{15}N$ values provide the basis of the ^{15}N natural abundance technique for estimating fixed-N contributions to terrestrial ecosystems [24]. However, attempting to trace fixed-N through ecosystems can be complicated by a myriad of fractionations, which are caused by numerous and often serial pathways of mineralization, nitrification, immobilization, and denitrification within the soil, plant root and mycorrhizal fungus mediation of soil N uptake [27].

Nutrient cycling processes in the soil are known to vary with plant community composition [28]. The processes associated with litter decomposition and enzymatic transformations of organic substrates in particular can vary because of differences in a variety of factors such as chemistry of the litter material, soil biota and soil chemistry associated with different plant species [29]. Both litter accumulation and stem flow can deliver N to epiphytes from various N pools [6], making the abundance and branch position of lichens on a tree significant. Lichens may play an important role in nutrient cycling in forest ecosystems, but the relative impact of lichens compared to other ecosystem components is not well understood. The effect of lichens on forest N-cycling could range from large in ecosystems where epiphytic lichens, especially cyanolichens are abundant, to insignificant where lichens are only present in low amounts [30].

In this study, we examine the importance of the vertical distributions of arboreal lichen biomasses [5] to the N-status of wet sub-boreal spruce and fir forest ecosystems of the central interior of BC, Canada. To do this, we measured the ^{15}N:^{14}N, %N of host tree (*i.e.*, conifer) foliage, host tree epiphytic lichen functional groups and organic and mineral soil horizons under the tree crowns at sites containing variable amounts of cyanolichen tripartite (*i.e.*, *L. pulmonaria)* and bipartite lichens from relatively low to high levels. We hypothesized that sub-boreal forest trees with high cyanolichen abundance should have higher needle %N reflecting the greater inputs of biologically fixed-N into these systems [8,31]. Knowledge obtained from this study should provide valuable information on the functional importance of epiphytic (N_2-fixing) cyanolichens to the N-status of wet sub-boreal conifer ecosystems of central BC.

2. Experimental Section

2.1. Study Area

The study area was described previously by Kobylinski and Fredeen [5]. In brief, this field study was carried out on the north side of the Fraser River near the town of Upper Fraser, BC, located approximately 70 km NE of Prince George, BC. All study sites were characterized by having relatively cool and moist summers and cold, snowy winters. Four mature sub-boreal forest sites were chosen based on having trees with predominantly high (High Cyano) or low (Low Cyano) epiphytic cyanolichen abundance and diversity, denoted as 'Upper Fraser' and 'Herrick' in previous work [5,32], respectively.

The two dominant tree species representing the vast majority of the canopy trees at all four sites were interior hybrid spruce (*Picea engelmannii* Parry ex Engelm. x *glauca* (Moench)) and subalpine fir (*Abies lasiocarpa* (Hook.) Nutt.) High Cyano sites were in "sub-boreal spruce" (SBS) 'wk1' and the 'vk' subzones (BC Biogeoclimatic Ecosystem Classification scheme) with a mean elevation of 680 m above sea level (a.s.l.), mean summer temperature of 11.8 ± 5.3 °C and relative humidity of 78% [8]. Low Cyano sites had lower average cyanolichen abundances and diversities and were located within 10–20 km of the High Cyano sites at a slightly higher mean elevation of 850 m a.s.l. and similar mean summer temperature and relative humidity of 10.8 ± 5.3 °C and 78%, respectively [8]. Soils at all sites were Orthic Humo-Ferric Podzols formed from sandy-colluvial materials at the High Cyano sites and from sandy–skeletal glaciofluvial materials at the Low Cyano sites. Average precipitation in the ecotonal study area is approximately 897 mm per year [8].

2.2. Canopy Access and Biomass Sampling

Sample tree selection, canopy access and biomass sampling procedures were previously described in Kobylinski and Fredeen [5]. In brief, trees were selected randomly from High Cyano and Low Cyano sites, with the exception of the fact that trees adjacent to Sitka alder (*Alnus viridis*) were excluded from the study given that annual N-inputs from Sitka alder can be substantial [26] and could have confounded ^{15}N natural abundance interpretation. All study trees were canopy trees and in excess of 22 m in height and 20 cm in DBH. Needles and lichens were sampled from each canopy height zone at the highest accessible point of each of the 32 study trees. Access into canopies was achieved through a single rope technique [33,34]. Selected trees were rigged, climbed and assessed vertically for epiphytic lichen biomass.

Lichen and needle sampling was described previously [5]. Briefly, epiphytic lichen functional group or species and host tree needle samples were collected over two summers: June to August 2008 and May to September 2009 at various heights within sample tree canopies. All cyanolichens in these mature forest tree sites were almost exclusively arboreal and on branches, with negligible cyanolichen occurring on the forest floor [35]. Bryophytes were essentially absent from all canopies, as is normal for these forest types [35]. Lichen biomasses were separated into five primary categories or functional groups based on biomass dominance by single species or functional group properties within different vertical canopy zones: *Alectoria sarmentosa* (Ach.) Ach., *Bryoria* Brodo & D. Hawksw. species, foliose chlorolichens, bipartite cyanolichens, and *Lobaria pulmonaria* (L.) Hoffm., the only tripartite cyanolichen at the site. The needle cohort was taken exclusively from the previous year's foliage to keep the age of the sampled needles constant. A processed and homogeneous sample for each functional group or needle was sent for isotopic analysis from the upper, middle and lower canopy height zones provided they were present in that zone.

2.3. Composite Soil Sampling

Soil samples were collected from the base of each sample tree between June and August 2009. We extracted soil cores using a soil auger (7.5 cm in diameter) by rotating the auger while applying downward force and lifting out the full blades. Soil core samples were extracted from a 1-m radius around each sample tree base in each of the cardinal directions. Samples were separated into three distinct layers: F-folic layer composed of organic matter rich in mycelia (~5 cm ± 2 cm deep), Ae-grayish surface soil layer (~2–10 cm thick); and Bf—yellowish brown to reddish brown subsoil layer (~10–20 cm). All layers were distinct and could be easily separated from each other; the soil auger was cleaned between cores. The three layers of the four subsamples were air dried for two days at room temperature (~22–25 °C), roots were removed and soil sieved through a 2-mm sieve. The F-layer samples were separated twice more (0.85 mm and 0.3 mm sieves) to remove small pieces of woody debris before composite samples were prepared. Composite samples were also made from the four Ae and Bf subsamples from each site. A dry weight of 2 g from each of the three layers of the four samples was mixed into one composite sample per soil layer (8 g dry weight), for a total of 96 soil samples.

2.4. Stable Isotope Analyses

Sample preparation of lichen, needle and soil samples took place at the University of Northern British Columbia (UNBC) before samples were shipped to the Stable Isotope lab at the University of Saskatchewan (U of S) in Saskatoon, Canada for analysis. Lichen and needle samples were oven dried for three days at 55 °C in paper bags. Samples were ball-milled (Retsch MM301, Hann, Germany) to a particle size of less than 250 µm and any fibrous matter or visible granules removed to improve precision of isotopic analysis. Ball-mill chambers were cleaned with deionized water and dried between samples. Samples (2 ± 0.2 mg) were stored in scintillation vials and encapsulated in 8 × 5-mm tin capsules (Catalog # D1008, Elemental Microanalysis Limited, Okehampton, UK) before combustion.

Soil, needle and lichen samples were analyzed using an elemental analyzer (Costech ECS4010, Valencia, CA, USA) coupled to a Delta V Advantage isotope ratio mass spectrometer with continuous flow (Conflo IV, Thermoscientific, Waltham, MA, USA) interface. The continuous flow gas isotope ratio mass spectrometer was used to measure % N, % C and the relative abundance of stable isotopes (^{15}N, ^{14}N, ^{13}C, and ^{12}C) in needle, lichen, and soil samples. Isotope ratio calibrations were performed with IAEA-N1, IAEA-N2, IAEA-NO-3, IAEA-CH6, and USGS-24 standards (International Atomic Energy Agency, 1995).

2.5. Natural Abundance Method

The δ values for ^{15}N and ^{13}C were calculated using equations 1 and 2, respectively.

$$(\delta\ ^{15}N\ (\text{‰ relative to atmospheric } N_2) = (R_{sample}/R_{standard} - 1) \times 1000) \quad (1)$$

$$(\delta\ ^{13}C\ [\text{‰ relative to V-PDB}] = (R_{sample}/R_{standard} - 1) \times 1000) \quad (2)$$

The value R is the ratio of heavy isotope (^{15}N and ^{13}C) to their lighter more abundant isotope.

2.6. Statistical Analyses

Data were analyzed using Sigma Plot 11 (Systat Software Inc., San Jose, CA, USA). Ordinary least square (OLS) multiple linear regression analyses were performed to address whether there were statistically significant differences in: (1) $\delta^{15}N$ and $\delta^{13}C$ between lichen functional groups and needles controlling for tree branch height, tree species, and lichen abundance; (2) $\delta^{15}N$ and $\delta^{13}C$ by tree branch height controlling for lichen functional groups and needles; (3) $\delta^{15}N$ and $\delta^{13}C$ between High and Low Cyano sites controlling for lichen functional groups, needles and soil; (4) $\delta^{15}N$ and $\delta^{13}C$ between tree species (spruce and fir) controlling for the lichen functional groups, needles and soil; and (5) foliar (needle) %N at sites with higher abundances of cyanolichen. The unstandardized coefficient (B) for OLS linear regression analyses provide an estimate of the change in the dependent variable associated with a one-unit change in the identified independent variable controlling for all other independent variables.

Six OLS regression models were created to predict: (i) $\delta^{15}N$ in lichen functional groups; (ii) $\delta^{15}N$ in needles; (iii) $\delta^{13}C$ in lichen functional groups; iv) $\delta^{13}C$ in needles; (v) $\delta^{15}N$ in soil; and (vi) $\delta^{13}C$ in soil. Correlation analyses were conducted between all the variables in each model to check for potential collinearity. Additional correlations were also conducted to look at the relationships between needle $\delta^{15}N$ and cyanolichen abundance, % N and cyanolichen abundance, and soil $\delta^{15}N$ and cyanolichen abundance.

Kolmogorov-Smirnov and Shapiro-Wilk tests were used to determine normality of data. Both tests showed lichen and needle $\delta^{15}N$ (KS = 0.191, $p < 0.001$ and W = 0.886, $p < 0.001$) and $\delta^{13}C$ (KS = 0.122, $p < 0.001$ and W = 0.949, $p < 0.001$) were not normally distributed. However, because of bimodal distributions of the dependent variable, data transformations were not used. The distribution of lichen and needle $\delta^{15}N$ and $\delta^{13}C$ (but not soil) showed some signs of heteroscedasticity and non-linearity across all independent variables.

3. Results

Nitrogen contents varied by over six-fold among lichen epiphytes sampled. At the extremes, *Alectoria sarmentosa* had the lowest % N (0.50 ± 0.13) and bipartite cyanolichens had the greatest %N (3.24 ± 0.52) on a dry weight basis (Figure 1). *Bryoria* spp. and foliose chlorolichens had very similar %N contents of 0.77 ± 0.18 and 0.73 ± 0.19, respectively. Needles had a higher %N content (1.11 ± 0.14) than hair lichens and foliose chlorolichens but not as high as *L. pulmonaria* (2.26 ± 0.30). The differences in mean %N between functional groups were significant ($F_{5,404} = 950.3, p < 0.001$: Tukey test) except for *Bryoria* spp. and foliose chlorolichens (Figure 1).

All lichen functional groups or species and conifer foliage (needle) had mostly unique $\delta^{15}N$ signatures relative to their %N (Figure 2). *Bryoria* spp. hair lichens at the top of the canopy had the lowest $\delta^{15}N$, while the hair lichen *Alectoria sarmentosa* found lower in the canopy had slightly lower %N and higher $\delta^{15}N$ than *Bryoria*. Host tree needles and foliose chlorolichens had $\delta^{15}N$ values that were intermediate between hair lichens and cyanolichens but similar %N. While *L. pulmonaria* and bipartite cyanolichens both had $\delta^{15}N$ values close to atmospheric N_2 ($\delta^{15}N = 0$), *L. pulmonaria* had 1.5-times less N on average (Figure 2). Mean $\delta^{15}N$ isotope values were slightly higher at Low Cyano sites for all lichen functional groups and host tree needles, but there was no significant difference in the $\delta^{15}N$ values or %N of lichens or needles on trees from High Cyano (black symbols) *versus* Low Cyano (white symbols) sites (Figure 2).

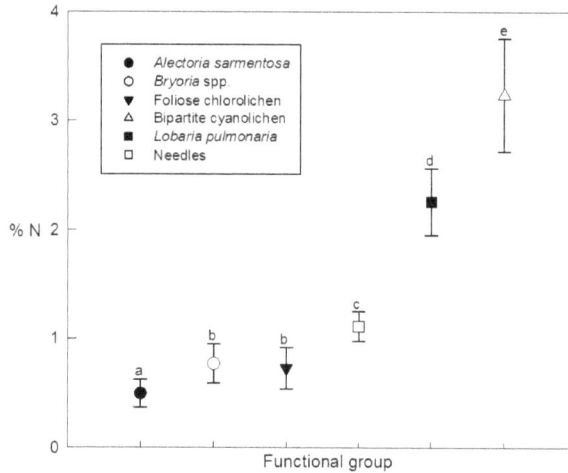

Figure 1. Mean %N (± SD) of lichen biomass in functional groups of lichen and foliage (needle) from 32 spruce and fir study trees with varying amounts of tree-level cyanolichen abundance. Different letters above error bars indicate significant ($p < 0.001$) differences between lichen functional groups (Tukey multiple mean comparison test).

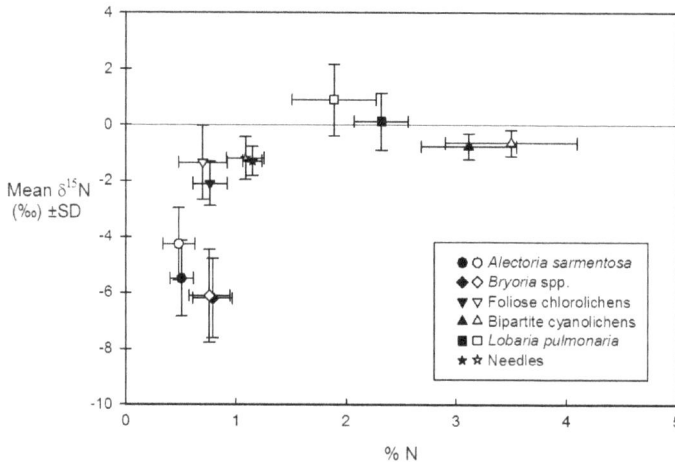

Figure 2. Mean (± SD) % N and $\delta^{15}N$ signatures of five epiphytic lichen and host tree foliage (needle) biomasses across four sites and two host tree species from sub-boreal spruce and fir forests in central interior BC. Half of the host trees were from high (High Cyano: black symbols) and the other half from low (Low Cyano: white symbols) cyanolichen abundance sites.

When we compared $\delta^{15}N$ of living conifer foliage (needle) on each of the 32 study trees to their actual cyanolichen abundance, there was a small, but ultimately not significant ($r = -0.16$, $p = 0.121$) negative correlation between mean needle $\delta^{15}N$ and increasing cyanolichen abundance (Figure 3a). Host tree foliage (needles) %N showed a significant but weak positive correlation to actual tree-level cyanolichen abundance ($r = 0.24$, $p = 0.019$; Figure 3b).

When lichen species, tree branch height, and tree species were controlled for, cyanolichen abundance in the canopy was a statistically significant but ineffective predictor of $\delta^{15}N$ (B < 0.001, $p < 0.001$, Table 1). The regression coefficients of the dependent variable $\delta^{15}N$ suggest that lichen species and cyanolichen abundance, but not tree species or tree branch height, were significant predictors of $\delta^{15}N$ in lichen functional groups. The regression analyses also show that *A. sarmentosa* (B = -3.657, $p < 0.001$), *Bryoria* spp. (B = -4.838, $p < 0.001$) and foliose chlorolichen (B = -0.482, $p = 0.002$) tree-level abundances were all negatively related to their $\delta^{15}N$ signatures. Conversely, bipartite cyanolichens (B = 0.561, $p < 0.001$) and *L. pulmonaria* (B = 1.608, $p < 0.001$) tree-level abundances were positively related to their $\delta^{15}N$ signatures. Both tree species (B = -0.166, $p = 0.095$) and tree branch height (B = 0.010, $p = 0.238$) were not significantly related to $\delta^{15}N$ levels in lichen functional groups (data not shown) and insignificant or with a very small coefficient for $\delta^{15}N$ in needles (B = 0.028, $p = 0.002$). Similarly, tree species (B = -0.146, $p = 0.267$) and cyanolichen abundance (data not shown) were not significant predictors of $\delta^{15}N$ in lichen or host-tree foliage.

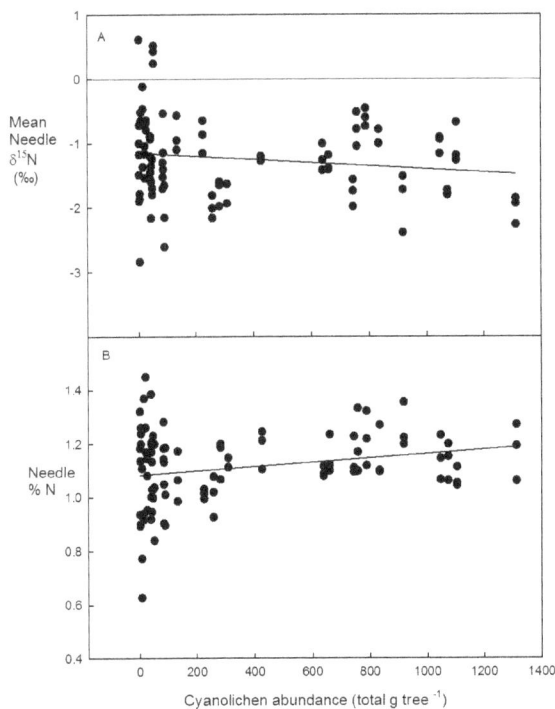

Figure 3. Mean (**a**) $\delta^{15}N$ and (**b**) %N of host tree foliage (needle) from low to high tree-level cyanolichen abundance (total g tree^{-1}).

Table 1. Linear regression (OLS) coefficients with standard error for the regression model predicting $\delta^{15}N$ of lichen functional groups (F = 220.501, $p < 0.001$, $R^2 = 0.817$) and host tree needles (F = 4.363, $p = 0.006$, $R^2 = 0.126$) based on tree species (spruce and fir), cyanolichen abundance, tree branch height, and lichen species or functional group.

	Lichen Functional Groups (Model i)		Needles (Model ii)	
	Unstandardized coefficient (B)	Std Error	Unstandardized coefficient (B)	Std Error
Intercept	−1.126 ***	0.165	−1.499 ***	0.171
Tree species (Spruce)	−0.166	0.099	−0.146	0.131
Tree branch height	0.010	0.008	0.028 **	0.009
Alectoria sarmentosa	−3.657 ***	0.168		
Bryoria spp.	−4.838 ***	0.183		
Foliose chlorolichen	−0.482 **	0.156		
Bipartite cyanolichen	0.561 ***	0.137		
Lobaria pulmonaria	1.608 ***	0.165		

*** $p < 0.001$; ** $p < 0.01$; * $p < 0.05$.

Soil horizon (F, Ae and Bf) $\delta^{15}N$ values were consistently higher at Low Cyano (means: F = 1.340, Ae = 5.130, Bf = 6.262) than at High Cyano (F = 0.782, Ae = 4.765, Bf = 5.383) sites, and decreased with increasing cyanolichen abundance in all three horizons (Figure 4A–4C), respectively. The abundances of cyanolichens and $\delta^{15}N$ were negatively correlated in all three soil horizons, with correlations being nearly significant for the uppermost horizons (F: $r = -0.341$, $p = 0.056$; Ae: $r = -0.263$, $p = 0.146$), and significant for the Bf horizon ($r = -0.356$, $p = 0.045$). Total %N in the F-horizon decreased with increasing tree-level cyanolichen abundance (Figure 4D), but was relatively unaffected in Ae and Bf horizons (Figure 4E and 4F).

Tree species and both upper mineral soil horizons were all significant predictors and all contributed to the overall relationship of both soil $\delta^{15}N$ and $\delta^{13}C$ (Table 2). Cyanolichen abundance was a significant predictor for both soil $\delta^{15}N$ and $\delta^{13}C$, but coefficients were <0.01 and therefore not relevant (data not shown). There was also a statistically significant difference in soil $\delta^{15}N$ (B = −0.525, $p < 0.001$) and soil $\delta^{13}C$ (B = −0.221, $p = 0.009$) between host tree species (spruce and fir).

When controlling for all independent variables, tree species and branch height were significant predictors of $\delta^{13}C$ of host-tree foliage (needles) and to lesser extents for lichen epiphytes (Table 3 and Figure 5A). Trends of increasing (less negative) $\delta^{13}C$ values with increasing branch height were evident for both *L. pulmonaria* and tree foliage (Figure 5A). There was evident clumping of $\delta^{13}C$ into three distinct groups (Figures 5A and 6). Specifically, $\delta^{13}C$ values of the lone tripartite cyanolichen (*L. pulmonaria*) were lower than host tree needles, which were in turn lower than hair lichens (*A. sarmentosa* and *Bryoria* spp.), bipartite cyanolichens and foliose chlorolichens.

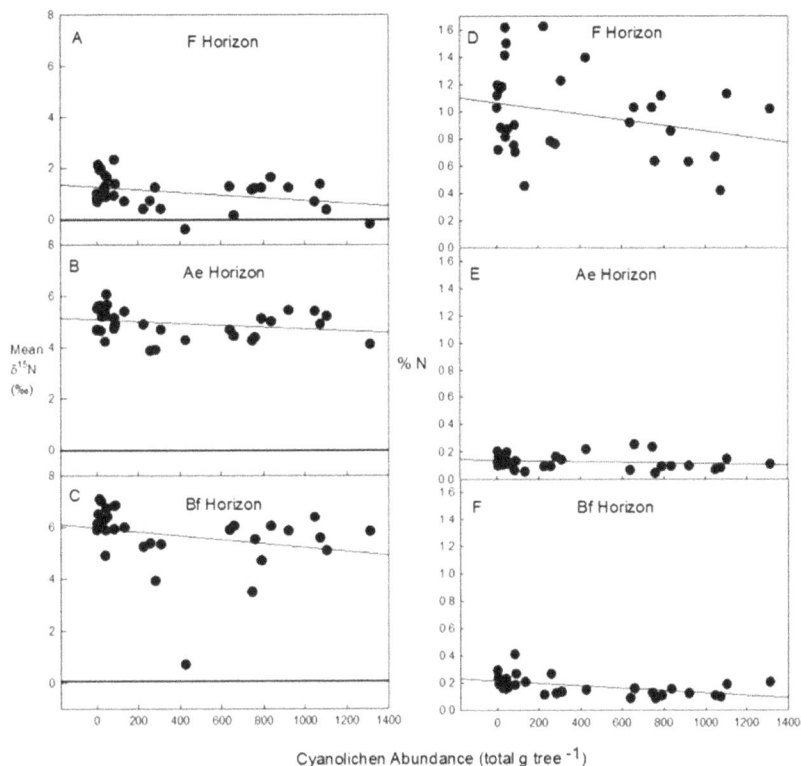

Figure 4. Mean $\delta^{15}N$ and %N of (A and B) the surface organic F horizons, (B and E) upper-most mineral (Ae) horizons and (C and F) lower mineral (Bf) horizons, respectively, beneath host trees with varying levels of cyanolichen abundance (total g tree^{-1}).

Table 2. Linear regression (OLS) coefficients with standard error for the regression model predicting $\delta^{15}N$ (F = 314.044, $p < 0.001$, $R^2 = 0.932$) and $\delta^{13}C$ (F = 117.897, $p < 0.001$, $R^2 = 0.838$) of soil based on tree species, and Ae and Bf soil horizons.

	$\delta^{15}N$		$\delta^{13}C$	
	Unstandardized coefficient (B)	Std. Error	Unstandardized coefficient (B)	Std. Error
Intercept	1.586 ***	0.140	−26.923 ***	0.094
Tree species	−0.525 ***	0.123	−0.221 **	0.062
Mineral surface soil (Ae)	3.887 ***	0.145	1.523 ***	0.097
Mineral subsoil (Bf)	4.777 ***	0.145	1.961 ***	0.097

*** $p < 0.001$; ** $p < 0.01$; * $p < 0.05$.

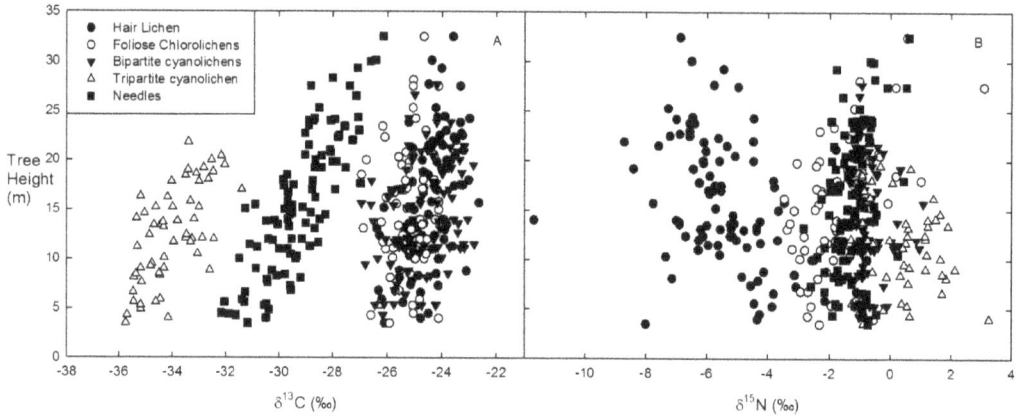

Figure 5. Variation in the natural abundances of (**A**) $\delta^{13}C$ and (**B**) $\delta^{15}N$ of five epiphytic lichen functional group and host tree foliage (needle) biomasses according to tree branch height for all sites.

Table 3. Linear regression (OLS) coefficients with standard error for the regression model predicting $\delta^{13}C$ in lichen functional groups (F = 711.517, $p < 0.001$, $R^2 = 0.942$) and needles (F = 79.545, $p < 0.001$, $R^2 = 0.724$) based on tree species (spruce), tree branch height and lichen functional groups.

	Lichen Functional Groups (Model iii)		Needles (Model iv)	
	Unstandardized coefficient (B)	Std. Error	Unstandardized coefficient (B)	Std. Error
Intercept	−30.523 ***	0.154	−31.868 ***	0.204
Tree species (Spruce)	0.048	0.089	0.622 ***	0.156
Tree branch height	0.093 ***	0.008	0.148 ***	0.011
Alectoria sarmentosa	4.899 ***	0.151		
Bryoria spp.	4.513 ***	0.164		
Foliose chlorolichen	3.969 ***	0.140		
Bipartite cyanolichen	4.324 ***	0.166		
Lobaria pulmonaria	−4.358 ***	0.149		

*** $p < 0.001$; ** $p < 0.01$; * $p < 0.05$.

Hair lichens *A. sarmentosa* and *Bryoria* spp. had uniformly negative $\delta^{15}N$ values that were also more negative than all other functional groups (Figure 5B), and trending lower with increase in branch height. The $\delta^{15}N$ of other lichen functional groups and host-tree foliage did not appear to be influenced by branch height.

Host tree needles had $\delta^{13}C$ values intermediate to *L. pulmonaria* and all other lichen functional groups (Figure 6), but at distinctly higher C concentrations than all lichen groups (Figure 7). Lichen functional groups, cyanolichen abundance, tree branch height, and the interaction between cyanolichens and site cyanolichen abundance (but not tree species) were significant predictors of $\delta^{13}C$ in lichen functional groups. The regression analyses showed that *A. sarmentosa* (B = 4.899, $p < 0.001$), *Bryoria* spp. (B = 4.513, $p < 0.001$), foliose chlorolichens (B = 3.969, $p < 0.001$), and bipartite cyanolichens (B = 4.324, $p < 0.001$) were positively related to $\delta^{13}C$ levels in lichen functional groups (Table 2). Conversely, *L. pulmonaria* (B = −4.358, $p < 0.001$) was negatively related to $\delta^{13}C$ levels in lichen functional groups. Both cyanolichen abundance (B = −0.001, $p < 0.001$) and tree branch height (B = 0.093, $p < 0.001$) were significantly related to $\delta^{13}C$ levels in lichen functional groups. There was also a statistically significant interaction between cyanolichens and site cyanolichen abundance and $\delta^{13}C$ increased slightly as site-level cyanolichen abundance increased (B = 0.001, $p = 0.005$, Table 2). There was no significant difference in $\delta^{13}C$ between sample tree species spruce and fir in lichen functional groups (B = 0.048, $p = 0.594$, Table 2). Alternatively, both tree species (B = 0.622, $p < 0.001$) and tree branch height (B = 0.148, $p < 0.001$), but not cyanolichen abundance (B < 0.001, $p = 0.942$), were significant predictors of $\delta^{13}C$ in needles (B = 0.028, $p = 0.002$).

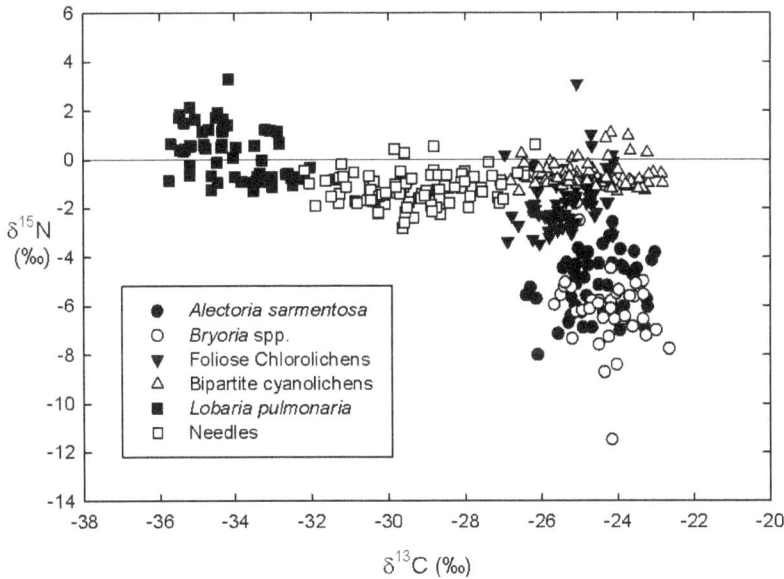

Figure 6. Dual natural abundance isotope plot of $\delta^{15}N$ *versus* $\delta^{13}C$ of the five epiphytic lichen functional group and host tree foliage (needle) biomasses across four sites and two host tree species.

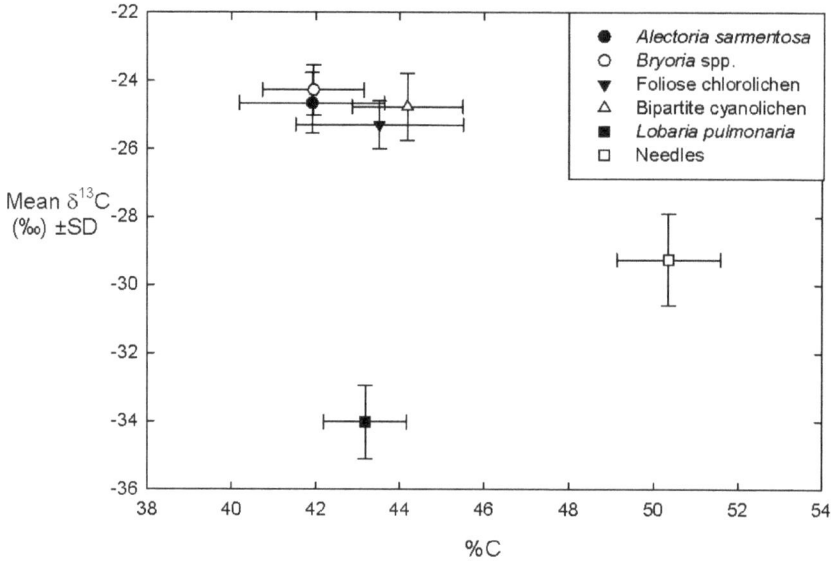

Figure 7. Mean (± SD) C concentrations *versus* $\delta^{13}C$ signals of five epiphytic lichen functional group and host tree foliage (needle) biomasses across four sites and two host tree species.

4. Discussion

Nitrogen-fixing cyanolichen epiphytes can be both diverse and abundant in conifer forests of central BC, and have been conjectured to be an important component of the N-cycle in these forest locations [4,8,32,36]. Although lichen N pool sizes and decomposition rates in these studies were indicative of enhanced mineral N flux rates, as observed in other forests types [7], direct links between cyanolichen abundance and improved host tree N status have been elusive. In our study, gradients in cyanolichen abundance across host trees (interior hybrid spruce and subalpine fir) in two generally high cyanolichen abundance sites and two generally low cyanolichen abundance sites [8,32] provided us with an opportunity to examine the potential for enhanced cyanolichen N-inputs into these forests in central BC. We hypothesized that foliar (needle) N contents would be positively correlated with cyanolichen abundance, and in fact this was found to be the case, albeit the relationship was not strong (Figure 3b). We further sought to more directly link increased host tree foliar N with cyanolichen N using the natural abundances of ^{15}N and ^{14}N of forest tree, lichen and soil components of these systems.

Our results corroborate previous findings of $\delta^{15}N$ values in lichen functional groups and or species. It is already known that cyanolichens that fix N_2 typically have $\delta^{15}N$ values close to atmospheric N_2, *i.e.*, 0‰ [37,38]. We also found $\delta^{15}N$ values close to 0‰ for both bipartite and tripartite (*L. pulmonaria*) cyanolichens in our study. By contrast, other non-N_2-fixing arboreal macrolichens (hair and chlorolichen) as well as host-tree foliage all had $\delta^{15}N$ values that were negative and less enriched in ^{15}N than either cyanolichen functional group. Hair lichens had the lowest $\delta^{15}N$, while chlorolichens and host tree foliage were intermediate. Interestingly,

A. sarmentosa, occurring lower in the canopy than *Bryoria* spp., and therefore overlapping with cyanolichen canopy zones [5], had $\delta^{15}N$ closer to atmospheric $\delta^{15}N$ (*i.e.*, 0) (Figure 2). This is consistent with *A. sarmentosa* receiving more leached-N from cyanolichens than *Bryoria* spp., and supported by the fact that an increase in the $\delta^{15}N$ of *A. sarmentosa* was observed in High Cyano sites, while relatively little change in $\delta^{15}N$ was observed for *Bryoria* spp. In both hair lichens, relatively low N contents and low $\delta^{15}N$ (~6 ‰ in Low Cyano sites) were consistent with low atmospheric N inputs in central BC [39] and the negative $\delta^{15}N$ of precipitation N measured in other studies [31].

Although biological N_2-fixation may represent a significant N input into ecosystems, both via leaching (less well documented) and decomposition, it may be difficult to identify because of its small relative isotopic effect against the background of ecosystem component $\delta^{15}N$ signatures [12]. Nitrogen isotope signatures of epiphytes varied across functional groups and with canopy position, but the reason for lichen $\delta^{15}N$ variability with canopy position in most studies remains unexplained. The $\delta^{15}N$ of hair lichens, with single sources of N (*i.e.*, atmospheric fixed-N), are relatively easy to explain relative to other lichens (cyanolichen and chlorolichen) lower in the canopy where inputs and outputs are more complex. At least 10 processes have been identified that can alter $\delta^{15}N$ values, none of which can currently be separated out in field studies [9]. Key explanations for $\delta^{15}N$ variability in lichens are: (i) a preference for uptake of the lighter ^{14}N isotope which can then lead to $\delta^{15}N$ enrichment of outputs [40]; (ii) fractionation of N isotopes in gaseous phase ammonia is greater than in the liquid phase of nitrate; and (iii) transfer of organic N which can result in increased ^{15}N depletion of the photobiont and less depletion in the mycobiont [41]. In general, our results were in agreement with previous work showing that lichens with a green alga as their photobiont (all lichens except bipartite cyanolichens) showed greater relative ^{15}N depletion [42] (Figure 1).

Soil $\delta^{15}N$ values increased with soil depth in our study, consistent with results of Gebauer and Schulze [43]. In keeping with our expectations, we also observed decreasing trends in $\delta^{15}N$ with increasing cyanolichen abundance in all soil horizons (Figure 4A–4E) equating to less positive soil horizon $\delta^{15}N$ values, more proximal to the $\delta^{15}N$ of cyanolichen biomasses which were uniformly close to zero. There are many reasons why soil $\delta^{15}N$ wouldn't necessarily be closely correlated with tree $\delta^{15}N$. First, measured $\delta^{15}N$ of soil pools may not represent the true isotopic composition of N available to plant roots, since most of the N in soils is bound in forms that are not immediately available to plants [25,44]. Second, only a few % of total soil N becomes available in a year [18], and symbiotic fungi (mycorrhizae) can alter the $\delta^{15}N$ of the N transferred from soil to host plant. Nevertheless, a similar downward trend in tree foliage $\delta^{15}N$ at our sites with increasing abundance of cyanolichen (Figure 3) was consistent with the downward trend in soil $\delta^{15}N$ (Figure 4A–4C). Explaining the increase in %N of host tree foliage with increased cyanolichen abundance is difficult to reconcile with total soil N. If greater cyanolichen inputs of N increased soil N concentrations, then a straightforward mass action of greater plant N uptake could have explained enhanced foliar %N. However, soil N decreased in all soil horizons, though not significantly, with higher cyanolichen abundance (Figure 4D–4F). Since total soil N does not in any way equate with amounts and forms of available inorganic N, it is possible that more readily available N fractions were more available at high abundance cyanolichen sites, even though total %N was not. Further work on soil N at these sites would be required to test this hypothesis.

Foliar (needle) $\delta^{15}N$ values in our host trees ranged between 0.6 and −2.8‰ and did not differ significantly between High and Low Cyano sites or tree branch height. However, there was a trend of decreasing foliar $\delta^{15}N$ with increasing cyanolichen abundance (Figure 3a). Gebauer and Schulze [43] reported lower, but overlapping, $\delta^{15}N$ values for conifer needles (between −2.5 and −4.1‰), which varied according to stand and age, with foliage from the healthiest site having the lowest $\delta^{15}N$. It is unclear why the range of $\delta^{15}N$ values observed by Gebauer & Schulze were more negative than ours, but one possibility would be the presumed greater atmospheric N inputs in these European forests when compared to the N-limited forests of central BC. Gebauer and Schulze [43] also reported a similar trend of $\delta^{13}C$ values of needles, which ranged between −26.2 and −32.2‰ and did not differ between lichen abundance sites but did change with canopy height. Similar to our study, needle $\delta^{13}C$ values were not significantly different between abundance sites (Table 3), were more negative in the lower canopy and increasing with branch height (Figure 5).

Lichen $\delta^{13}C$ values were previously found to vary widely over a large range of habitats and species [45]. Our lichen functional groups and tree foliage (needles) fell into three distinct groups with respect to $\delta^{13}C$ values: the tripartite (and only cephalodic) cyanolichen *L. pulmonaria* with the lowest $\delta^{13}C$ values (−36 to −31), host tree foliage with intermediate $\delta^{13}C$ values (−33 to −26), and all other lichens including bipartite cyanolichens, hair lichens and chlorolichens with the least negative $\delta^{13}C$ values (−28 to −23). Our values are entirely consistent with those of Maguas *et al.* [46], namely that tripartite lichen associations typically have the lowest and bipartite lichens the highest $\delta^{13}C$. Carbon isotope discrimination in lichens [42] has been attributed to the species of photobiont present. Carbon concentrations also clearly distinguished between conifer foliage (mean %C of ~50.5%) and all lichen functional groups (means of %C ranging from ~42%–45%). This range of values are in close agreement with %C values observed previously for epiphytic lichens in sub-boreal BC [8] and conifer foliage [47], respectively.

Interestingly, we observed positive relationships between tree branch height and $\delta^{13}C$ values for both *L. pulmonaria* and foliage. Possible causes that may be responsible for the observed height-specific differences in these $\delta^{13}C$ values are CO_2 source, light level and factors influencing diffusion resistance such as water availability. Carbon dioxide source influences the $\delta^{13}C$ value of lichens [48,49], and lichens growing in the canopy close to the forest floor (*i.e.*, *L. pulmonaria*) would more readily assimilate CO_2 enriched by soil respiration [46,50,51]. Light levels also influence photosynthesis and can alter the CO_2 gradient inside the lichen thallus. Inorganic carbon acquisition by lichen photobionts fluctuate greatly with moisture and light availability [52], and lichen hydration is known to affect $\delta^{13}C$ values. Increased $\delta^{13}C$ values have been reported for lichens in drier habitats [45] and on thinner branches, both being directly and indirectly associated with desiccation stress, respectively [53]. Water stress would be expected to increase with branch height. Therefore, differences in $\delta^{13}C$ likely exist due to biological traits (e.g., foliage (stomatal) *versus* lichen, cephalodic *versus* non-cephalodic lichen), different microclimatic conditions in the canopy and/or the uptake of isotopically different CO_2 sources. Irrespective of mechanism and ecological significance, the $\delta^{13}C$—in concert with the $\delta^{15}N$—are clearly useful in chemically distinguishing among epiphyte functional group and host tree foliage.

5. Conclusions

Stable isotope natural abundance techniques have been important tools for ecophysiology and ecosystem research for many decades now [10,18], but have infrequently been applied to determining the contributions that arboreal cyanolichens make to forest N-cycling. Despite the narrow geographic and temporal scope of this study, the results from our study support previous work at our high and low cyanolichen abundance sites [8,32], suggesting that trees with a high abundance of arboreal cyanolichens provide greater N-inputs into N-limited conifer forests of central BC. High cyanolichen abundance, typically coupled with higher overall arboreal lichen diversity, is usually constrained to mature and old-growth stands in drier parts of the central interior of BC. However, the reasons for the variability in cyanolichen abundance at the tree and site level, and the contributions that cyanolichen abundances make to tree and forest growth remain uncertain and are the topics of current and future research at our forest research sites. Low-input, extensive forestry practiced in central British Columbia could benefit from this increased understanding.

Acknowledgements

The authors are grateful for Jocelyn Campbell's guidance and assistance with field site selection and project design. Kevin Jordan provided expert professional help and support in accessing canopies in year one of this study. Thanks are extended to Myles Stocki for stable isotope analyses and to the many graduate student field assistants for help in gathering samples in the canopy. This project was primarily funded by a Discovery Grant from the Natural Sciences and Engineering Research Council of Canada (NSERC 194405; ALF).

Author Contributions

Ania Kobylinski is the lead author and researcher. She climbed nearly half of the experimental trees, conducted the field sampling, prepared samples for analyses, and collated and analyzed data. This work comprised one chapter in her M.Sc. thesis. Art Fredeen conceived of the fundamental research questions, acquired the research funding, helped in decision-making regarding sampling, analyses and data processing, and helped in the editing and/or writing of the thesis chapter and manuscript.

Conflicts of Interest

The authors declare no conflict of interest.

References

1. Holub, S.M.; Lathja, K. The fate and retention of organic and inorganic ^{15}N-nitrogen in an old-growth forests soil in western Oregon. *Ecosystems* **2004**, *7*, 368–380.
2. Brockley, R.P.; Simpson D.G. Effects of Intensive Fertilization on the Foliar Nutrition and Growth of Young Lodgepole Pine and Spruce Forests in the Interior of British Columbia (E.P. 886.13): Establishment and Progress Report. Available online: https://www.for.gov.bc.ca/hfd/pubs/Docs/tr/tr018.pdf (accessed on 27 July 2015).
3. Coxson, D.S.; Nadkarni, N.M. Ecological Roles of Epiphytes in Nutrient Cycles of Forest Ecosystems. In *Forest Canopies*; Academic Press: San Diego, California, CA, USA, 1995.
4. Campbell, J.; Fredeen, A.L. Contrasting the abundance, nitrogen, and carbon of epiphytic macrolichen species between host trees and soil types in a sub-boreal forest. *Can. J. Bot.* **2007**, *85*, 31–42.
5. Kobylinski, A.; Fredeen, A.L. Vertical distribution and nitrogen content of epiphytic macrolichen functional groups in sub-boreal forests of central British Columbia. *For. Ecol. Manag.* **2014**, *329*, 118–128.
6. Bergstrom, D.M.; Tweedie, C.E. A conceptual model for integrative studies of epiphytes: Nitrogen utilisation, a case study. *Aust. J. Bot.* **1998**, *46*, 273–280.
7. Knops, J.M.H.; Nash, T.H.I.; Schlesinger, W.H. The influence of epiphytic lichens on the nutrient cycling of a blue oak woodland. *Ecol. Monogr.* **1997**, *66*, 75–82.
8. Campbell, J.; Fredeen, A.L.; Prescott, C.E. Decomposition and nutrient release from four epiphytic lichen litters in sub-boreal spruce forests. *Can. J. Bot.* **2010**, *40*, 1473–1484.
9. Robinson, D. δ^{15}N as an integrator of the nitrogen cycle. *Trends Ecol. Evol.* **2001**, *16*, 153–162.
10. Dawson, T.E.; Mambelli, S.; Plamboeck, A.H.; Templer, P.H.; Tu, K.P. Stable isotopes in plant ecology. *Annu. Rev. Ecol. Syst.* **2002**, *33*, 507–559.
11. Boddey, R.M.; Peoples, M.B.; Palmer, B.; Dart, P.J. Use of the ^{15}N natural abundance technique to quantify biological nitrogen fixation by woody perennials. *Nutr. Cycl. Agroecosyst.* **2000**, *57*, 235–270.
12. Vitousek, P.M.; Shearer, G.; Kohl, D.H. Foliar ^{15}N natural abundance in Hawaiian rainforest: Patterns and possible mechanisms. *Oecologia* **1989**, *78*, 383–388.
13. Junk, G.; Svec, H.J. The absolute abundance of the nitrogen isotopes in the atmosphere and compressed gas from various sources. *Geochim. Cosmochim. Acta* **1958**, *14*, 234–243.
14. Fogel, M.L.; Wooller, M.J.; Cheeseman, J.; Smallwood, B.J.; Roberts, Q.; Romero, I.; Meyers, M.J. Unusually negative nitrogen isotopic compositions (δ^{15}N) of mangroves and lichens in an oligotrophic, microbially-influenced ecosystem. *Biogeosciences* **2008**, *5*, 1693–1704.
15. Huiskes, A.H.L.; Boschker, H.T.S.; Lud, D.; Moerdijk-Poortvliet, T.C.W. Stable isotope ratios as a tool for assessing changes in carbon and nutrient sources in antarctic terrestrial ecosystems. *Plant Ecol.* **2006**, *182*, 79–86.

16. Amundson, R.; Austin, A.T.; Schurur, E.A.G.; Yoo, K.; Matzek, V.; Kendall, C.; Uebersax, A.; Brenner, D.; Baisden, W.T. Global patterns of the isotopic composition of soil and plant nitrogen. *Glob. Biogeological Cycles* **2003**, *17*, 1–10.

17. Wallander, H.; Göransson, H.; Rosengren, U. Production, standing biomass and natural abundance of 15N and 13C in ectomycorrhizal mycelia collected at different soil depths in two forest types. *Oecologia* **2004**, *139*, 89–97.

18. Högberg, P. Tansley Review No. 95 ^{15}N natural abundance in soil-plant systems. *New Phytol.* **1997**, *137*, 179–203.

19. Högberg, P.; Högberg, M.N.; Quist, M.E.; Ekblad, A.; Nasholm, T. Nitrogen isotope fractionation during nitrogen uptake by ectomycorrhizal and non-mycorrhizal Pinus sylvestris. *New Phytol.* **1999**, *142*, 569–576.

20. Peterson, B.J.; Fry, B. Stable isotopes in ecosystem studies. *Annu. Rev. Ecol. Syst.* **1987**, *18*, 293–320.

21. Mariotti, A.; Pierre, D.; Vedy, J.; Bruckert, S.; Guillemot, J. The abundance of natural nitrogen-15 in the organic matter of soils along an altitudinal gradient. *Catena* **1980**, *7*, 293–300.

22. Nadelhoffer, K.J.; Fry, B. Controls on natural nitrogen-15 and carbon-13 abundances in forest soil organic matter. *Soil Sci. Soc. Am. J.* **1988**, *52*, 1633–1640.

23. Billings, S.A.; Richter, D.D. Changes in stable isotopic signatures of soil nitrogen and carbon during 40 years of forest development. *Oecologia* **2006**, *148*, 325–333.

24. Shearer, G.B.; Kohl, D.H. N2-fixation in field settings: Estimations based on natural ^{15}N abundance. *Aust. J. Plant Physiol.* **1986**, *13*, 699–756.

25. Binkley, D.; Hart, S.C. Components of nitrogen availability assessments in forest soils. *Adv. Soil Sci.* **1989**, *10*, 57–112.

26. Sanborn, P.; Preston, C.; Brockley, R. N2 fixation by Sitka alder in a young lodgepole pine stand in central interior British Columbia, Canada. *For. Ecol. Manag.* **2002**, *167*, 223–231.

27. Lajtha, K.; Schlesinger, W.H. Plant response to variations in nitrogen availability in a desert shrubland community. *Biogeochemistry* **1986**, *2*, 29–37.

28. Ehrenfeld, J.G.; Ravit, B.; Elgerma, K. Feedback in the plant–soil system. *Annu. Rev. Ecol. Syst.* **2005**, *30*, 1–41.

29. Aerts, R. Climate, leaf litter chemistry and leaf litter decomposition in terrestrial ecosystems: A triangular relationship. *Oikos* **1997**, *79*, 439–449.

30. Pike, L.H. The importance of epiphytic lichens in mineral cycling. *Bryol.* **1978**, *81*, 247–257.

31. Freyer, H.D. Seasonal variation of ^{15}N/^{14}N ratios in atmospheric nitrate species. *Tellus* **1991**, *43*, 30–44.

32. Campbell, J.; Bradfield, G.E.; Prescott, C.E.; Fredeen, A.L. The influence of overstorey *Populus* on epiphytic lichens in subboreal spruce forests of British Columbia. *Can. J. For. Res.* **2010**, *40*, 143–154.

33. Denison, W.C. Life in tall trees. *Sci. Am.* **1973**, *228*, 74–80.

34. Perry, D.; Wiliams, J. The tropical rain forest canopy: A method providing total access. *Biotropica* **1981**, *13*, 283–285.

35. Botting, R.S.; Fredeen, A.L. Contrasting terrestrial moss, lichen and liverwort diversity and abundance between old-growth and young second-growth sub-boreal spruce forest in central British Columbia. *Can. J. Bot.* **2006**, *84*, 120–132.

36. Campbell, J.; Fredeen, A.L. *Lobaria pulmonaria* abundance as an indicator of macrolichen diversity in interior cedar hemlock forests of East-Central British Columbia. *Can. J. Bot.* **2004**, *82*, 970–982.

37. Macko, S.A.; Fogel, M.L.; Hare, P.E.; Hoering, T.C. Isotopic fractionation of nitrogen and carbon in the synthesis of amino acids by microorganisms. *Chem. Geol.* **1987**, *65*, 79–92.

38. Virginia, R.A.; Delwiche, C.C. Natural ^{15}N abundance of presumed N_2-fixing and non-N_2-fixing plants from selected ecosystems. *Oecologia* **1982**, *54*, 317–325.

39. Hope, G. The Soil Ecosystem of an ESSF Forest and Its Response to a Range of Harvesting Disturbances (Extension Note 53). Available online: http://www.for.gov.bc.ca/hfd/pubs/ Docs/En /En53.pdf (accessed on 11 February 2014).

40. Wania, R.; Hietz, P.; Wanek, W. Natural ^{15}N abundance of epiphytes depends on the position within the forest canopy: Source signals and isotope fractionation. *Plant Cell Environ.* **2002**, *25*, 581–589.

41. Hobbie, E.A.; Hobbie, J.E. Natural abundance of ^{15}N in nitrogen-limited forests and tundra can estimate nitrogen cycling through mycorrhizal fungi: A review. *Ecosystems* **2008**, *11*, 815–830.

42. Riera, P. δ^{13}C and δ^{15}N comparisons among different co-occurring lichen species from littoral rocky substrata. *Lichenologist* **2005**, *37*, 93–95.

43. Gebauer, G.; Schulze, E.D. Carbon and nitrogen isotope ratios in different compartments of a healthy and a declining *Picea abies* forest in the Fichtelgebirge, NE Bavaria. *Oecologia* **1991**, *87*, 198–207.

44. Jansson, S.L. Tracer studies on nitrogen transformations in soil with special attention to mineralization-immobilization relationships. *Ann. R. Agri. Coll. Swed.* **1958**, *24*, 101–361.

45. Batts, J.E.; Calder, L.J.; Batts, B.D. Utilizing stable isotope abundances of lichens to monitor environmental change. *Chem. Geol.* **2004**, *204*, 345–368.

46. Máguas, C.; Griffiths, H.; Broadmeadow, M.S.J. Gas exchange and carbon isotope discrimination in lichens: Evidence for interactions between CO_2-concentrating mechanisms and diffusion limitation. *Planta* **1995**, *196*, 95–102.

47. Thomas, S.C.; Martin, A.R. Carbon content of tree tissues: A synthesis. *Forests* **2012**, *3*, 332–352.

48. Farquhar, G.D.; Ehleringer, J.R. Carbon isotope discrimination and photosynthesis. *Annu. Rev. Plant Physiol. Plant Mol. Biol.* **1989**, *40*, 503–537.

49. Lakatos, M.; Hartard, B.; Maguas, C. The stable isotopes of δ^{13}C and δ^{18}O of lichens can be used as tracers of microenvironmental carbon and water sources. In *Stable Isotopes as Indicators of Ecological Change (Terrestrial Ecology)*; Dawson, T.E., Siegwolf, R.T.W., Eds.; Elsevier: Oxford, UK, 2007; pp. 73–88.

50. Broadmeadow, M.S.J.; Griffiths, H.; Maxwell, C.; Borland, A.M. The carbon isotope ratio of plant organic material reflects temporal and spatial variations in CO_2 within tropical forest formations in Trinidad. *Oecologia* **1992**, *89*, 435–441.

51. Máguas, C.; Griffiths, H.; Ehleringer, J.R.; Serodio, J. Characterizations of photobiont associations in lichens using carbon isotope discrimination techniques. In *Stable Isotopes and Plant Carbon-Water Relations*; Academic Press London: New York, NY, USA, 1993; pp. 201–212.

52. Palmqvist, K. Tansley review No. 117 Carbon economy in lichens. *New Phytol.* **2000**, *148*, 11–36.

53. Hietz, P.; Wanek, W.; Wania, R.; Nadkarni, N.M. Nitrogen-15 natural abundance in a montane cloud forest canopy as an indicator of nitrogen cycling and epiphyte nutrition. *Oecologia* **2002**, *131*, 350–355.

Growth and Nutrient Status of Foliage as Affected by Tree Species and Fertilization in a Fire-Disturbed Urban Forest

Choonsig Kim, Jaeyeob Jeong, Jae-Hyun Park and Ho-Seop Ma

Abstract: The aim of the present study was to evaluate the growth and macronutrient (C, N, P, K) status in the foliage of four tree species (LT: *Liriodendron tulipifera* L.; PY: *Prunus yedoensis* Matsumura; QA: *Quercus acutissima* Carruth; PT: *Pinus thunbergii* Parl.) in response to fertilization with different nutrient ratios in a fire-disturbed urban forest located in BongDaesan (Mt.), Korea. Two fertilizers ($N_3P_8K_1$ = 113:300:37 kg·ha^{-1}·year^{-1}; $N_6P_4K_1$ = 226:150:37 ha^{-1}·year^{-1}) in four planting sites were applied in April 2013 and March 2014. The growth and nutrient responses of the foliage were monitored six times for two years. Foliar growth and nutrient concentrations were not significantly different ($p > 0.05$) in response to different doses of N or P fertilizer, but the foliage showed increased N and P concentrations and content after fertilization compared with the control ($N_0P_0K_0$). Foliar C and K concentrations were little affected by fertilization. Foliar nutrient concentrations and contents were significantly higher in PY and LT than in PT. The results suggest that the foliar N and P concentration could be used as a parameter to assess the nutrient environments of tree species restored in a fire-disturbed urban forest.

Reprinted from *Forests*. Cite as: Kim, C.; Jeong, J.; Park, J.-H.; Ma, H.-S. Growth and Nutrient Status of Foliage as Affected by Tree Species and Fertilization in a Fire-Disturbed Urban Forest. *Forests* **2015**, *6*, 2199-2213.

1. Introduction

Foliage analysis has received considerable research attention because the nutrient concentrations of foliage have been accepted as adequate indicators of growth and soil fertility at sites in forest stands [1–3]. Generally, the nutrient responses of foliage are commonly used as a tool to assess the nutrient requirements and deficiencies [4].

Urban forests play an important role in enhancing ecosystem service with many demands, such as recreational service, esthetics and biodiversity [5]. However, many urban forest ecosystems face challenging natural and anthropogenic influences, such as forest fire and air pollutions [6]. For example, soils disturbed through forest fire exhibit a myriad of nutritional problems, such as nitrogen (N) and phosphorus (P) deficiency through increased nutrient leaching, surface runoff, and soil erosion [7–8]. The loss of plant nutrients and destabilization of soils in fire-disturbed urban forests might inhibit the rooting and growth of newly planted tree species [9]. In addition, forest management practices, such as nutrient additions, are required to supply sufficient nutrients to optimize the growth of newly planted tree species in urban forests burned by forest fire.

Fertilization stimulates tree growth through positive effects on leaf area and foliar photosynthetic rate [10,11] with increased nutrient concentrations in the living components of trees [12]. However, the status of foliar nutrients was dependent on the type [13] and dose [14] of the fertilizer, tree species [12,15] and many environmental resource factors, such as soil properties and water supply [1,16]. Foliar growth and nutrient status might be important factors in determining responses to fertilization because foliar nutrient concentrations are sensitive to soil nutrient conditions, although the concentrations were much more dependent on species than on sites [1,13,14]. In addition, the application of a suitable fertilizer ratio after foliage analysis, considering the soil environmental conditions and tree growth characteristics, is one of the most effective ways to reduce cost and fertilizer waste.

Although several studies have investigated the changes in soil properties following fires in forest ecosystems in Korea [12,14], limited information is available on urban forest landscapes. In addition, the quantification of foliar nutrient status following fertilization is critical for the actual and potential vegetation restoration of a fire-disturbed urban forest. The aim of the present study was to examine the growth and nutrient responses of foliage based on the compound ratio of fertilizer from four tree species planted in a fire-disturbed urban forest. We hypothesized that the growth and nutrient concentrations of foliage may correspond to the fertilization with different nutrient ratios by tree species-related differences.

2. Experimental Section

2.1. Site

The study site was located in BongDaesan (Mt.) of the Ulsan Metropolitan city, located in southeastern Korea (Figure 1). This mountain was a frequent forest fire area (37 times from 2003 to 2011), primarily resulting from arson. The dominant forest soils are a slightly dry brown forest soil (mostly Entisols or Inceptisols, United States Soil Classification System) derived from granite parent rocks. The annual average precipitation and temperature in this area are 1277 mm·year^{-1} and 14.1 °C, respectively.

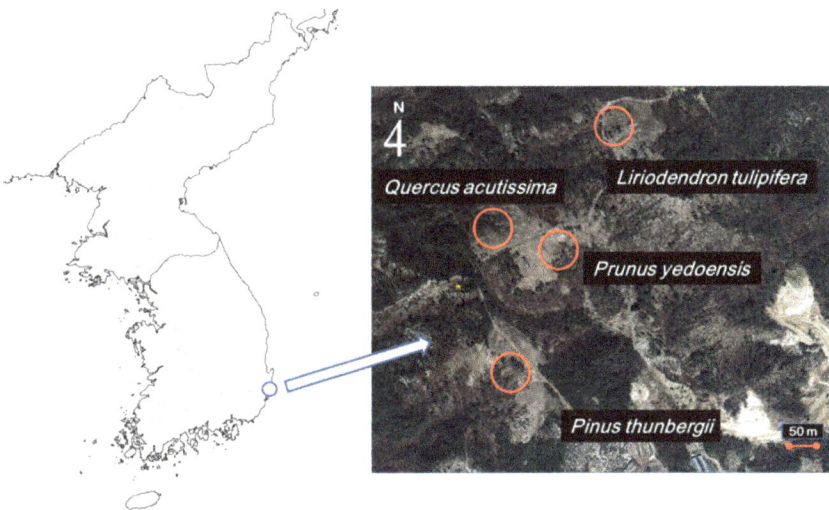

Figure 1. Location of tree planting sites in a fire-disturbed urban forest.

2.2. Methods

This study comprised a completely randomized design with 2 blocks involving total 24 plots (two fertilizer ($N_3P_8K_1$ and $N_6P_4K_1$) and one control ($N_0P_0K_0$) treatments × four different tree planting species (*Liriodendron tulipifera* L. (LT); *Prunus yedoensis* Matsumura (PY); *Quercus acutissima* Carruth (QA); and *Pinus thunbergii* Parl. (PT)) × two blocks) in a fire-disturbed urban forest. The fire resulted from arson in the spring of 2008. In the spring of 2009, the two-year-old (1-1) seedlings of four tree species were planted under a similar site environmental condition on gentle slopes (5–15°) within a close situated distance among tree planting sites from burned forests (Figure 1). Experimental plots consisting of three deciduous hardwood (LT, PY, QA) and one evergreen coniferous plantations (PT) were located adjacent to each other (Figure 1), as sampling bias due to differing site components can be reduced by using a sampling scheme with identical designs for adjacent sites [17]. The stand characteristics and soil properties before the fertilization are shown in Tables 1 and 2, respectively. The soil properties of the study site were similar to a severely burned forest soil in Korea [14].

Table 1. Stand characteristics before fertilization in the study site.

Tree Species	Location	Elevation (m)	Stand Density (trees·ha^{-1})	DBH [†] (mm)	Height (m)
Liriodendron tulipifera	35°32'34.92" N 129°26'22.92" E	154	1666 (160)	28.6 (2.8)	2.87 (0.14)
Prunus yedoensis	35°32'26.48" N 129°26'16.22" E	121	1267 (160)	49.4 (2.3)	4.10 (0.13)
Quercus acutissima	35°32'26.77" N 125°26'15.15" E	118	1533 (66)	26.4 (1.1)	2.22 (0.10)
Pinus thunbergii	35°32'16.36" N 125°26'17.56" E	98	1600 (206)	27.3 (3.7)	2.31 (0.11)

Standard errors in parenthesis. [†] DBH: diameter at breast height (1.2 m).

Table 2. Soil property before fertilization in the study site.

Tree Species	B.D. [†] g·cm^{-3}	Sand	Silt	Clay %	C	N	P mg·kg^{-1}	K$^+$	Ca^{2+} cmol$_c$·kg^{-1}	Mg^{2+}
Liriodendron tulipifera	0.90 (0.01)	49 (4.7)	40 (4.0)	10 (0.7)	3.1 (0.50)	0.15 (0.02)	6.0 (2.0)	0.16 (0.02)	1.60 (0.28)	0.50 (0.09)
Prunus yedoensis	0.89 (0.01)	47 (1.3)	43 (0.7)	10 (1.2)	3.6 (0.70)	0.17 (0.02)	1.2 (0.5)	0.29 (0.11)	1.54 (0.39)	0.56 (0.13)
Quercus acutissima	0.88 (0.02)	43 (1.8)	47 (2.9)	10 (1.2)	1.9 (0.04)	0.08 (0.01)	9.9 (0.7)	0.09 (0.01)	0.43 (0.06)	0.19 (0.02)
Pinus thunbergii	0.91 (0.01)	63 (8.7)	28 (8.1)	9 (0.7)	1.9 (0.01)	0.07 (0.01)	6.5 (2.6)	0.08 (0.02)	0.38 (0.11)	0.16 (0.03)

Standard errors in parenthesis. [†] B.D.: bulk density.

Treatment plots (plot size = 5 × 5 m) of each tree species were randomly assigned with a 1 m buffer zone between each plot on the same facing slopes and aspects to minimize spatial variation in soil properties. The combination of fertilizer ratios ($N_3P_8K_1$ (113 kg N ha^{-1}·year^{-1}, 300 kg P ha^{-1}·year^{-1}, 37 kg K ha^{-1}·year^{-1}) and $N_6P_4K_1$ (226 kg N ha^{-1}·year^{-1}, 150 kg P ha^{-1}·year^{-1}, 37 kg K ha^{-1}·year^{-1})) was based on the guideline of fertilization after forest fire in Korean forests [9]. Urea, fused superphosphate, and potassium chloride fertilizers were used as sources of N, P, and potassium (K), respectively. Fertilizers were applied for two years (in April 2013 and March 2014) by hand across each plot.

Fresh foliar (current-year-old) samples were collected at six times (27 June, 23 August, and 17 October, 2013 and 26 June, 22 August, and 18 October, 2014) for two years with pruners from the mid-crown of three trees per each treatment plot (Figure 2). The foliar samples, in plastic zipper bags, were transported to the laboratory and foliage was separated from twigs or small branches. For each treatment, three repetitions of 10 leaves were counted and weighted. Leaf area was measured by fresh foliar samples by using scanned leaf meter (CI-202 area meter CID, Inc., Camas, WA, USA). The specific leaf area of the foliage was determined as the leaf area (cm^2) and dry weight (g) of the ratio [16]. The foliar samples were oven-dried at 65 °C for 48 h, and the dried samples were ground in a Wiley mill and passed through a 40-mesh stainless steel sieve. Foliar carbon (C) and N concentrations from the ground materials were determined using an elemental analyzer (Thermo Scientific, Flash 2000, Italy). Foliar P and K concentrations were determined through dry ashing 0.5 g of dry foliage at 470 °C for 4 h, digesting the ash with 3 mL of concentrated 5 M HCl, diluting the digest with 0.25 mL of concentrated HNO_3 and 3 mL concentrated 5 M HCl, and measuring the concentrations via ICP (Perkin Elmer Optima 5300DV).

Figure 2. Morphological characteristics of foliage by fertilizer compound ratios of tree species collected on August 2014: (**a**) *Liriodendron tulipifera* L.; (**b**) *Prunus yedoensis* Matsumura; (**c**) *Quercus acutissima* Carruth; and (**d**) *Pinus thunbergii* Parl.

2.3. Statistical Analysis

All data were analyzed using the PROC MIXED procedure of SAS [18] to determine the significance of main fixed effects (tree species (S), fertilizer treatment (F), sampling month (M)) and their interactions (S × F, S × M, F × M, S × F × M), whereas the sampling years of foliage were considered a random effect. The following model was used to describe the data analysis (Equation (1)).

$$Y_{ijk} = u + S_i + F_j + M_k + (SF)_{ij} + (SM)_{ik} + (FM)_{jk} + (SFM)_{ijk} + e_{ijk} \tag{1}$$

where u is the overall mean effect, S is the different tree species (i = 1, 2, 3, 4), F is the fertilizer treatment (j = 1, 2, 3), and M is the sampling month (k = 1, 2, 3). When significant differences were observed, the treatment means were compared using Tukey's test at $p < 0.05$.

3. Results and Discussion

3.1. Results

3.1.1. Growth Response

Foliar growth responses (leaf area, dry weight, and specific leaf area) were significantly ($p < 0.05$) affected by tree species and fertilization (Table 3). The foliar leaf area and dry weight showed a significant two-factor interaction between tree species and fertilization (Figure 3). The leaf area of LT, PY and PT increased after the fertilization, but the leaf area of QA was not affected by fertilization (Figure 3). The foliar dry weight of LT and QA was significantly higher in the $N_6P_4K_1$ treatment than in the $N_0P_0K_0$ treatment (Figure 3).

Table 3. P-value from results of three-way ANOVA on tree species, fertilization and sampling month on growth and nutrient responses of foliage from tree species planted in a fire-disturbed urban forest. Bold values denote significance at $p < 0.05$.

Component	Leaf Area	Specific Leaf Area	Dry Weight	Nutrient Concentration				C/N Ratio	Nutrient Content			
				C	N	P	K		C	N	P	K
Species (S)	**<0.001**	**<0.001**	**<0.001**	**<0.001**	**<0.001**	**<0.001**	**<0.001**	**<0.001**	**<0.001**	**<0.001**	**<0.001**	**<0.001**
Fertilizer (F)	**<0.001**	**0.049**	**0.002**	0.444	**<0.001**	**<0.001**	0.666	**<0.001**	**0.003**	**<0.001**	**<0.001**	0.158
Month (M)	**0.003**	0.309	**<0.001**	**<0.001**	**<0.001**	**<0.001**	**<0.001**	**<0.001**	**<0.001**	0.066	**<0.001**	**<0.001**
S × F	**0.002**	0.545	**0.018**	0.231	**0.032**	0.264	0.986	**0.037**	**0.030**	**0.007**	**0.003**	0.707
S × M	0.234	**0.012**	0.269	**<0.001**	**0.016**	0.770	0.126	0.424	0.187	0.278	**0.002**	**<0.001**
F × M	0.861	0.981	0.903	0.419	0.614	0.275	0.699	0.834	0.917	0.945	0.807	0.994
S × F × M	0.999	0.685	0.625	0.830	0.978	0.999	0.867	0.940	0.607	0.950	0.659	0.963

A significant main effect of the foliar leaf area and dry weight on the sampling month, which was lower in June than in August and October (Figure 3), was observed. The specific leaf area was marginally affected by fertilization ($p = 0.049$), with a significant two-factor interaction between tree species and sampling month (Figure 4).

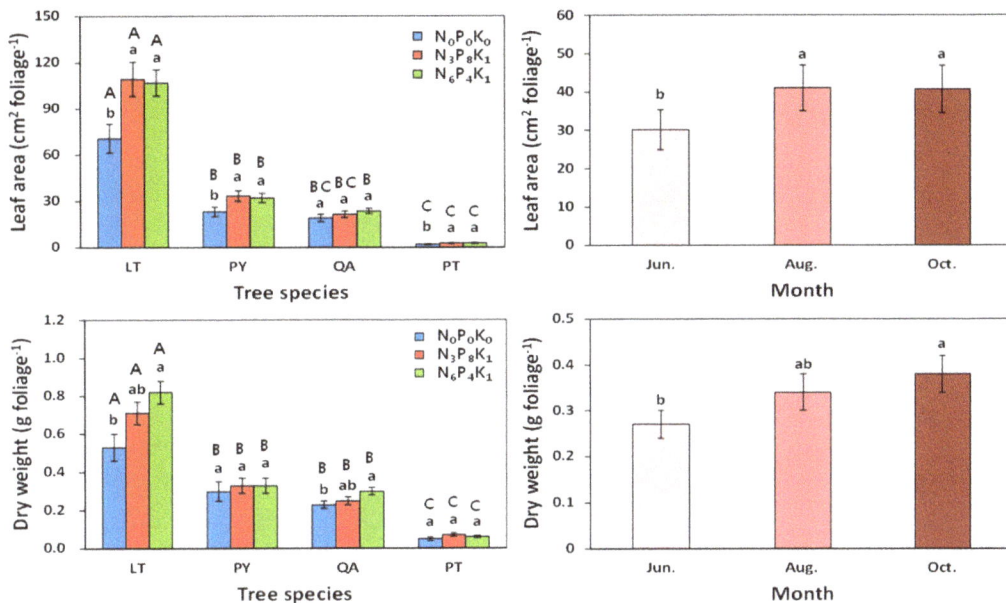

Figure 3. Leaf area and dry weight of foliage by tree species and sampling month in a fire-disturbed urban forest (mean ± standard error). Treatment means with the same lower case letter among fertilizer or sampling month treatments and treatment means with the same upper case letter among tree species are not significantly different at $p < 0.05$. LT: *Liriodendron tulipifera* L.; PY: *Prunus yedoensis* Matsumura; QA: *Quercus acutissima* Carruth; and PT: *Pinus thunbergii* Parl.

3.1.2. Nutrient Responses

There was no significant fertilization effect of the foliar C concentration, while the C concentration showed a two-factor interaction between tree species and sampling month (Table 3, Figure 5). The foliar C content among tree species was highest in LT, followed by PY, QA and PT, while the content of LT and QA was significantly higher in the $N_6P_4K_1$ treatments compared with the $N_0P_0K_0$ treatments. The foliar N concentration significantly differed with the sampling month (Figure 5). The foliar N concentration of four tree species was lower in October or in the $N_0P_0K_0$ treatment than in June or in the fertilizer treatments (Figure 5), but the concentration did not significantly differ between the different N ratio of the fertilizer, such as the $N_3P_8K_1$ and $N_6P_4K_1$ treatments. The foliar N content was similar to foliage C content of the four tree species (Figure 5).

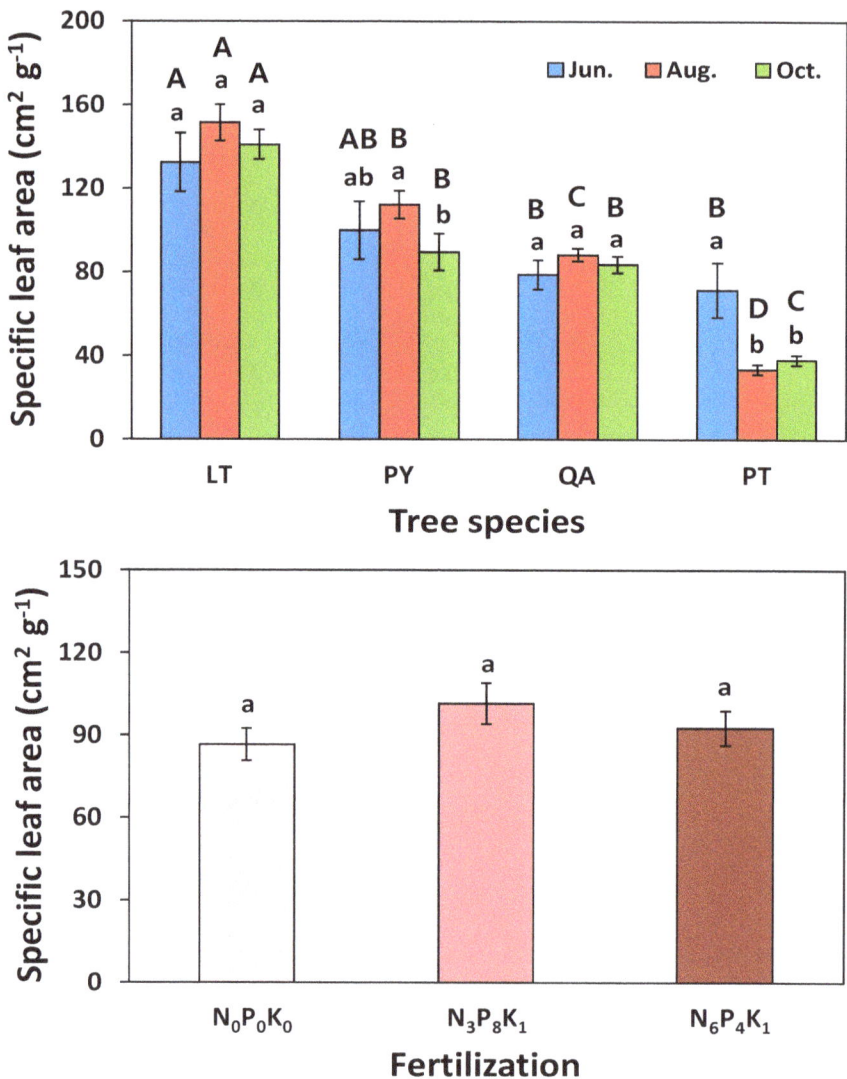

Figure 4. Specific leaf area of foliage by tree species and fertilization in a fire-disturbed urban forest (mean ± standard error). Treatment means with the same lower case letter among sampling month or fertilizer treatments and treatment means with the same upper case letter among tree species are not significantly different at $p < 0.05$. LT: *Liriodendron tulipifera* L.; PY: *Prunus yedoensis* Matsumura; QA: *Quercus acutissima* Carruth; and PT: *Pinus thunbergii* Parl.

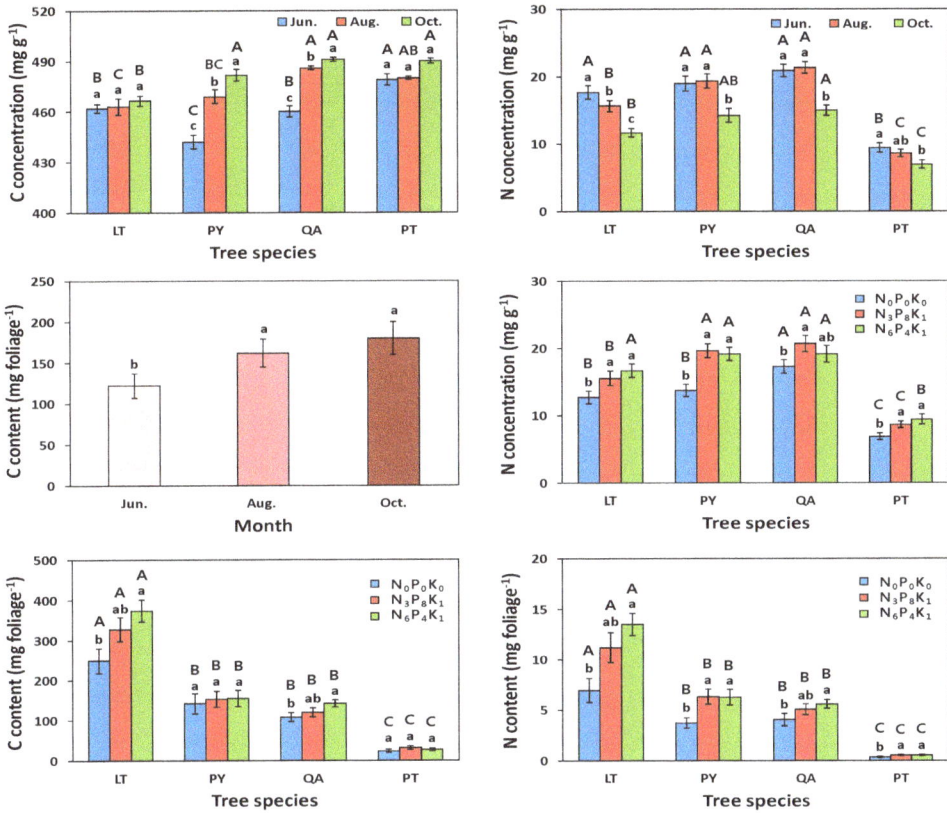

Figure 5. Carbon and nitrogen concentration and content of foliage by tree species and sampling month in a fire-disturbed urban forest (mean ± standard error). Treatment means with the same lower case letter among sampling month or fertilizer treatments and treatment means with the same upper case letter among tree species are not significantly different at $p < 0.05$. LT: *Liriodendron tulipifera* L.; PY: *Prunus yedoensis* Matsumura; QA: *Quercus acutissima* Carruth; and PT: *Pinus thunbergii* Parl.

A significant two-factor interaction between the tree species and fertilizer treatments on the C/N ratio of foliage was observed (Figure 6). The C/N ratio in all tree species was significantly lower in the fertilizer than in the $N_0P_0K_0$ treatments, regardless of the compound ratio of fertilizer, while the C/N ratio was significantly higher in conifers (PT) than in broadleaf tree species (LT, PY, and QA). The C/N ratio was also highest in October, followed by August and June (Figure 6).

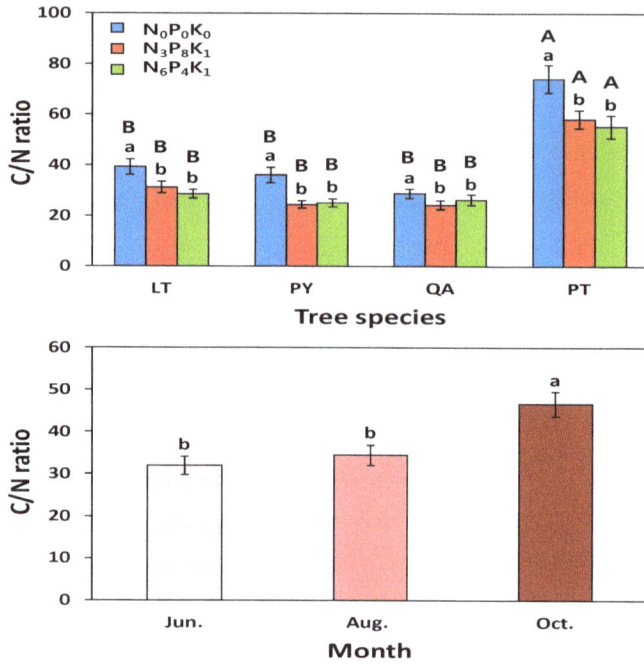

Figure 6. Carbon and nitrogen ratio of foliage by tree species and sampling month in a fire-disturbed urban forest (mean ± standard error). Treatment means with the same lower case letter among fertilizer or sampling month treatments and treatment means with the same upper case letter among tree species are not significantly different at $p < 0.05$. LT: *Liriodendron tulipifera* L.; PY: *Prunus yedoensis* Matsumura; QA: *Quercus acutissima* Carruth; and PT: *Pinus thunbergii* Parl.

The foliar P concentration was significantly affected by tree species, fertilization and sampling month with no two- or three-way interaction (Table 3, Figure 7). The foliar P concentration was significantly higher in PY than in other tree species. The foliage of PY showed a significantly high P concentration compared with the other three tree species (Figure 7). The foliar P concentration was highest in August, followed by October and June (Figure 7). There was a significant two-factor interaction between the tree species and fertilizer treatments or between the sampling month and tree species on foliar P content (Figure 7). In contrast to the foliar P concentrations, the foliar K concentration and content were little affected by fertilization (Table 3), while the concentration and content were higher in October than in June, except for PT. The foliar K concentration was significantly higher in PY than in the other three tree species (Figure 8).

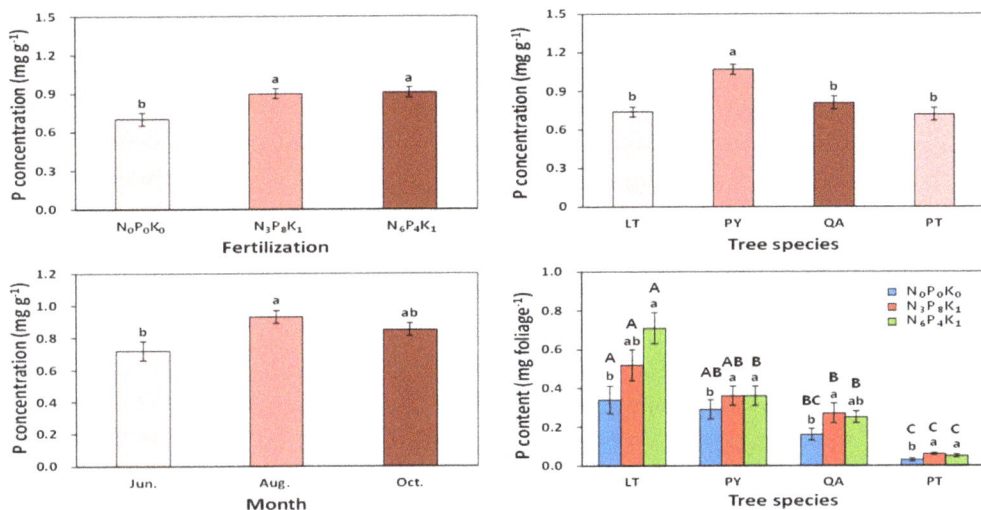

Figure 7. Phosphorus concentration and content of foliage by fertilization, tree species, and sampling month in a fire-disturbed urban forest (mean ± standard error). Treatment means with the same lower case letter among fertilizer, tree species or sampling month treatments and treatment means with the same upper case letter among tree species are not significantly different at $p < 0.05$. LT: *Liriodendron tulipifera* L.; PY: *Prunus yedoensis* Matsumura; QA: *Quercus acutissima* Carruth; and PT: *Pinus thunbergii* Parl.

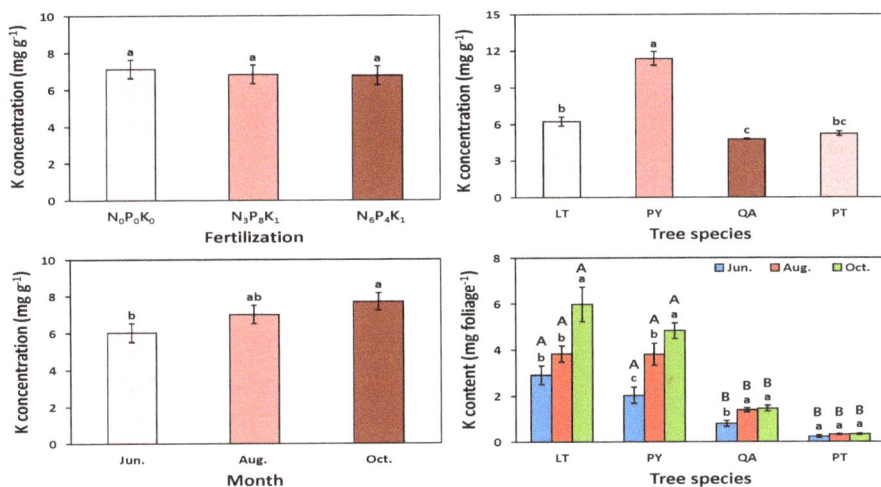

Figure 8. Potassium concentration and content of foliage by tree species, fertilization and sampling month in a fire-disturbed urban forest (mean ± standard error). Treatment means with the same lower case letter among fertilizer, tree species or sampling month treatments and treatment means with the same upper case letter among tree species are not significantly different at $p < 0.05$. LT: *Liriodendron tulipifera* L.; PY: *Prunus yedoensis* Matsumura; QA: *Quercus acutissima* Carruth; and PT: *Pinus thunbergii* Parl.

3.2. Discussion

3.2.1. Growth Responses

Because nutrient availability is generally considered the major resource factor limiting growth in many forest tree species [19], foliage growth after fertilization would have greater leaf area and dry weight than foliage grown without fertilization. Although there was a significant tree species effect on foliage growth because of the morphological difference among tree species (Figure 2), fertilization induced different growth responses among tree species planted in a fire-disturbed urban forest. Most tree species showed increased leaf area following fertilization, while the dry weight and specific leaf area were not affected by fertilization, except for the foliar dry weight in LT and QA. This result might reflect the different C allocation patterns during the growth and development of foliage in different tree species [20]. For example, N fertilization increased the foliage biomass, rather than the dry weight of the foliage in eucalyptus plantations [21]. However, the lack of a significant fertilizer compound ratio on the morphological growth responses of foliage might reflect other micronutrient imbalances resulting from increased N fertilization [11] or multi-factorial influences that occur during the leaf life-span, such as light environment, nutrients, temperature and water supply [22]. Seasonal patterns of foliar leaf area and dry weight could be attributed to foliage maturation or resource allocation between woody (roots, stem wood, and branches) and photosynthetic compounds [20]. The high leaf area and dry weight in August and October could reflect the maturation of foliage, whereas the low leaf area in June indicated that the foliage in the four tree species was not fully elongated during this season.

3.2.2. Nutrient Responses

The foliar C concentration was a poor indicator of the fertilizer response in a fire-disturbed urban forest ($p = 0.444$). Similarly, other studies reported that the inter- and intra-specific variations of the C concentration in tree species were determined through genetic and environmental factors [23,24], rather than resource factors, such as nutrient availability [25]. For example, genetic differences among tree species might result in different C concentration of foliage, which was significantly higher in PT than in LT (Figure 5). In addition, the foliage C concentration was generally increased in the heavy litter fall season (October) compared with the growing season (June–August) because of low mineral concentrations of foliage by nutrient resorption (Figure 5). The foliar C content reflected differences in the foliar dry weight induced after fertilization rather than the foliar C concentration among tree species and reflected differences in foliage maturation associated with the translocation of carbohydrates and other cellular materials to active growing tissues [15].

As expected, the foliar N concentration and content among tree species were lower in conifers (PT) than in broadleaf tree species (LT, PY, and QA). The foliar N concentration and content of the four tree species increased following fertilization, while insignificant increases in foliage N concentration occurred with different doses of N fertilizer ($N_3P_8K_1$ and $N_6P_4K_1$). A high foliar N concentration and content with fertilization likely reflects the increased uptake of available N in soil depths [15,19], as tree species with high N availability in mineral soil produce high foliage N concentration [26]. The low foliar N concentration and content observed under $N_0P_0K_0$ treatment suggested the N deficiency and limited tree growth in a fire-disturbed urban forest. However, a similar foliar N concentration of the fertilizer compound ratio might result from luxury N consumption from a high dose of N fertilizer ($N_6P_4K_1$) compared with $N_3P_8K_1$ fertilizer, although the foliar N concentration after fertilization was regulated through factors, such as the combined effect of the available soil N status, tree growth, and climate factors. The foliar N concentration and content were affected by sampling months because of the resorption of foliar N in October or the dilution effect due to the increased relative accumulation of carbohydrates with increasing dry weight [4]. Similarly, the high C/N ratio in October reflected foliar N resorption before heavy litter fall [27]. Also, C/N ratio was higher under nutrient deficient conditions ($N_0P_0K_0$) compared with trees grown under improved nutrient conditions ($N_6P_4K_1$ and $N_3P_8K_1$).

The foliar P concentration and content of the four tree species were increased after fertilization, suggesting that the four tree species were P deficient in the fire-disturbed urban forest. Similarly, the foliage of eucalypt forests showed an increased P concentration following P fertilizer application [28], reflecting enhanced uptake and mineralization in the rhizosphere. However, the foliar P concentration was not responsible for the increased dose of the P in the fertilizer ($N_6P_4K_1$ and $N_3P_8K_1$), indicating that much of the P uptake at a high dose might be associated with luxury consumption. The difference of foliage P content among tree species is similar to those observed in dry weight of foliage.

The K fertilizer had minor effects on the foliar K concentration and content of the tree species except of PY, suggesting that this nutrient might not play an important role in limiting growth with N or P applications. Similar foliar K concentrations following fertilization could be affected through a diluting effect via increased foliar dry weight with N or P supply levels compared with the $N_0P_0K_0$ treatment. Additionally, the K in foliage could be increasingly leached after rain from the tree canopy in the different fertilizer treatments because higher fluxes of K from throughfall were observed at more fertile sites compared with sites with a poorer nutrient status [27]. However, the significant high K concentration and content of foliage in PY could be influenced by high soil K availability (0.29 cmolc·kg^{-1}) compared with other tree planting sites of 0.08–0.16 cmolc·kg^{-1} (Table 2). This result supports that the K concentration in sweetgum (*Liquidambar styraciflua* L.) foliage was associated with the inherent soil chemical properties rather than fertilization [29].

4. Conclusions

The growth characteristics, such as leaf area, dry weight and specific leaf area in foliage, were not affected by fertilizer compound ratios, whereas fertilization produced high foliar N or P concentrations of four tree species planted in a fire-disturbed urban forest. In contrast to N and P, foliar C and K concentrations and contents in tree species were little affected by fertilization. The foliar N, P, and K concentrations substantially decreased during senescence (October) compared with the growing season (June, August), except for the increased foliar C concentration. There was no clear effect on the foliar nutrient status from different compound ratios (e.g., $N_3P_8K_1$ and $N_6P_4K_1$) of fertilizer on different tree species because of the luxury consumption at high N and P doses. These results suggest that a new compound ratio of fertilizer is needed to optimize the vegetation growth of trees planted in fire-disturbed urban forests because of tree species differences on foliage N and P concentration and content following fertilization.

Acknowledgments

This study was carried out with the support of "Forest Science & Technology Projects (Project No. S211212L030320)" provided by Korea Forest Service.

Author Contributions

Choonsig Kim led the writing; Jaeyeob Jeong performed the experiments; and Jae-Hyun Park and Ho-Seob Ma analyzed the data.

Conflicts of Interest

The authors declare no conflict of interest.

References

1. Weetman, G.F.; Wells, C.G. Plant analyses as an aid in fertilizing forests. In *Soil Testing and Plant Analysis*, 3rd ed.; Westerman, R.L., Ed.; Soil Science Society of America: Madison, WI, USA, 1990; pp. 659–690.

2. Barron-Gafford, G.A.; Will, R.E.; Burkes, C.; Shiver, B.; Teskey, R.O. Nutrient concentrations and content, their relation to stem growth, of intensively managed *Pinus taeda* and *Pinus elliottii* stands of different planting densities. *For. Sci.* **2003**, *49*, 291–300.

3. Tausz, M.; Trummer, W.; Wonisch, A.; Goessler, W.; Grill, D.; Jiménez, M.S.; Morales, D. A survey of foliar mineral nutrient concentrations of *Pinus canariensis* at field plots in Tenerife. *For. Ecol. Manag.* **2004**, *189*, 49–55.

4. Silfverberg, K.; Moilanen, M. Long-term nutrient status of PK fertilized Scots pine stands on drained peatlands in North-Central Finland. *Suo* **2008**, *59*, 71–88.

5. Pakaki, D.E.; Carreir, M.M.; Cherrier, J.; Grulke, N.E.; Jennings, V.; Princetl, S.; Pouyat, R.V.; Whitlow, T.H.; Zipperer, W.C. Coupling biogeochemical cycles in urban environments: Ecosystem services, green solutions, and misconceptions. *Front. Ecol. Environ.* **2011**, *9*, 27–36.

6. Escobedo, F.J.; Nowak, D.J. Spatial heterogeneity and air pollution removal by an urban forest. *Landsc. Urban Plan.* **2009**, *90*, 102–110.

7. Kim, C.; Koo, K.S.; Byun, J.K.; Jeong, J.H. Post-fire effects on soil properties in red pine (*Pinus densiflora*) stands. *Forest Sci. Tech.* **2005**, *1*, 1–7.

8. Xue, L. Li, Q.; Chen, H. Effects of a wildfire on selected physical, chemical and biochemical soil properties in a *Pinus massoniana* forest in south China. *Forests* **2014**, *5*, 2947–2966.

9. Kim, Y.S.; Byun, J.K.; Kim, C.; Park, B.B.; Kim, Y.K.; Bae, S.W. Growth response of *Pinus densiflora* seedlings to different fertilizer compound ratios in a recently burned area in the eastern coast of Korea. *Landsc. Ecol. Eng.* **2014**, *10*, 241–247.

10. Garrison, M.T.; Moore, J.A.; Shaw, T.M.; Mika, P.G. Foliar nutrient and tree growth response of mixed-conifer stands to three fertilization treatments in northeast Oregon and north central Washington. *For. Ecol. Manag.* **2000**, *132*, 183–198.

11. Amponsah, I.G.; Comeau, P.G.; Brockley, R.P.; Lieffers, V.J. Effects of repeated fertilization on needle longevity, foliar nutrition, effective leaf area index, and growth characteristics of lodgepole pine in interior British Columbia, Canada. *Can. J. For. Res.* **2005**, *35*, 440–451.

12. Miller, J.H.; Allen, H.L.; Zutter, B.R.; Zedaker, S.M.; Newbold, R.A. Soil and pine foliage nutrient responses 15 years after competing-vegetation control and their correlation with growth for 13 loblolly pine plantations in the southern United States. *Can. J. For. Res.* **2006**, *36*, 2412–2425.

13. Kim, C.; Ju, N.K.; Lee, H.Y.; Lee, K.S. Effects of growth, carbon and nitrogen response of foliage in a red pine stand. *Korean J. Soil Sci. Fertil.* **2013**, *46*, 9–15.

14. Kim, C.; Byun, J.K.; Park, J.H.; Ma, H.S. Litter fall and nutrient status of green leaves and leaf litter at various compound ratios of fertilizer in sawtooth oak stands, Korea. *Ann. For. Res.* **2013**, *56*, 339–350.

15. Mugasha, A.G.; Pluth, D.J.; Macdonald, S.E. Effects of fertilization on seasonal patterns of foliar mass and nutrients of tamarack and black spruce on undrained and drained minerotrophic peatland sites. *For. Ecol. Manag.* **1999**, *116*, 13–31.

16. Bauer, G.; Schulze, E-D.; Mund, M. Nutrient contents and concentrations in relation to growth of *Picea abies* and *Fagus sylvatica* along a European transect. *Tree Physiol.* **1997**, *17*, 777–786.

17. Oksanen, L. Logic of experiments in ecology: Is pseudoreplication a pseudoissue? *Oikos* **2002**, *94*, 27–38.

18. SAS Institute Inc. *SAS/STAT Statistical Software*; Version 9.1; SAS Institute Inc.: Cary, NC, USA, 2003.

19. Binkley, D.; Fisher, R. Nutrient management. In *Ecology and Management of Forest Soils*, 4th ed.; Wiley-Blackwell: Hoboken, NJ, USA, 2013; pp. 254–275.

20. Gough, C.M.; Seiler, J.R.; Maier, C.A. Short-term effects of fertilization on loblolly pine (*Pinus taeda* L.) physiology. *Plant Cell Environ.* **2004**, *27*, 876–886.

21. Laclau, J.P.; Almeida, J.S.R.; Gonçalves, J.L.M.; Saint-André, L.; Ventura, M.; Ranger, R.; Moreira, R.M.; Nouvellon, Y. Influence of nitrogen and potassium fertilization on leaf lifespan and allocation of above-ground growth in *Eucalyptus* plantations. *Tree Physio.* **2009**, *29*, 114–124.

22. Wilson, P.J.; Thompson, K.; Hodgson, J.G. Specific leaf area and leaf dry matter content as alternative predictors of plant strategies. *New Phytol.* **1999**, *143*, 155–162.

23. Bert, D.; Danjon, D. Carbon concentration variations in the roots, stem and crown of mature *Pinus pinaster* (Ait.). *For. Ecol. Manag.* **2006**, *222*, 279–295.

24. Zhang, Q.; Wang, C.; Wang, X.; Quan, X. Carbon concentration variability of 10 Chinese temperate tree species. *For. Ecol. Manag.* **2009**, *258*, 722–727.

25. Jeong, J.; Park, J.-H.; Kim, J.-I.; Lim, J.-T.; Lee, S.-R.; Kim, C. Effects of container volumes and fertilization on red (*Pinus densiflora*) and black pine (*Pinus thunbergii*) seedlings growth. *Forest Sci. Tech.* **2010**, *6*, 80–86.

26. Sariyildiz, T.; Anderson, J.M. Variation in the chemical composition of green leaves and leaf litters from three deciduous tree species growing on different soil types. *For. Ecol. Manag.* **2005**, *210*, 303–319.

27. Hagon-Thorn, A.; Varnagiryte, I.; Nihlgard, B.; Armolaitis, K. Autumn nutrient resorption and losses in four deciduous forest tree species. *For. Ecol. Manag.* **2006**, *228*, 33–39.

28. O'connell, A.M.; Grove, T.S. Influence of nitrogen and phosphorus fertilizers on amount and nutrient content of litterfall in a regrowth eucalypt forest. *New For.* **1993**, *7*, 33–47.

29. Scott, A.; Burger, J.A.; Kaczmarek, D.J.; Kane, M.B. Growth and nutrition response of young sweetgum plantations to repeated nitrogen fertilizer on two site types. *Biomass Bioenerg.* **2004**, *27*, 313–325.

Impact of Nitrogen Fertilization on Forest Carbon Sequestration and Water Loss in a Chronosequence of Three Douglas-Fir Stands in the Pacific Northwest

Xianming Dou, Baozhang Chen, T. Andrew Black, Rachhpal S. Jassal and Mingliang Che

Abstract: To examine the effect of nitrogen (N) fertilization on forest carbon (C) sequestration and water loss, we used an artificial neural network model to estimate C fluxes and evapotranspiration (ET) in response to N fertilization during four post-fertilization years in a Pacific Northwest chronosequence of three Douglas-fir stands aged 61, 22 and 10 years old in 2010 (DF49, HDF88 and HDF00, respectively). Results showed that N fertilization increased gross primary productivity (GPP) for all three sites in all four years with the largest absolute increase at HDF00 followed by HDF88. Ecosystem respiration increased in all four years at HDF00, but decreased over the last three years at HDF88 and over all four years at DF49. As a result, fertilization increased the net ecosystem productivity of all three stands with the largest increase at HDF88, followed by DF49. Fertilization had no discernible effect on ET in any of the stands. Consequently, fertilization increased water use efficiency (WUE) in all four post-fertilization years at all three sites and also increased light use efficiency (LUE) of all the stands, especially HDF00. Our results suggest that the effects of fertilization on forest C sequestration and water loss may be associated with stand age and fertilization; the two younger stands appeared to be more efficient than the older stand with respect to GPP, WUE and LUE.

Reprinted from *Forests*. Cite as: Dou, X.; Chen, B.; Black, T.A.; Jassal, R.S.; Che, M. Impact of Nitrogen Fertilization on Forest Carbon Sequestration and Water Loss in a Chronosequence of Three Douglas-Fir Stands in the Pacific Northwest. *Forests* **2015**, *6*, 1897-1921.

1. Introduction

The factors affecting the physiological processes controlling the amount of terrestrial carbon (C) sequestered in terrestrial ecosystems mainly include atmospheric CO_2 concentration, climatic variability, land use change and nitrogen (N) fixation [1–3]. In terrestrial ecosystems, the important processes of determining whether an ecosystem is a C sink or source are C fixation and release through photosynthesis and respiration, respectively. For N-limited terrestrial ecosystems, additional N supply can affect these processes, and in turn affect the strength of C sinks or sources [4–9].

Research has shown that the response of C exchange to N fertilization in terrestrial ecosystems is positive or negative depending on the ecosystem N status [10–13]. Many researchers have studied the effects of N enrichment on net ecosystem productivity (NEP) in the forest ecosystem, but it remains uncertain as to how much N addition contributes to the magnitude of terrestrial NEP [1,14,15].

Many studies have found that N addition increases NEP [16–21]. Magnani *et al.* [22] found NEP was highly correlated with N deposition in forest ecosystems and the response of NEP to N in wet deposition was approximately 725 kg C kg^{-1} N. However, Sutton *et al.* [23] found that the response of NEP to N was about 50–75 kg C kg^{-1} N based on a model re-analysis across 22 European forest sites and considering the impacts of climatological differences among stands. de Vries *et al.* [24] also found that the response of NEP to N would be 30–70 kg C kg^{-1} N based on a multi-factor analysis of European forest measurements at nearly 400 intensively monitored forest plots in Europe. The results of both Sutton *et al.* [23] and de Vries *et al.* [24] are much smaller than the estimates of Magnani *et al.* [22]. Therefore, there is still some controversy over the magnitude and sustainability of NEP resulting from N enrichment and its potential mechanisms [25–27].

However, other studies have reported that N addition may cause a relatively small increase in NEP [12,28], have no effect [29], or have a negative effect by altering plant and microbial communities, including threatened and endangered species [30]. Nadelhoffer *et al.* [31] pointed out that the influence of N enrichment on ecosystem C storage should be minor in forest ecosystems in the northern temperate zone. Currie *et al.* [32] stated that small increases in C storage occurring primarily in living and dead wood may result from elevated N deposition over the next few decades. Using meta-analysis from global N addition experiments, Liu and Greaver [33] concluded that the increases in short-term C sequestration below ground caused by N enrichment are due to increased C storage in the surface organic layer, but it appeared to be difficult to predict the response of C storage to N enrichment through a long-term experiment. Harpole *et al.* [34] reported that the ability of grassland ecosystems to sequester C late in the growing season was decreased by N enrichment because of increased growing season water use and earlier leaf senescence.

Furthermore, water and C cycles in terrestrial ecosystems are closely coupled. A global sensitivity analysis for an integrated forest ecosystem model [35] also showed the intrinsic coupling between water and the C cycle through stomatal conductance, which affect both water and C fluxes between the biosphere and atmosphere. Nutrients may not only affect productivity and foliar biomass but are also associated with evapotranspiration (ET) in forests and other ecosystems [36]. Increased productivity and foliar biomass induced by increased nutrient availability are probably responsible for the increase in ET resulting from increased rainfall interception and transpiration. As a consequence, more attention should be paid to water use efficiency (WUE), defined as the ratio of C gain (usually gross primary productivity, GPP) to water loss, *i.e.*, ET [37,38]. Studies have reported that by increasing WUE, N addition could enhance plant productivity [39–41]. On the other hand, other researchers have found that N enrichment has no impact on WUE [42–44] or a negative effect on WUE [45,46]. Consequently, it is difficult to account for the differences in N-induced effects on WUE [47]. Furthermore, fertilization may increase NEP through an increase in GPP, resulting from an increase in resource capture and resource use efficiency, e.g., increase in absorbed photosynthetically active radiation (APAR), light use efficiency (LUE, the ratio of GPP to APAR), and/or ecosystem carbon-use efficiency (CUEe, the ratio of NEP to GPP) [5]. Although LUE models have been extensively used to estimate regional or global GPP through remote sensing techniques [48–50], the biophysical controls on LUE in terrestrial ecosystems, especially soil N, remain poorly understood, especially in Douglas-fir stands on the west coast of Canada.

The three different-aged Douglas-fir stands in this study (DF49, established in 1949, HDF88, established in 1988, and HDF00, established in 2000) are in the Pacific Northwest coastal forest region of Canada. Forests in this region between Oregon and Alaska cover approximately 10^5 km^2

and play an important role in the global C cycle. In this region, due to their considerable distances from heavy industrial activity, there is very little atmospheric N deposition, and N deficiency generally occurs in soils [51]. Fertilization using a single dose of urea in early 2007 in all three of these Douglas-fir stands and on-going long-term measurements of C fluxes using the eddy-covariance (EC) technique offer an opportunity to examine the effects of N fertilization on C and water fluxes. Previous studies on the effects of N fertilization have focused on C fluxes at DF49 in the first post-fertilization year using a process-based model [52], and C fluxes, ET and WUE in all three stands in the two post-fertilization years using an empirical model (multiple linear regression, MLR) [53].

In this study, we have used the measured C fluxes (viz., GPP, ecosystem respiration (R), NEP ($=GPP - R$)), ET and climatic variables in the three stands from 1998 to 2010. The objectives of this study are threefold: (1) to estimate the effects of N fertilization in the three stands on the C fluxes, ET, WUE and LUE using two common data-driven models, MLR and an artificial neural network (ANN); (2) to examine the variation of the N fertilization effects during the 4-year post-fertilization period; and (3) to gain insight into the mechanisms of N fertilization controlling the C fluxes, ET, WUE and LUE in the three stands.

2. Materials and Methods

2.1. Site Descriptions

Three stands were located on the east coast of Vancouver Island, BC, between Campbell River and Denman Island. In 2010, the ages of the stands were 61, 22 and 10 years old, which spanned the typical ages of stands in this area. The three stands were dominated by Douglas-fir (*Pseudotsuga menziesii* (Mirb.) Franco) with relatively similar stand densities, soil, topography, elevation, and biogeoclimatic classification [54]. On the whole, stand age and the corresponding stand structural characteristics were considered to be largely responsible for the differences among these stands in C uptake and ET. The oldest stand (49°52'7.8" N, 125°20'6.3" W, flux tower location), DF49, was planted with Douglas-fir seedlings in 1949 and occupied an area of 130 ha. The stand comprised 80% Douglas-fir, 17% western red cedar (*Thuja plicata*), and 3% western hemlock (*Tsuga heterophylla*). The mean annual temperature and precipitation measured during 1998–2010 in the stand are 8.1 °C and 1350 mm, respectively. The 22-year-old (pole-sapling) stand (49°32'10.49" N, 124°54'7.18" W), HDF88, 110 ha in size, was situated about 30 km southwest of DF49. The stand comprised 75% Douglas-fir, 21% western red cedar and 4% grand fir (*Abies grandis*). The mean annual temperature and rainfall in the stand, since measurements began in 2002, are 9.1 °C and 1579 mm, respectively. The youngest stand (49°52'1.08" N, 125°16'43.80" W), HDF00, was located about 2 km southeast of DF49. This 32-ha stand was harvested in the winter of 1999/2000 and in the following spring was planted with one-year-old seedlings (93% Douglas-fir, 7% western red cedar). Much of the leaf area at this site was due to the growth of pioneer and understory species from the previous stand. The mean annual temperature and rainfall at the site are 8.5 °C and 1321 mm, respectively. Additional details on stand history, vegetation and soil can be found in Humphreys *et al.* [54], Chen *et al.* [55], Krishnan *et al.* [56] and Morgenstern *et al.* [57].

2.2. Stand Fertilization

Urea fertilizer was spread aerially at 200 kg·N·ha^{-1} over about 110 ha at DF49 and 20 ha at HDF88, which included most of the respective tower flux footprints, on 13 January and 17 February 2007, respectively. Because of the young age of the planted trees and competing understory at HDF00, 80 g urea per tree (~60 kg·N·ha^{-1}) was manually applied along the tree drip line of trees located in the 5-ha tower flux footprint on 13–14 February 2007. The fertilized areas around the respective towers accounted for more than 80% of the EC fluxes measured during unstable atmospheric conditions, which occurred during the daytime and nighttime [55].

2.3. EC and Climate Measurements

The EC technique has been used to measure continuous half-hourly CO_2 and water vapor fluxes since 2000, 2001 and 1997 at DF49, HDF88 and HDF00, respectively. Climate variables that were also measured included precipitation, net radiation (R_n), photosynthetically active radiation (PAR), wind speed, air temperature (T_a) and humidity, soil temperature (T_s) and volumetric soil moisture content (θ). Details on EC instrument characteristics, measurements and calculation procedures for these stands are described in Humphreys *et al.* [54], Chen *et al.* [55] and Morgenstern *et al.* [57]. Half-hourly EC-measured CO_2 fluxes above the stand were corrected by adding the estimated rate of change in CO_2 storage in the air column below the EC sensor height to obtain the net ecosystem exchange (NEE). NEP was obtained as NEP = −NEE [54].

The methodology of gap-filling for EC-measured NEP and its separation into GPP and R was adopted from the Fluxnet-Canada Research Network protocol [55,58]. Missing nighttime values of R were obtained using an exponential relationship between NEE and T_s (Q_{10} model) using measured NEE values corresponding to friction velocity values greater than a threshold value. This model was also used to obtain daytime values of R. GPP during the growing season was obtained as the sum of daytime-measured NEP and daytime R and was set to zero during the cold season. An empirical light-response relationship was used to fill gaps in GPP. The details on NEP gap-filling and its separation into GPP and R as well as the uncertainties in the annual sums of NEP, GPP and R can be found in Chen *et al.* [59].

2.4. Artificial Neural Networks

An ANN model is characterized by flexible mathematical structures, which can be used to investigate complicated non-linear relationships between inputs and outputs [60]. For training of time series prediction, the ANN model is conducted by propagating the input data and then back-propagating the error by means of self-adjustment of the weights using least squares residuals so that the simulated outputs best approximate the target outputs (observed data). The performance of an ANN model as a purely empirical non-linear regression model is commonly affected by the quality of the training dataset, network architecture and network training [61]. Additional mathematical background on ANN models can be found in White [62], Jain *et al.* [63], Zhang *et al.* [64] and Basheer and Hajmeer [65]. The back-propagation (BP) algorithm used in this study for network training has been successfully applied in ecosystem C flux estimation [66–68] because it is regarded as the most widely applied algorithm in the ANN literature [69,70]. A three-layer BP neural network (BPNN) was employed with one hidden layer because Funahashi [71] and Cybenko [72] demonstrated that even with one hidden layer, a BPNN can approximate any continuous multivariate function with reasonable precision. In the calibration phase, the optimization approach applied was based on Levenberg–Marquardt [73], which is known to outperform the simple gradient descent and other conjugate gradient methods. In this three-layer BPNN, a hyperbolic tangent sigmoid transfer function (tansig) at hidden layer and a linear transfer function (purelin) at output layer was used. Aiming at obtaining the hidden node number and avoiding over-fitting during the training period, we used a trial and error method to select the optimal solution by altering the number of hidden nodes.

2.5. Comparison of ANN and MLR Models

Six environmental variables that included T_s at the 5-cm depth, T_a, downwelling PAR (Q), vapor pressure deficit (D), R_n, and θ in the 0–30 cm layer, and stand age were selected to train the ANN model using pre-fertilization flux measurements. The particular environmental variables that worked best for GPP, R and ET were chosen as inputs after trying several ANN networks. GPP, R and ET were simulated independently using different input variables for the ANN model. Specifically, three different variable groups, (T_a, θ, Q, D, stand age), (T_a, θ, T_s, stand age) and (T_a, θ, R_n, D, stand age), were then used to train the ANN model and predict monthly GPP, R and ET, respectively, for the three sites assuming the stands had not been fertilized. To ensure high precision in the period of model prediction, we used multi-year monthly values of climate variables, stand age and EC-measured C fluxes and ET before 2005 to train the ANN model and then verified the trained model with measurements in 2005 and 2006. When we were convinced that the optimized ANN model successfully simulated the multiyear seasonal variations in C fluxes and ET, the input values for the post-fertilization period were transferred into the trained model to predict the GPP, R and ET for 2007 to 2010. The resulting differences between the measurements and predictions were used to discern the impact of fertilization.

At the beginning of the training, inputs and outputs were normalized between 0 and 1, which is a common preprocessing method for variables with different values. Moreover, we also used an MLR model to simulate C fluxes and ET with the same variables to assess the advantages of the ANN modeling approach over traditional regression modeling techniques. Similar to the ANN model, we used a variable group (e.g., T_a, θ, Q, D, stand age) as the independent variables to establish an MLR equation and predict the dependent variable (e.g., GPP). Because of the differences in age among these three stands, it is important to note that both ANN and MLR models were run separately for each stand. The performances of the models were evaluated by comparing predicted with measured values in the calibration and verification periods. In this study, computer programming and data analysis were performed using MATLAB (version 7.13, R2011b).

3. Results

3.1. Environmental Variables

Since all three stands were exposed to similar weather conditions (warm dry summers and wet cool winters), we describe only the seasonal and annual variations in environmental variables at DF49. The seasonal patterns of T_a, T_s, θ, D, Q and R_n during 2007–2010 were similar to those measured before fertilization from 1998 to 2006 (Figure 1). Compared with the 9-year means during the pre-fertilization period, the annual average values of T_a during 2007–2010 were lower by about 0.7, 1.1, 0.4 and 0.8 °C, respectively, and the annual average values of T_s were higher by about 0.3, 0.1, 0.1 and 0.6 °C, respectively (Figure 1a,b). Annual values of total Q in 2007, 2008 and 2010 were lower by about 9%, 3% and 6%, respectively, suggesting possible growth limitation in those years, but were higher by about 8% in 2009 (Figure 1f). Annual average values of θ in the 0–30 cm soil layer in the growing season (May–September) in the post-fertilization period during 2007–2010 were 0.22, 0.20, 0.19 and 0.20 $m^3 \cdot m^{-3}$, respectively, which were higher, especially in the noticeably wet year of 2007, compared with the pre-fertilization 9-year mean (0.16 $m^3 m^{-3}$) (Figure 1c).

3.2. Seasonal and Interannual Variations of Observed C Fluxes and ET before and after Fertilization

The seasonal variations in the monthly EC-measured C fluxes and ET in the three stands before and after fertilization are presented in Figure 2. Monthly GPP values during the 4 post-fertilization years were similar to the means during the 9 non-fertilized years (Figure 2a). The monthly values of R during the 4 post-fertilization years were close to the 9-year pre-fertilization means for Jan–May but lower for the later months except for June and July in 2009 (Figure 2b). Therefore, the monthly NEP values during the 4 post-fertilization years were greater than the 9-year pre-fertilization average values for March–December, especially in June, August and September (Figure 2c).

Similarly, for HDF88, monthly NEP values during the 4 post-fertilization years at HDF88 were generally greater than the pre-fertilization average values for 2002 to 2006 (Figure 2g). In the case of HDF00, compared with the 2001–2006 means, the largest increases in monthly values of NEP during 2007–2010 occurred in June, July and August (Figure 2k). The monthly GPP values during 2007–2010 at HDF88 and HDF00 were higher than the mean values for the pre-fertilization periods (Figure 2e,i). The monthly R values for HDF00 were also generally higher than the monthly means in the pre-fertilization period (Figure 2j). Furthermore, at DF49, the maximum NEP occurred in April and May (Figure 2c), while maximum NEP at HDF88 and HDF00 occurred in June and July (Figure 2g,k).

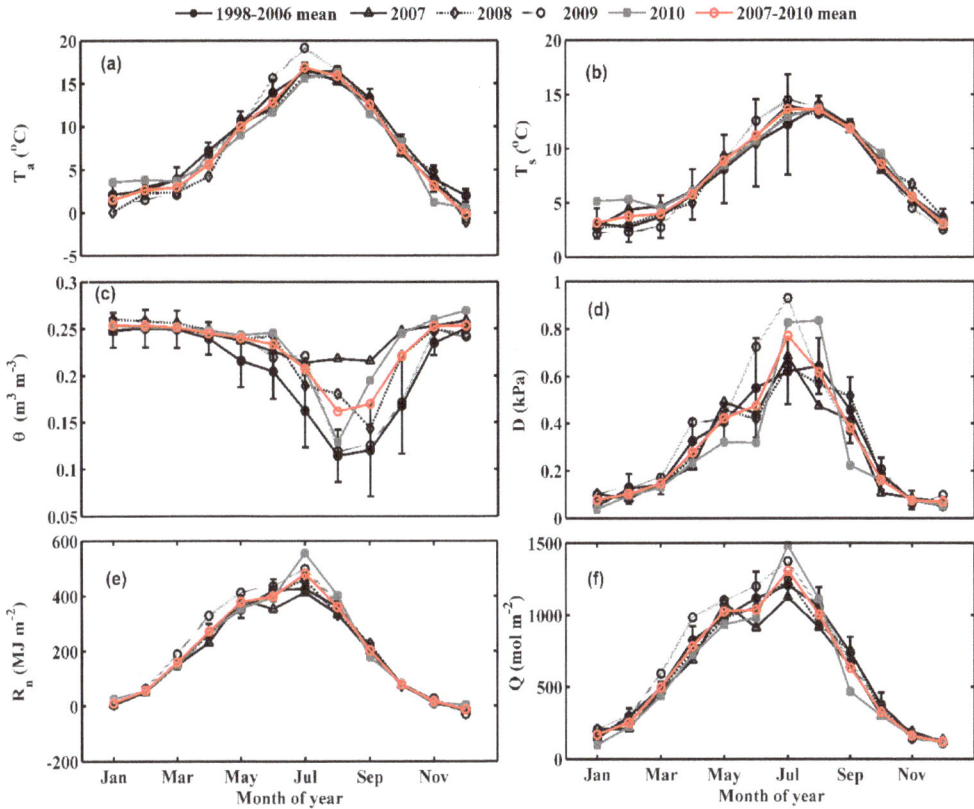

Figure 1. Comparisons of mean monthly values of environmental variables observed at DF49 for the pre-fertilization (1998–2006) and post-fertilization years (2007–2010). The error bars are ± 1 standard deviation (SD) for all the pre-fertilization data. (**a**) air temperature above the canopy (T_a); (**b**) 5-cm depth soil temperature (T_s); (**c**) 0–30-cm soil water content (θ); (**d**) water vapor pressure deficit (D); (**e**) total downwelling photosynthetically active radiation (Q) and (**f**) net radiation (R_n).

Most of the monthly ET values during 2007–2010 at DF49 and HDF88 were close to the monthly means in the pre-fertilization period (Figure 2d,h). However, at HDF00, monthly ET values for May–July during 2009 to 2010 were significantly higher than the means for the pre-fertilization period (Figure 2l). Moreover, for DF49, the monthly values of ET during January to April in 2010 were higher than the previous 9-year average values (Figure 2d).

To better examine the interannual variations of EC-measured C fluxes, ET, WUE and LUE, the annual measured values are compared in Figure 3. The results indicate that EC-measured NEP during the post-fertilization four years increased in all three stands (Figure 3c). Annual NEP at DF49 remained stable over post-fertilization years with values about 200 g C m^{-2} year^{-1} higher than in the pre-fertilization period, despite the post-fertilization years being cooler and wetter than the pre-fertilization years (Figure 1a,c). However, at HDF88, annual NEP consistently increased with increasing stand age (Figure 3c).

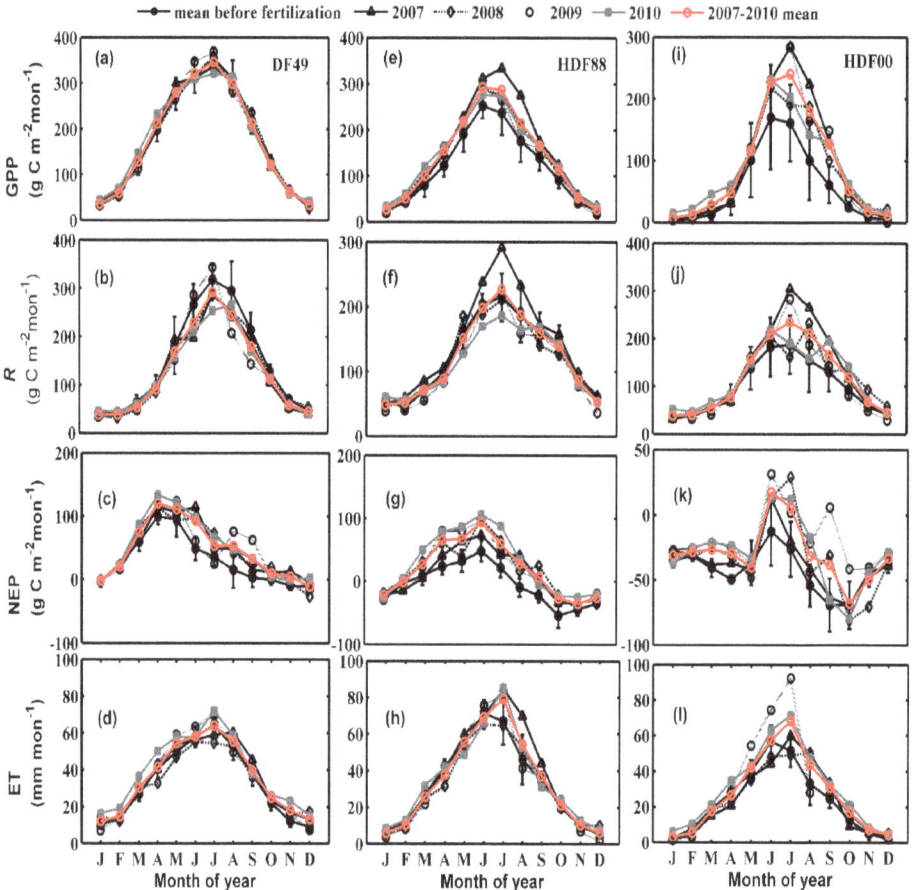

Figure 2. Comparisons of mean monthly values of eddy-covariance tower measured gross primary productivity (GPP), ecosystem respiration (R), net ecosystem productivity (NEP) and evapotranspiration (ET) in all three stands for the pre-fertilization and post-fertilization years (2007–2010). The means of measured carbon component fluxes and ET at DF49, HDF88 and HDF00 are shown for 1998 to 2006, 2002 to 2006 and 2001 to 2006, respectively. (**a–d**) for DF49; (**e–h**) for HDF88; and (**i–l**) for HDF00. The error bars are ± 1 SD for all the pre-fertilization data.

The interannual variations in GPP and R at DF49 since 1998 show that R decreased while GPP showed little change during 2007–2010 (Figure 3a,b). However, for the other two stands, annual R increased in the first post-fertilization year and decreased in the next three years (Figure 3b). In addition, compared with the pre-fertilization (9-years) mean CUEe (17%) in DF49, annual average values of CUEe during 2007–2010 were higher by about 9%, 8%, 9% and 12%, respectively.

Annual ET during the post-fertilization period did not show an apparent difference from the previous 9-year means in all three stands (Figure 3d). Annual values of WUE varied greatly among the three stands (Figure 3e). Annual WUE of DF49 was the highest, ranging from 4.5 to 5.4 g·C (kg water)$^{-1}$, and steadily decreased during 2008–2010. Annual LUE values in DF49 were also the highest among the three stands, ranging from 0.0216 to 0.0254 mol·C·mol^{-1} photons (Figure 3f).

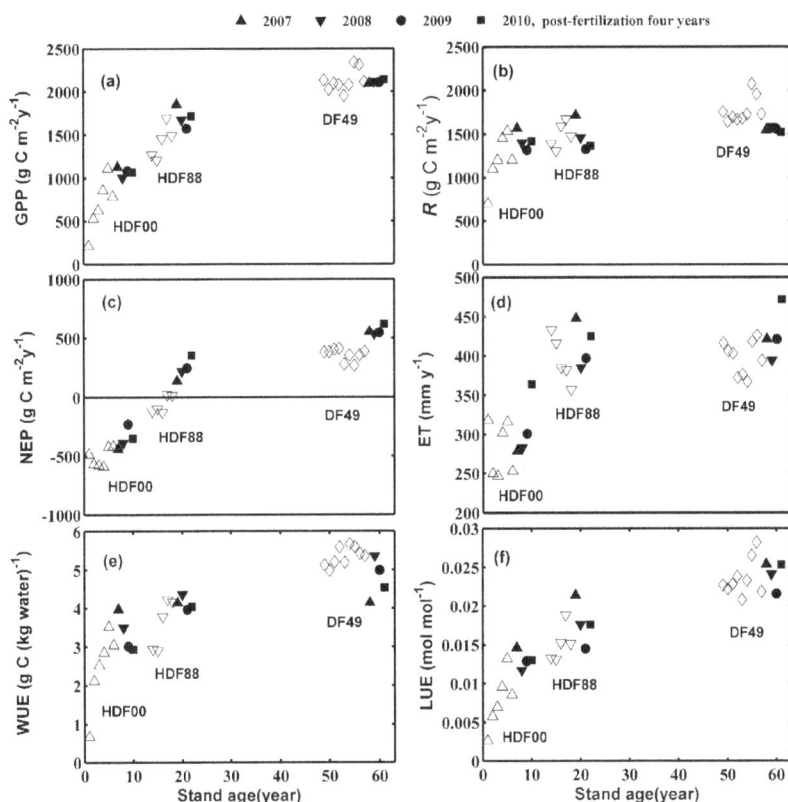

Figure 3. Interannual variations of eddy-covariance tower measured carbon fluxes, evapotranspiration (ET), water-use efficiency (WUE) and light-use efficiency (LUE) in the three Douglas-fir stands. (**a**) gross primary productivity (GPP); (**b**) ecosystem respiration (R); (**c**) net ecosystem productivity (NEP); (**d**) ET; (**e**) WUE and (**f**) LUE. Note: The open symbols in all three stands indicates the annual pre-fertilization values.

3.3. Verification of the ANN Model and Its Comparison to the MLR Model

The differences between EC-measured, ANN- and MLR-modeled monthly GPP, R, and NEP in the three stands are compared in Figure 4 and Tables 1–3. ANN simulations for the model calibration period (1998–2004) at DF49 showed strong agreement with the observed values and explained about 99%, 99% and 96% of the variance in monthly GPP, R, and NEP, respectively (Root mean square error (RMSE) = 2, 8, and 8 g·C·m^{-2}·month^{-1}, respectively). These results indicate that by optimizing the ANN parameters, the trained ANN model achieved reasonable simulation of monthly GPP, R and NEP at DF49. The linear regression analysis between ANN-modeled and measured values for the model verification period (2005–2006) for DF49 showed R^2 = 0.99, 0.99 and 0.98 and RMSE = 3, 6 and 6 $g \cdot C \cdot m^{-2} \cdot month^{-1}$ for GPP, R, and NEP, respectively (Table 1).

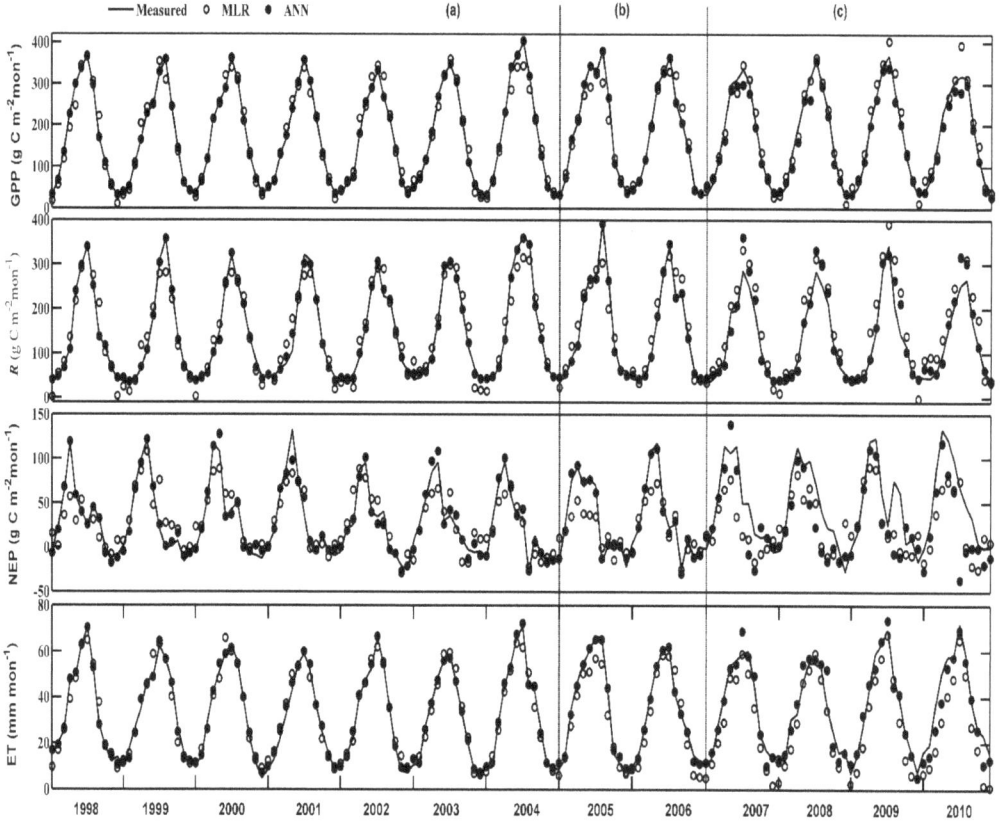

Figure 4. Artificial neural network (ANN) and multiple linear regression (MLR) simulated, and eddy-covariance tower measured monthly gross primary productivity (GPP), ecosystem respiration (R), net ecosystem productivity (NEP) and evapotranspiration (ET) at DF49 from 1998 to 2010. (**a**) for the ANN and MLR calibration years 1998–2004; (**b**) for the model validation years 2005–2006; and (**c**) for the four post-fertilization years 2007–2010.

MLR simulations for DF49 for the model calibration period, however, explained about 97%, 92% and 76% of the variance of monthly GPP, R, and NEP, respectively (Figure 4a and Table 1), which are somewhat lower than for the ANN simulations. The linear regression analysis showed RMSE = 21, 29, and 19 $g \cdot C \cdot m^{-2} \cdot month^{-1}$ for GPP, R and NEP, respectively, which are significantly higher than those for the comparison of ANN-simulated and EC-measured values. In addition, for the model verification period, linear regression between MLR-simulated and observed values showed RMSE = 30, 34 and 24 g·C·m^{-2} month^{-1} for GPP, R, and NEP, respectively, which are all significantly higher than those for the calibration period.

For the other two stands, the performance of the ANN model was also superior to that of the MLR model in the estimates of GPP, R and NEP (Table 1). Regarding ET, similar results were observed in the calibration and verification periods in all three stands (Table 1). As a consequence, we focused on describing and discussing the results of ANN modeling in the prediction period to ascertain the responses of C fluxes and ET to fertilization.

3.4. Effects of N addition on C exchange Fluxes, ET, WUE and LUE in a Chronosequence

In order to provide insights into the influence of fertilization on the C fluxes, ET as well as WUE and LUE during the period of prediction, monthly measured and ANN-modeled values, and their differences are shown for the four post-fertilization years (2007–2010) in Figures 5 and 6. The annual effects of N fertilization during this period are summarized in Tables 2 and 3. Figure 7 compares the annual measured and ANN modeled values of the C fluxes and ET, and shows ± 1 standard deviation estimated using the optimized model parameters.

Over the four post-fertilization years at DF49, most of the ANN modeled monthly values of GPP were lower than the measured values (Table 1 and Figure 5a), but were higher for R (Table 1 and Figure 5d) and consequently, modeled monthly values of NEP were lower than the observed values (Table 1 and Figure 5g).

At the annual time scale over the four years, N fertilization at DF49 caused a 6% to 11% increase in annual GPP and a 6% to 10% decrease in R, resulting in an 84% to 164% increase in NEP (Table 2). For HDF88, N fertilization caused an 18% to 29% increase in annual GPP, an 11% increase in R in 2007, and a 6% to 14% decrease during 2008–2010, resulting in a 208% to 321% increase in NEP (Table 2). For HDF00, N fertilization caused a 43% to 70% increase in annual GPP and an 8% to 21% increase in R, resulting in a 27% to 50% increase in NEP (Table 2). Overall, N fertilization had a positive impact on GPP for all three stands with the response being the highest at HDF00, followed by HDF88 (Table 2).

No consistent effects on annual ET were evident in the four post-fertilization years in all three stands (Table 3). For annual WUE, N fertilization of DF49 caused an increase of 0.30 to 0.67 g·C (kg·water)$^{-1}$ during 2007 to 2009, but a decrease of 0.21 g·C (kg·water)$^{-1}$ in 2010. N fertilization caused increases of 0.59 to 1.05 C (kg·water)$^{-1}$ and 0.76 to 1.75 C (kg·water)$^{-1}$ in annual WUE during 2007 to 2010 at HDF88 and HDF00, respectively (Table 3). For annual LUE, N fertilization had positive impacts for all three stands, especially at HDF00 with the greatest response ranging from 0.0039 to 0.0057 mol·C·mol^{-1} photons (Table 3).

Table 1. Comparisons of artificial neural network (ANN) and multiple linear regression (MLR) model performances among the calibration, verification and prediction periods for monthly gross primary productivity (GPP, $g \cdot C \cdot m^{-2} \cdot year^{-1}$), ecosystem respiration (R, $g \cdot C \cdot m^{-2} \cdot year^{-1}$) and evapotranspiration (ET, $mm \cdot year^{-1}$) in the three Douglas-fir stands.

Flux	Stand	Method	Calibration			Verification			Prediction		
			Regression equation	RMSE	R^2	Regression equation	RMSE	R^2	Regression equation	RMSE	R^2
GPP	DF49	MVMLR	$y = 0.97x + 5.94$	21	0.97	$y = 0.88x + 18.18$	30	0.94	$y = 1.01x + 6.1$	22	0.97
		MVANN	$y = x + 0.07$	2	0.99	$y = 0.99x - 0.35$	3	0.99	$y = 0.86x + 15.42$	27	0.97
	HDF88	MVMLR	$y = 0.92x + 8.65$	21	0.92	$y = 0.84x - 2.36$	36	0.91	$y = 0.82x + 16.47$	33	0.88
		MVANN	$y = x + 0.42$	5	0.99	$y = 0.97x + 2.22$	6	0.99	$y = 0.78x + 6.51$	40	0.90
	HDF00	MVMLR	$y = 0.82x + 8.23$	24	0.82	$y = 0.36x + 48.16$	62	0.55	$y = 0.43x + 68.56$	59	0.61
		MVANN	$y = 0.98x + 0.24$	7	0.98	$y = 0.98x - 1.25$	6	0.99	$y = 0.65x - 0.42$	43	0.96
R	DF49	MVMLR	$y = 0.92x + 11.45$	29	0.92	$y = 0.89x + 22.65$	34	0.90	$y = 1.2x - 17.4$	33	0.94
		MVANN	$y = 0.99x + 1.31$	8	0.99	$y = x + 0.03$	6	0.99	$y = 1.02x - 5.39$	25	0.96
	HDF88	MVMLR	$y = 0.92x + 9.71$	16	0.92	$y = 0.73x + 26.58$	25	0.92	$y = 0.59x + 43.8$	33	0.78
		MVANN	$y = x - 0.07$	3	0.99	$y = 0.98x + 1.63$	2	0.99	$y = 0.96x + 9.18$	23	0.89
	HDF00	MVMLR	$y = 0.84x + 14.57$	23	0.84	$y = 0.66x + 58.33$	40	0.80	$y = 0.81x + 59.39$	42	0.76
		MVANN	$y = 0.99x + 2.53$	3	0.99	$y = 1.01x - 0.98$	3	0.99	$y = 0.89x - 1.03$	28	0.91
NEP	DF49	MVMLR	$y = 0.71x + 8.56$	19	0.76	$y = 0.55x + 4.65$	24	0.82	$y = 0.59x - 2.14$	35	0.64
		MVANN	$y = 0.99x + 0.63$	8	0.96	$y = 0.99x - 0.54$	6	0.98	$y = 0.67x - 9.36$	39	0.54
	HDF88	MVMLR	$y = 0.93x - 0.65$	12	0.86	$y = 1.01x - 13.44$	23	0.81	$y = 1.08x - 5.52$	24	0.80
		MVANN	$y = 0.99x + 0.62$	4	0.98	$y = 0.96x - 1.33$	5	0.98	$y = 0.53x - 21$	41	0.51
	HDF00	MVMLR	$y = 0.67x - 15.4$	17	0.34	$y = -0.04x - 58.5$	42	0.01	$y = -0.21x - 55.3$	53	0.02
		MVANN	$y = 0.99x + 1.21$	8	0.82	$y = 0.89x - 6.47$	7	0.87	$y = 0.31x - 32.2$	42	0.06
ET	DF49	MVMLR	$y = 0.96x + 1.26$	4	0.96	$y = 0.91x - 0.82$	6	0.94	$y = 1.05x - 10.3$	5	0.93
		MVANN	$y = 0.99x + 0.34$	1	0.99	$y = 0.98x + 0.36$	1	0.99	$y = 1.04x - 2.1$	6	0.9
	HDF88	MVMLR	$y = 0.93x + 2.06$	6	0.93	$y = 1.24x - 14.9$	13	0.90	$y = 1.09x - 20.64$	20	0.91
		MVANN	$y = 0.99x + 0.21$	1	0.99	$y = 0.99x - 0.08$	1	0.99	$y = x + 0.56$	5	0.95
	HDF00	MVMLR	$y = 0.92x + 1.78$	5	0.92	$y = 0.91x - 2.6$	9	0.88	$y = 0.8x - 5.49$	16	0.76
		MVANN	$y = 0.99x - 0.03$	1	0.99	$y = x + 0.23$	1	0.99	$y = 0.83x + 3.62$	9	0.84

Note: The regressions among calibration, verification and prediction periods were obtained using the monthly EC-measured values for the C component fluxes as the independent variable and the monthly values estimated using the ANN and MLR models as the dependent variable. MVMLR indicates measured *versus* MLR modeled; MVANN indicates measured *versus* ANN modeled; three different-aged Douglas-fir stands: DF49, HDF88 and HDF00, established in 1949, 1988, and 2000, respectively.

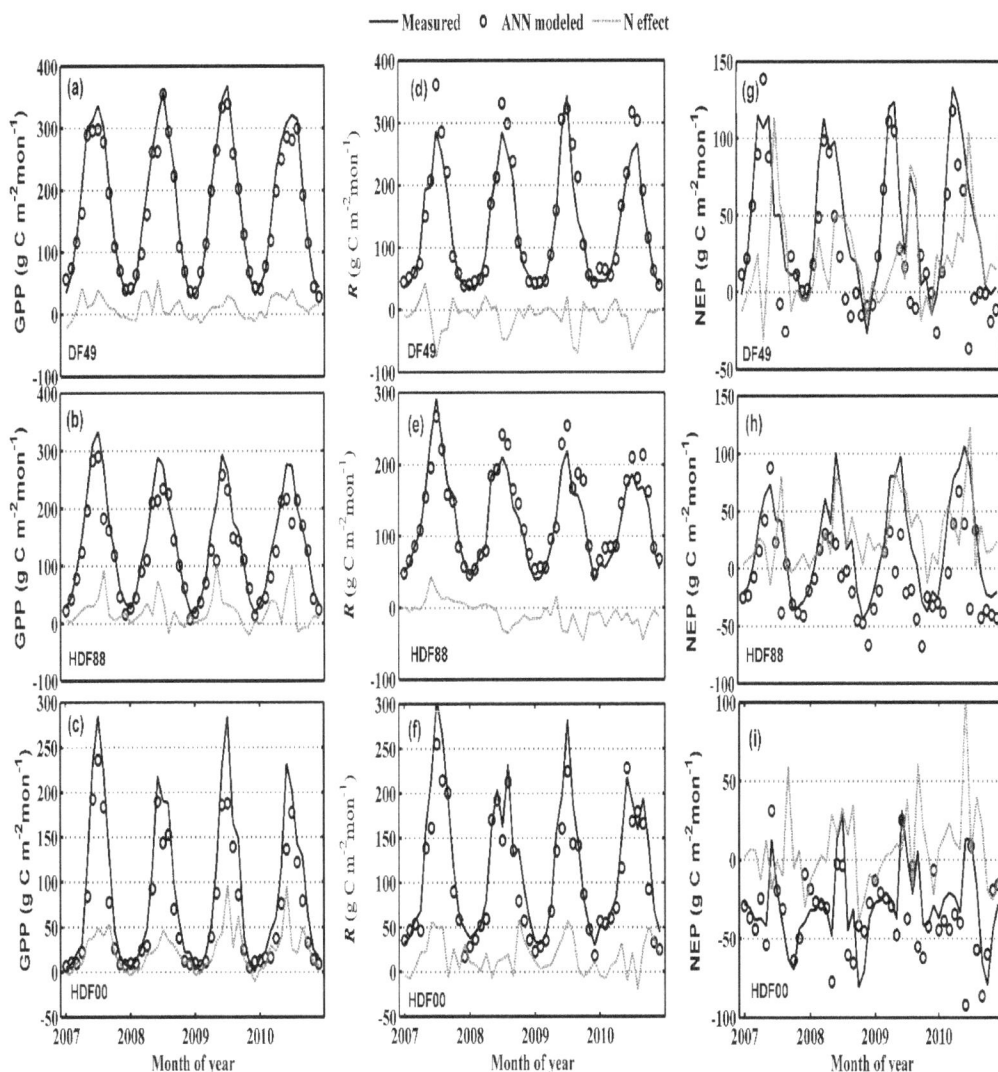

Figure 5. Artificial neural network (ANN) simulated and eddy-covariance tower measured monthly gross primary productivity (GPP), ecosystem respiration (R) and net ecosystem productivity (NEP) in all three stands for the four post-fertilization years 2007–2010. (**a–c**) for GPP; (**d–f**) for R; and (**g–i**) for NEP. The effects of N fertilization on the carbon (C) component fluxes were estimated as the differences between the measured C fluxes and their corresponding modeled values.

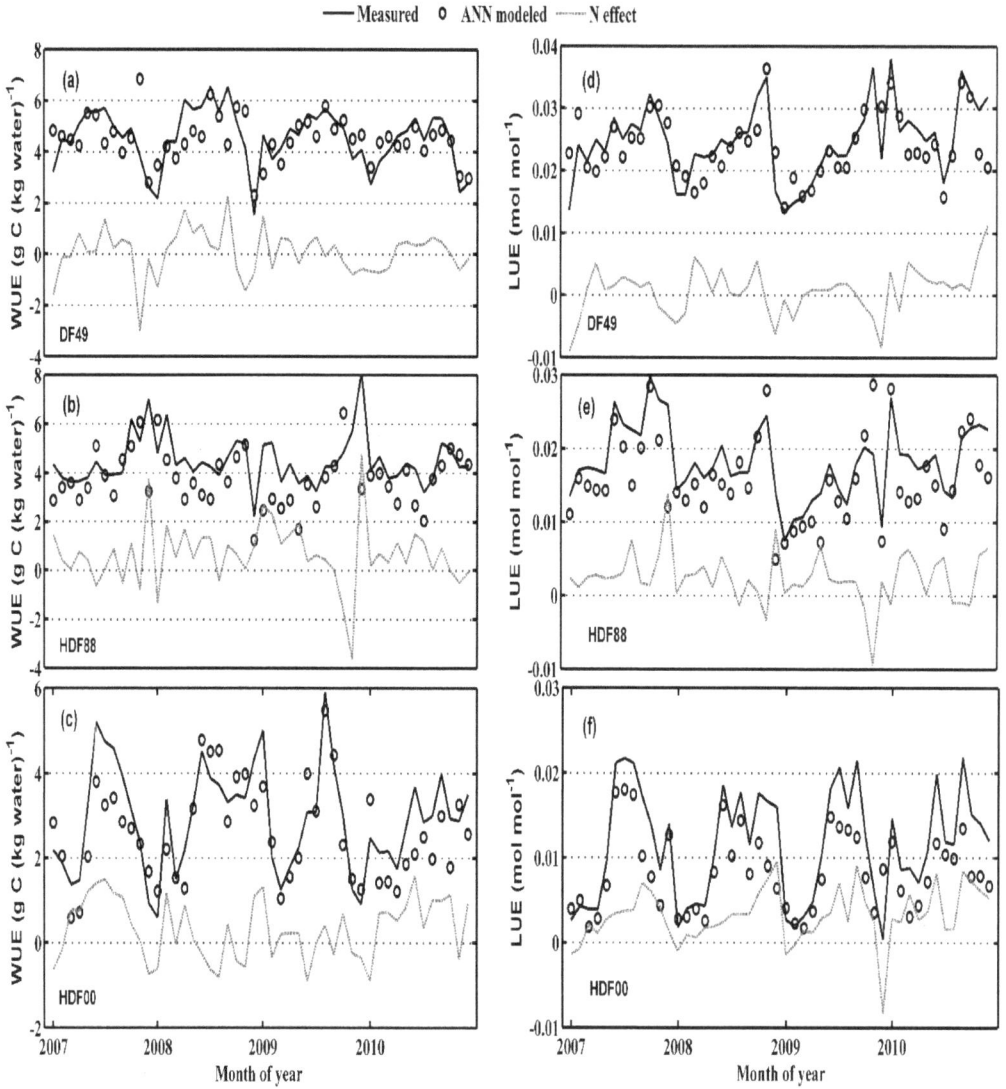

Figure 6. Artificial neural network (ANN) simulated and eddy-covariance tower measured monthly water-use efficiency (WUE) and light-use efficiency (LUE) in all three stands for the four post-fertilization years 2007–2010. (**a**–**c**) for WUE and (**d**–**f**) for LUE. The effects of N fertilization on WUE and LUE were estimated as the differences between the measured WUE and LUE values and their corresponding modeled values.

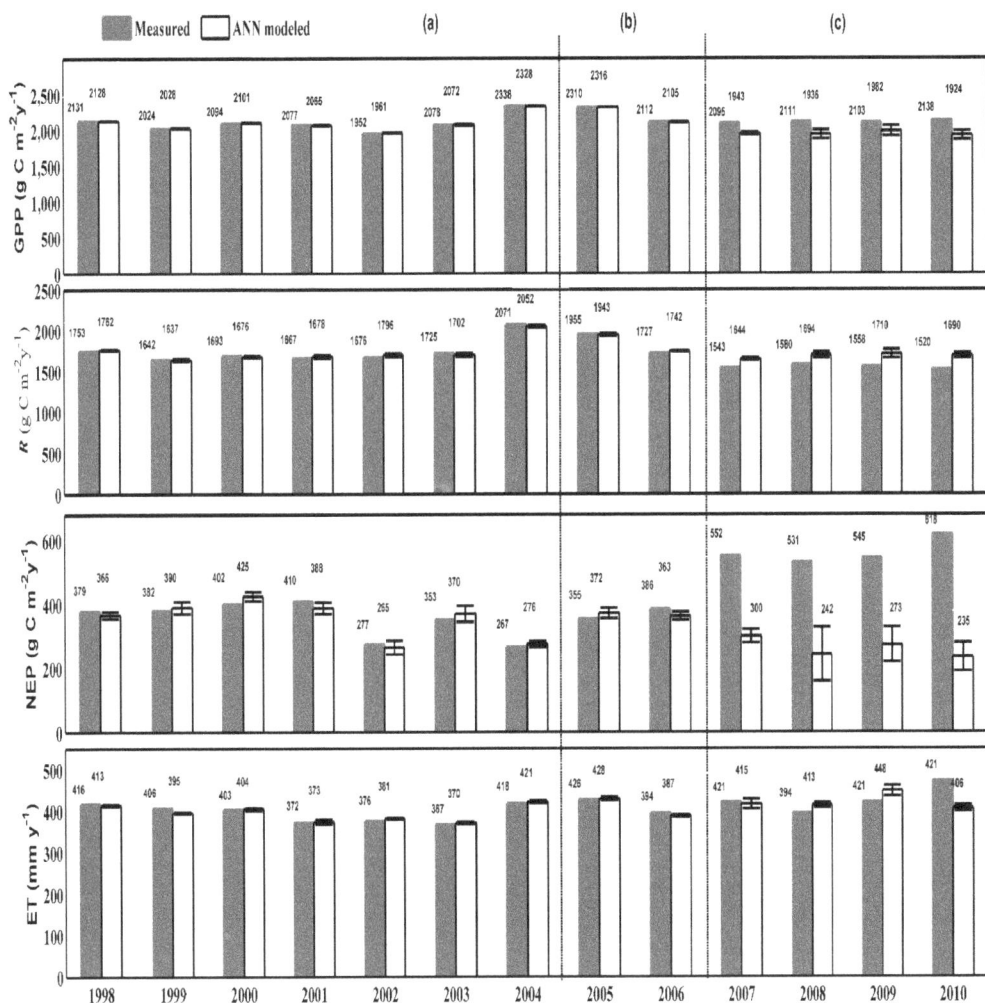

Figure 7. Comparisons of measured and artificial neural network (ANN) simulated annual gross primary productivity (GPP), ecosystem respiration (R), net ecosystem productivity (NEP) and evapotranspiration (ET) at DF49 from 1998 to 2010. (**a**) for the ANN calibration years 1998–2004; (**b**) for the model validation years 2005–2006; and (**c**) for the four post-fertilization years 2007–2010. The error bars with ± 1 SD are the standard errors for ANN-modeled annual values estimated using the optimized model parameters.

Table 2. Effects of nitrogen fertilization on annual gross primary productivity (GPP, g·C·m⁻²·year⁻¹), ecosystem respiration (R, g·C·m⁻²·year⁻¹) and net ecosystem productivity (NEP, g·C·m⁻²·year⁻¹) for the four post-fertilization years 2007–2010 in the three Douglas-fir stands.

Stand	Flux	2007			2008			2009			2010			Mean (2007–2010)		
		Measured	Modeled	N effect	Measured	Modeled	N effect	Measured	Modeled	N effect	Measured	Modeled	N effect	Measured	Modeled	N effect
DF49	GPP	2095	1943	+152	2111	1936	+175	2103	1982	+121	2138	1924	+214	2112	1924	+188
	R	1543	1644	−101	1580	1694	−114	1558	1710	−152	1520	1690	−170	1550	1690	−140
	NEP	552	300	+253	531	242	+289	545	273	+272	618	235	+383	562	235	+327
HDF88	GPP	1851	1435	+416	1675	1406	+269	1572	1316	+256	1716	1456	+260	1704	1456	+248
	R	1714	1546	+168	1457	1548	−91	1326	1499	−173	1365	1583	−218	1466	1583	−117
	NEP	137	−112	+249	218	−142	+360	246	−183	+429	351	−159	+478	238	−159	+397
HDF00	GPP	1126	691	+435	1005	665	+340	1083	759	+324	1068	629	+439	1071	629	+442
	R	1566	1297	+269	1396	1278	+118	1314	1216	+98	1418	1258	+160	1424	1258	+166
	NEP	−440	−606	+166	−391	−613	+222	−231	−457	+226	−350	−629	+279	−353	−629	+276

Note: Modeled, artificial neural network (ANN) modeled without N fertilization; N effect, measured − ANN modeled. Positive values indicate positive effects caused by N fertilization, visa versa; three different-aged Douglas-fir stands: DF49, HDF88 and HDF00, established in 1949, 1988, and 2000, respectively.

Table 3. Effects of nitrogen fertilization on annual evapotranspiration (ET, mm·y⁻¹), water use efficiency (WUE, g·C (kg·water)⁻¹) and light use efficiency (LUE, mol·C·mol⁻¹ photons) during the four post-fertilization years 2007–2010 in the three Douglas-fir stands.

Stand		2007			2008			2009			2010			Mean (2007–2010)		
		Measured	Modeled	N effect	Measured	Modeled	N effect	Measured	Modeled	N effect	Measured	Modeled	N effect	Measured	Modeled	N effect
DF49	ET	421	415	+6	394	413	−19	421	448	−27	472	406	+66	427	420.5	+6.5
	WUE	4.98	4.68	+0.3	5.36	4.69	+0.67	5	4.42	+0.58	4.53	4.74	−0.21	4.9675	4.6325	+0.335
	LUE	0.0254	0.0235	+0.0019	0.0241	0.0221	+0.002	0.0216	0.0204	+0.0012	0.0253	0.0228	+0.0025	0.0241	0.0222	+0.0019
HDF88	ET	448	405	+43	385	426	−41	397	413	−16	425	437	−12	413.75	420.25	−6.5
	WUE	4.13	3.54	+0.59	4.35	3.3	+1.05	3.96	3.19	+0.77	4.04	3.33	+0.71	4.12	3.34	+0.78
	LUE	0.0214	0.0166	+0.0048	0.0176	0.0148	+0.0028	0.0145	0.0121	+0.0024	0.0176	0.015	+0.0026	0.017775	0.014625	+0.00315
HDF00	ET	283	312	−29	288	244	+44	360	338	+22	364	431	−67	323.75	331.25	−7.5
	WUE	3.97	2.22	+1.75	3.49	2.73	+0.76	3.01	2.24	+0.77	2.93	1.46	+1.47	3.35	2.1625	+1.1875
	LUE	0.0146	0.0089	+0.0057	0.0117	0.0077	+0.004	0.0129	0.009	+0.0039	0.013	0.0077	+0.0053	0.01305	0.008325	+0.004725

Note: Modeled, artificial neural network (ANN) modeled without N fertilization; N effect, measured − ANN modeled. Positive values indicate positive effects caused by N fertilization, visa versa; three different-aged Douglas-fir stands: DF49, HDF88 and HDF00, established in 1949, 1988, and 2000, respectively.

4. Discussion

4.1. Effects of N Fertilization on Gross Primary Productivity

N fertilization generally results in GPP and net primary productivity (NPP) increases in terrestrial ecosystems by promoting growth and biomass accumulation in terrestrial plants [7,21,25,74,75]. Our results showed that N fertilization generally increased GPP in all three stands, especially in the younger stands, HDF00 and HDF88. N addition resulted in a GPP increase of 8% at DF49 in the first post-fertilization year, which is in excellent agreement with an 8% increase obtained by Chen et al. [52] using a model-data synthesis approach and a 10% increase calculated by Jassal et al. [53] using a simple empirical model (Table 4). This is lower than the 14% increase estimated by Grant et al. [76] using a process-based model. In addition, our results, which showed that N fertilization led to an increase in GPP in the three stands during the first two years after fertilization, are consistent in sign with the findings by Jassal et al. [53], although the magnitude of the increases in GPP are somewhat different. This difference may partly result from the reported uncertainty due to the use of the empirical relationships used for gap-filling, resulting in an uncertainty of ± 50 g·C·m^{-2}·year^{-1} in EC-measured GPP, and ± 75 and ± 25 g·C·m^{-2}·year^{-1} for measured R and NEP, respectively [54]. Overall, N fertilization had a positive effect on GPP and the effectiveness of fertilization varied with stand age.

4.2. Effects of N Fertilization on Ecosystem Respiration

Our results indicated that N fertilization of DF49 reduced R by approximately 101 g·C·m^{-2}·year^{-1} in the first post-fertilization year, which is in good agreement with the 93 g·C·m^{-2}·year^{-1} decrease calculated by Chen, et al. [52] (Table 4). During the next three years (2008–2010), modeled R remained greater than measured R by 114 to 170 g·C·m^{-2}·year^{-1}, implying that N fertilization suppressed R. However, the increase in R induced by N fertilization at HDF00 during the first two years after fertilization agrees with the findings reported in Jassal et al. [53]. Our ANN modeling indicated that this positive effect on R at HDF00 continued for the third and fourth post-fertilization years (2009 to 2010) (Table 2). It should be noted that Jassal et al [53] partitioned NEE by estimating daytime R using the relationship between daytime NEP and PAR rather than by using the relationship between nighttime NEE and T_s. Such variable effects of N addition on R in different stands suggest the need to investigate the mechanisms.

4.3. Effects of N fertilization on Net Ecosystem Production

N fertilization significantly increased NEP during the four post-fertilization years in all the three stands, especially HDF88, followed by DF49, largely due to N-induced GPP increases and R decreases. These results are consistent with those reported by Chen, et al. [52] in the first post-fertilization year, and are also consistent in sign with the results obtained by Jassal et al. [53] both at DF49 and HDF88 in the first two post-fertilization years (Table 4), although the magnitude of the increase in NEP was less than that in our study. However, at HDF00, a substantial increase in NEP in the four post-fertilization years was primarily due to GPP increasing more than R (Table 2).

The C:N response [kg C (sequestered) kg^{-1} N added] has been widely used to estimate the response of C sequestration to N addition [22–24,77]. Our results suggest that C:N response varied with stand age. In the first post-fertilization year, DF49 and HDF88 had approximately the same C:N response (13 kg C kg^{-1} N, followed by HDF00 (8 kg C kg^{-1} N). Our results are consistent with those of Jassal et al. [53], de Vries et al. [24], Högberg [26] and Sutton et al. [23], which are significantly smaller than the assessment made by Magnani et al. [22], with an extremely high C:N response (approximately 725 kg C kg^{-1} N) to wet deposition. Furthermore, it seems that the 60 kg·N·ha^{-1} application to individual trees at HDF00 was less efficient with respect to NEP than the larger application (200 kg N ha^{-1}) at DF49 and HDF88.

Table 4. N-induced changes (= measured – modeled) in annual gross primary productivity (GPP, g·C·m^{-2}·year^{-1}), ecosystem respiration (R, g·C·m^{-2}·year^{-1}) and net ecosystem productivity (NEP, g·C·m^{-2}·year^{-1}) from different study results in the three West Coast Douglas-fir stands during 2007 to 2010.

Stand	Flux	Study	2007	2008	2009	2010
DF49	NEP	This study (ANN)	253	289	272	383
		Jassal et al. [53]	168	78		
		Chen et al. [52]	250			
	GPP	This study (ANN)	152	175	121	214
		Jassal et al. [53]	203	139		
		Chen et al. [52]	157			
	R	This study (ANN)	−101	−114	−152	−170
		Jassal et al. [53]	35	61		
		Chen et al. [52]	−93			
HDF88	NEP	This study (ANN)	249	360	429	478
		Jassal et al. [53]	182	179		
	GPP	This study (ANN)	416	269	256	260
		Jassal et al. [53]	615	366		
	R	This study (ANN)	168	−91	−173	−218
		Jassal et al. [53]	433	187		
HDF00	NEP	This study (ANN)	166	222	226	279
		Jassal et al. [53]	−3	−82		
	GPP	This study (ANN)	435	340	324	439
		Jassal et al. [53]	305	329		
	R	This study (ANN)	269	118	98	160
		Jassal et al. [53]	308	411		

Three different-aged Douglas-fir stands: DF49, HDF88 and HDF00, established in 1949, 1988, and 2000, respectively.

4.4. Effects of N fertilization on ET, WUE and LUE

N application showed no consistent effect on ET among the three stands. N fertilization slightly decreased ET at HDF88 in the last three years and had no clear effect on ET at HDF00 over the four post-fertilization years (Table 3). On the other hand, increased nutrient availability (e.g., N fertilization) had significantly positive effects on WUE at HDF88 and HDF00 due to its strong positive effect on GPP compared with its small effect on ET, but had no significant effect at DF49. Ripullone *et al.* [47] found that N addition increased both WUE and biomass production. They concluded that this was mainly due to the positive response of photosynthesis to N availability with it having no impact on either stomatal conductance or transpiration. Consequently, improved tree nutrition (e.g., N fertilization) could cause an increase in forest productivity [78] without leading to an increase in demand for water.

Furthermore, N fertilization caused an increase in the 4-year mean WUE of 0.34 g·C (kg·water)$^{-1}$ for DF49, implying a relatively conservative response to N fertilization, and increases of 0.78 and 1.19 g·C (kg·water)$^{-1}$ for HDF88 and HDF00, respectively, likely due to their younger stand ages and less developed root system [56]. Wu *et al.* [79] suggested that low or appropriate N application might improve the survivability of *Sophora davidii* seedlings under water deficit conditions, resulting in an increase in WUE and stimulating growth and biomass production. Our results indicated that the supply of 60 kg N ha^{-1} applied to individual trees at HDF00 may be more efficient in terms of WUE compared with the higher application of 200 kg N ha^{-1} at HDF88 and DF49, especially in the latter stand.

N fertilization in our study had positive impacts on LUE in all three stands in the four post-fertilization years, especially at HDF00 followed by HDF88 with the smallest increase occurring in 2009 in all three stands accompanied by the highest annual PAR. This suggests that there was a reduction in the increase in LUE induced by N fertilization under high PAR, which has also been reported by Ibrom *et al.* [80] and Schwalm *et al.* [81]. Furthermore, N fertilization caused a 0.002·mol C·mol^{-1} photon increase in the 4-year mean LUE over the post-fertilization period for DF49, implying a relatively conservative plant productivity response to N fertilization, while higher increases of 0.0032 and 0.0047 mol·C·mol^{-1} photons for HDF88 and HDF00, respectively, were likely due to their younger stand age and rapid stand development. Overall, stand development had a significant impact on plant productivity, LUE and WUE as well as nutrient demand (e.g., N fertilization) [82].

4.5. Modeling Uncertainty and Limitations

Several studies have found that there is observable interannual variability of C component fluxes and ET in these stands, which to a large extent, results from interannual variability in climate [42,54,56,59]. We used both ANN and MLR approaches for removing the impacts of climatic variability on C fluxes and ET in the three stands during the post-fertilization period. In addition, a significant effect of stand development on C fluxes in the two younger stands (HDF88 and HDF00) was demonstrated in Figure 3. Our results show that pre-fertilization monthly C fluxes and ET affected by climate and stand age were better described using the ANN approach than the MLR approach.

The determination of BPNN model parameters including the hidden layer node number, momentum coefficient and learning rate and its shortcomings, such as the slow rate of convergence and local minimum, have considerably restricted its application and have led to uncertainties in the estimate of the N effect in this study. Nevertheless, we believe this study contributes significantly to quantifying the effects of N fertilization on C fluxes and ET and reducing the uncertainty in model simulation. Considering the limitations of the BPNN model, in order to further improve the estimate of the N effect, more importance should be attached to a hybrid BPNN model with optimized algorithms such as the genetic algorithm and particle swarm optimization.

5. Conclusions

1. Pre-fertilization monthly C fluxes and ET were better described using the ANN approach than the MLR approach.

2. N fertilization increased GPP in all three Douglas-fir stands over the four post-fertilization years with the greatest increase in the youngest stand (HDF00) followed by HDF88, suggesting that the effectiveness of fertilization on GPP may be associated with stand age. In addition, fertilization decreased R over the four years after fertilization at DF49, and over the last three years at HDF88, but increased R over the four years at HDF00. As a result, N fertilization increased NEP in all three stands, with the greatest increase in HDF88 and followed by DF49.

3. N fertilization had no discernible effects on ET among the three stands. This result led to a significant increase in WUE in the two younger stands (HDF88 and HDF00) but a small increase in only the first three years for the oldest stand (DF49).

4. N fertilization increased LUE in all three stands, especially the youngest stand (HDF00).

Acknowledgments

This research was supported a research grant (41271116) funded by the National Science Foundation of China and an NSERC Discovery Grant (6123-2012) to TAB. Thanks to TimberWest Forest Company and Island Timberlands LP for logistical support and to Agrium Inc. for providing financial support and the urea fertilizer. We greatly appreciate the assistance of Dominic Lessard, Zoran Nesic and Rick Ketler with the climate and EC observations, and Christian Brümmer, Nick Grant and Praveena Krishnan for data preprocessing.

Author Contributions

Baozhang Chen and T. Andrew Black, conceived and designed the paper. Xianming Dou performed the data analyses and ANN modeling. Xianming Dou and Baozhang Chen wrote the paper. Rachhpal S. Jassal conducted the fertilization experiments. Mingliang Che prepared the data for ANN modeling. All coauthors contributed to manuscript revision and approved the final manuscript.

Conflicts of Interest

The authors declare no conflict of interest.

References

1.   Houghton, R.A. Terrestrial carbon sinks–uncertain. *Biologist* **2002**, *49*, 155–160.

2. Piao, S.; Ciais, P.; Friedlingstein, P.; de Noblet-Ducoudré, N.; Cadule, P.; Viovy, N.; Wang, T. Spatiotemporal patterns of terrestrial carbon cycle during the 20th century. *Glob. Biogeochem. Cycles* **2009**, *23*, doi:10.1029/2008GB003339.

3. Tian, H.; Melillo, J.; Lu, C.; Kicklighter, D.; Liu, M.; Ren, W.; Xu, X.; Chen, G.; Zhang, C.; Pan, S. China's terrestrial carbon balance: Contributions from multiple global change factors. *Glob. Biogeochem. Cycles* **2011**, *25*, doi:10.1029/2010GB003838.

4. Cameron, D.R.; Oijen, M.V.; Werner, C.; Butterbach-Bahl, K.; Grote, R.; Haas, E.; Heuvelink, G.B.M.; Kiese, R.; Kros, J.; Kuhnert, M. Environmental change impacts on the C-and N-cycle of european forests: A model comparison study. *Biogeosciences* **2013**, *10*, 1751–1773.

5. Fernández-Martínez, M.; Vicca, S.; Janssens, I.A.; Sardans, J.; Luyssaert, S.; Campioli, M.; Chapin, F.S., III; Ciais, P.; Malhi, Y.; Obersteiner, M. Nutrient availability as the key regulator of global forest carbon balance. *Nat. Clim. Chang.* **2014**, *4*, 471–476.

6. Hilker, T.; Lepine, L.; Coops, N.C.; Jassal, R.S.; Black, T.A.; Wulder, M.A.; Ollinger, S.; Tsui, O.; Day, M. Assessing the impact of *N*-fertilization on biochemical composition and biomass of a Douglas-fir canopy—A remote sensing approach. *Agric. For. Meteorol.* **2012**, *153*, 124–133.

7. Högberg, P. Environmental science: Nitrogen impacts on forest carbon. *Nature* **2007**, *447*, 781–782.

8. Reich, P.B.; Hobbie, S.E.; Lee, T.; Ellsworth, D.S.; West, J.B.; Tilman, D.; Knops, J.M.H.; Naeem, S.; Trost, J. Nitrogen limitation constrains sustainability of ecosystem response to CO_2. *Nature* **2006**, *440*, 922–925.

9. Yang, Y.; Luo, Y.; Finzi, A.C. Carbon and nitrogen dynamics during forest stand development: A global synthesis. *New Phytol.* **2011**, *190*, 977–989.

10. Aber, J.; McDowell, W.; Nadelhoffer, K.; Magill, A.; Berntson, G.; Kamakea, M.; McNulty, S.; Currie, W.; Rustad, L.; Fernandez, I. Nitrogen saturation in temperate forest ecosystems. *Bioscience* **1998**, *48*, 921–934.

11. Du, Z.; Wang, W.; Zeng, W.; Zeng, H. Nitrogen deposition enhances carbon sequestration by plantations in northern China. *PLoS ONE* **2014**, *9*, doi:10.1371/journal.pone.0087975.

12. Krause, K.; Cherubini, P.; Bugmann, H.; Schleppi, P. Growth enhancement of picea abies trees under long-term, low-dose n addition is due to morphological more than to physiological changes. *Tree Physiol.* **2012**, *32*, 1471–1481.

13. Magill, A.H.; Aber, J.D.; Currie, W.S.; Nadelhoffer, K.J.; Martin, M.E.; McDowell, W.H.; Melillo, J.M.; Steudler, P. Ecosystem response to 15 years of chronic nitrogen additions at the harvard forest lter, Massachusetts, USA. *For. Ecol. Manag.* **2004**, *196*, 7–28.

14. Reay, D.S.; Dentener, F.; Smith, P.; Grace, J.; Feely, R.A. Global nitrogen deposition and carbon sinks. *Nat. Geosci.* **2008**, *1*, 430–437.

15. Fleischer, K.; Rebel, K.T.; Molen, V.D.M.K.; Erisman, J.W.; Wassen, M.J.; Loon, E.E.; Montagnani, L.; Gough, C.M.; Herbst, M.; Janssens, I.A. The contribution of nitrogen deposition to the photosynthetic capacity of forests. *Glob. Biogeochem. Cycles* **2013**, *27*, 187–199.

16. De Vries, W.; Du, E.; Butterbach-Bahl, K. Short and long-term impacts of nitrogen deposition on carbon sequestration by forest ecosystems. *Curr. Opin. Environ. Sust.* **2014**, *9*, 90–104.

17. Berg, B.; Matzner, E. Effect of n deposition on decomposition of plant litter and soil organic matter in forest systems. *Environ. Rev.* **1997**, *5*, 1–25.

18. Franklin, O.; Högberg, P.; Ekblad, A.; Ågren, G.I. Pine forest floor carbon accumulation in response to N and PK additions: Bomb ^{14}C modelling and respiration studies. *Ecosystems* **2003**, *6*, 644–658.

19. Leggett, Z.H.; Kelting, D.L. Fertilization effects on carbon pools in loblolly pine plantations on two upland sites. *Soil Sci. Soc. Am. J.* **2006**, *70*, 279–286.

20. Olsson, P.; Linder, S.; Giesler, R.; Högberg, P. Fertilization of boreal forest reduces both autotrophic and heterotrophic soil respiration. *Glob. Chang. Biol.* **2005**, *11*, 1745–1753.

21. Xia, J.; Wan, S. Global response patterns of terrestrial plant species to nitrogen addition. *New Phytol.* **2008**, *179*, 428–439.

22. Magnani, F.; Mencuccini, M.; Borghetti, M.; Berbigier, P.; Berninger, F.; Delzon, S.; Grelle, A.; Hari, P.; Jarvis, P.G.; Kolari, P. The human footprint in the carbon cycle of temperate and boreal forests. *Nature* **2007**, *447*, 849–851.

23. Sutton, M.A.; Simpson, D.; Levy, P.E.; Smith, R.I.; Reis, S.; van Oijen, M.; de Vries, W.I.M. Uncertainties in the relationship between atmospheric nitrogen deposition and forest carbon sequestration. *Glob. Chang. Biol.* **2008**, *14*, 2057–2063.

24. de Vries, W.; Solberg, S.; Dobbertin, M.; Sterba, H.; Laubhahn, D.; Reinds, G.J.; Nabuurs, G.J.; Gundersen, P.; Sutton, M.A. Ecologically implausible carbon response? *Nature* **2008**, *451*, E1–E3.

25. Dezi, S.; Medlyn, B.E.; Tonon, G.; Magnani, F. The effect of nitrogen deposition on forest carbon sequestration: A model-based analysis. *Glob. Chang. Biol.* **2010**, *16*, 1470–1486.

26. Högberg, P. What is the quantitative relation between nitrogen deposition and forest carbon sequestration? *Glob. Chang. Biol.* **2012**, *18*, 1–2.

27. Janssens, I.A.; Dieleman, W.; Luyssaert, S.; Subke, J.A.; Reichstein, M.; Ceulemans, R.; Ciais, P.; Dolman, A.J.; Grace, J.; Matteucci, G.; *et al.* Reduction of forest soil respiration in response to nitrogen deposition. *Nat. Geosci.* **2010**, *3*, 315–322.

28. Morell, F.J.; Lampurlanés, J.; Álvaro-Fuentes, J.; Cantero-Martínez, C. Yield and water use efficiency of barley in a semiarid mediterranean agroecosystem: Long-term effects of tillage and N fertilization. *Soil Till. Res.* **2011**, *117*, 76–84

29. Körner, C. Biosphere responses to CO_2 enrichment. *Ecol. Appl.* **2000**, *10*, 1590–1619.

30. Fenn, M.E.; Baron, J.S.; Allen, E.B.; Rueth, H.M.; Nydick, K.R.; Geiser, L.; Bowman, W.D.; Sickman, J.O.; Meixner, T.; Johnson, D.W. Ecological effects of nitrogen deposition in the western united states. *Bioscience* **2003**, *53*, 404–420.

31. Nadelhoffer, K.J.; Emmett, B.A.; Gundersen, P.; Kjønaas, O.J.; Koopmans, C.J.; Schleppi, P.; Tietema, A.; Wright, R.F. Nitrogen deposition makes a minor contribution to carbon sequestration in temperate forests. *Nature* **1999**, *398*, 145–148.

32. Currie, W.S.; Nadelhoffer, K.J.; Aber, J.D. Redistributions of highlight turnover and replenishment of mineral soil organic n as a long-term control on forest c balance. *For. Ecol. Manag.* **2004**, *196*, 109–127.

33. Liu, L.; Greaver, T.L. A global perspective on belowground carbon dynamics under nitrogen enrichment. *Ecol. Lett.* **2010**, *13*, 819–828.

34. Harpole, W.S.; Potts, D.L.; Suding, K.N. Ecosystem responses to water and nitrogen amendment in a California grassland. *Glob. Chang. Biol.* **2007**, *13*, 2341–2348.

35. Tian, S.; Youssef, M.A.; Amatya, D.M.; Vance, E.D. Global sensitivity analysis of drainmod-forest, an integrated forest ecosystem model. *Hydrol. Process.* **2014**, *28*, 4389–4410.

36. Felzer, B.S.; Cronin, T.W.; Melillo, J.M.; Kicklighter, D.W.; Schlosser, C.A.; Dangal, S.R.S. Nitrogen effect on carbon-water coupling in forests, grasslands, and shrublands in the arid western united states. *J. Geophys. Res.* **2011**, *116*, doi:10.1029/2010jg001621.

37. Hu, Z.; Yu, G.; Fu, Y.; Sun, X.; Li, Y.; Shi, P.; Wang, Y.; Zheng, Z. Effects of vegetation control on ecosystem water use efficiency within and among four grassland ecosystems in China. *Glob. Chang. Biol.* **2008**, *14*, 1609–1619.

38. Ito, A.; Inatomi, M. Water-use efficiency of the terrestrial biosphere: A model analysis focusing on interactions between the global carbon and water cycles. *J. Hydrometeorol.* **2012**, *13*, 681–694.

39. Brix, H.; Mitchell, A.K. Thinning and nitrogen fertilization effects on soil and tree water stress in a douglas-fir stand. *Can. J. For. Res.* **1986**, *16*, 1334–1338.

40. Dordas, C.A.; Sioulas, C. Safflower yield, chlorophyll content, photosynthesis, and water use efficiency response to nitrogen fertilization under rainfed conditions. *Ind. Crop. Prod.* **2008**, *27*, 75–85.

41. Livingston, N.J.; Guy, R.D.; Sun, Z.J.; Ethier, G.J. The effects of nitrogen stress on the stable carbon isotope composition, productivity and water use efficiency of white spruce (*Picea glauca* (moench) voss) seedlings. *Plant Cell Environ.* **1999**, *22*, 281–289.

42. Jassal, R.S.; Black, T.A.; Spittlehouse, D.L.; Brümmer, C.; Nesic, Z. Evapotranspiration and water use efficiency in different-aged Pacific Northwest Douglas-fir stands. *Agric. For. Meteorol.* **2009**, *149*, 1168–1178.

43. Korol, R.L.; Kirschbaum, M.U.F.; Farquhar, G.D.; Jeffreys, M. Effects of water status and soil fertility on the c-isotope signature in *Pinus radiata*. *Tree Physiol.* **1999**, *19*, 551–562.

44. Mitchell, A.K.; Hinckley, T.M. Effects of foliar nitrogen concentration on photosynthesis and water use efficiency in Douglas-fir. *Tree Physiol.* **1993**, *12*, 403–410.

45. Castellanos, M.T.; Tarquis, A.M.; Ribas, F.; Cabello, M.J.; Arce, A.; Cartagena, M.C. Nitrogen fertigation: An integrated agronomic and environmental study. *Agric. Water Manag.* **2013**, *120*, 46–55.

46. Claussen, W. Growth, water use efficiency, and proline content of hydroponically grown tomato plants as affected by nitrogen source and nutrient concentration. *Plant Soil* **2002**, *247*, 199–209.

47. Ripullone, F.; Lauteri, M.; Grassi, G.; Amato, M.; Borghetti, M. Variation in nitrogen supply changes water-use efficiency of pseudotsuga menziesii and populus x euroamericana; a comparison of three approaches to determine water-use efficiency. *Tree Physiol.* **2004**, *24*, 671–679.

48. Running, S.W.; Nemani, R.R.; Heinsch, F.A.; Zhao, M.; Reeves, M.; Hashimoto, H. A continuous satellite-derived measure of global terrestrial primary production. *Bioscience* **2004**, *54*, 547–560.

49. Yuan, W.; Liu, S.; Yu, G.; Bonnefond, J.M.; Chen, J.; Davis, K.; Desai, A.R.; Goldstein, A.H.; Gianelle, D.; Rossi, F. Global estimates of evapotranspiration and gross primary production based on modis and global meteorology data. *Remote Sens. Environ.* **2010**, *114*, 1416–1431.

50. Zhao, M.; Heinsch, F.A.; Nemani, R.R.; Running, S.W. Improvements of the modis terrestrial gross and net primary production global data set. *Remote Sens. Environ.* **2005**, *95*, 164–176.

51. Hanley, D.P.; Chappell, H.N.; Nadelhoffer, E.H. *Fertilizing Coastal Douglas-fir Forests*; Washington State University: Pullman, WA, USA, 1991.

52. Chen, B.; Coops, N.C.; Black, T.A.; Jassal, R.S.; Chen, J.M.; Johnson, M. Modeling to discern nitrogen fertilization impacts on carbon sequestration in a Pacific Northwest Douglas-fir forest in the first post-fertilization year. *Glob. Chang. Biol.* **2011**, *17*, 1442–1460.

53. Jassal, R.S.; Black, T.A.; Cai, T.; Ethier, G.; Pepin, S.; Brümmer, C.; Nesic, Z.; Spittlehouse, D.L.; Trofymow, J.A. Impact of nitrogen fertilization on carbon and water balances in a chronosequence of three Douglas-fir stands in the pacific northwest. *Agric. For. Meteorol.* **2010**, *150*, 208–218.

54. Humphreys, E.R.; Black, T.A.; Morgenstern, K.; Cai, T.; Drewitt, G.B.; Nesic, Z.; Trofymow, J.A. Carbon dioxide fluxes in coastal Douglas-fir stands at different stages of development after clearcut harvesting. *Agric. For. Meteorol.* **2006**, *140*, 6–22.

55. Chen, B.; Black, T.A.; Coops, N.C.; Hilker, T.; Trofymow, J.A.; Morgenstern, K. Assessing tower flux footprint climatology and scaling between remotely sensed and eddy covariance measurements. *Bound. Layer Meteorol.* **2009**, *130*, 137–167.

56. Krishnan, P.; Black, T.A.; Jassal, R.S.; Chen, B.; Nesic, Z. Interannual variability of the carbon balance of three different-aged Douglas-fir stands in the pacific northwest. *J. Geophys. Res.* **2009**, *114*, doi:10.1029/2008JG000912.

57. Morgenstern, K.; Black, T.A.; Humphreys, E.R.; Griffis, T.J.; Drewitt, G.B.; Cai, T.; Nesic, Z.; Spittlehouse, D.L.; Livingston, N.J. Sensitivity and uncertainty of the carbon balance of a pacific northwest Douglas-fir forest during an el niño/la niña cycle. *Agric. For. Meteorol.* **2004**, *123*, 201–219.

58. Barr, A.G.; Black, T.A.; Hogg, E.H.; Griffis, T.J.; Morgenstern, K.; Kljun, N.; Theede, A.; Nesic, Z. Climatic controls on the carbon and water balances of a boreal aspen forest, 1994–2003. *Glob. Chang. Biol.* **2007**, *13*, 561–576.

59. Chen, B.; Black, T.A.; Coops, N.C.; Krishnan, P.; Jassal, R.; Bruemmer, C.; Nesic, Z. Seasonal controls on interannual variability in carbon dioxide exchange of a near-end-of rotation Douglas-fir stand in the pacific northwest, 1997–2006. *Glob. Chang. Biol.* **2009**, *15*, 1962–1981.

60. Melesse, A.M.; Hanley, R.S. Artificial neural network application for multi-ecosystem carbon flux simulation. *Ecol. Model.* **2005**, *189*, 305–314.

61. Moffat, A.M.; Papale, D.; Reichstein, M.; Hollinger, D.Y.; Richardson, A.D.; Barr, A.G.; Beckstein, C.; Braswell, B.H.; Churkina, G.; Desai, A.R. Comprehensive comparison of gap-filling techniques for eddy covariance net carbon fluxes. *Agric. For. Meteorol.* **2007**, *147*, 209–232.

62. White, H. Learning in artificial neural networks: A statistical perspective. *Neural Comput.* **1989**, *1*, 425–464.

63. Jain, A.K.; Mao, J.; Mohiuddin, K.M. Artificial neural networks: A tutorial. *Computer* **1996**, *29*, 31–44.

64. Zhang, G.; Patuwo, B.E.; Hu, M.Y. Forecasting with artificial neural networks: The state of the art. *Int. J. Forecast.* **1998**, *14*, 35–62.

65. Basheer, I.A.; Hajmeer, M. Artificial neural networks: Fundamentals, computing, design, and application. *J. Microbiol. Methods* **2000**, *43*, 3–31.

66. Papale, D.; Valentini, R. A new assessment of european forests carbon exchanges by eddy fluxes and artificial neural network spatialization. *Glob. Chang. Biol.* **2003**, *9*, 525–535.

67. Ooba, M.; Hirano, T.; Mogami, J.I.; Hirata, R.; Fujinuma, Y. Comparisons of gap-filling methods for carbon flux dataset: A combination of a genetic algorithm and an artificial neural network. *Ecol. Model.* **2006**, *198*, 473–486.

68. Richardson, A.D.; Mahecha, M.D.; Falge, E.; Kattge, J.; Moffat, A.M.; Papale, D.; Reichstein, M.; Stauch, V.J.; Braswell, B.H.; Churkina, G. Statistical properties of random CO_2 flux measurement uncertainty inferred from model residuals. *Agric. For. Meteorol.* **2008**, *148*, 38–50.

69. Rumelhart, D.E.; Hintont, G.E.; Williams, R.J. Learning representations by back-propagating errors. *Nature* **1986**, *323*, 533–536.

70. Maier, H.R.; Dandy, G.C. Neural networks for the prediction and forecasting of water resources variables: A review of modelling issues and applications. *Environ. Model. Softw.* **2000**, *15*, 101–124.

71. Funahashi, K.I. On the approximate realization of continuous mappings by neural networks. *Neural Netw.* **1989**, *2*, 183–192.

72. Cybenko, G. Approximation by superpositions of a sigmoidal function. *Math. Control Signal Syst.* **1989**, *2*, 303–314.

73. Moré, J. The levenberg-marquardt algorithm: Implementation and theory. In *Numerical Analysis*; Watson, G.A., Ed.; Springer: Berlin Heidelberg, Germany, 1978; pp. 105–116.

74. Thomas, R.Q.; Canham, C.D.; Weathers, K.C.; Goodale, C.L. Increased tree carbon storage in response to nitrogen deposition in the us. *Nat. Geosci.* **2010**, *3*, 13–17.

75. Niu, S.; Wu, M.; Han, Y.I.; Xia, J.; Zhang, Z.; Yang, H.; Wan, S. Nitrogen effects on net ecosystem carbon exchange in a temperate steppe. *Glob. Chang. Biol.* **2010**, *16*, 144–155.

76. Grant, R.F.; Black, T.A.; Jassal, R.S.; Bruemmer, C. Changes in net ecosystem productivity and greenhouse gas exchange with fertilization of Douglas-fir: Mathematical modeling in ecosys. *J. Geophys. Res.* **2010**, *115*, doi:10.1029/2009jg001094.

77. Hyvönen, R.; Persson, T.; Andersson, S.; Olsson, B.; Ågren, G.I.; Linder, S. Impact of long-term nitrogen addition on carbon stocks in trees and soils in northern Europe. *Biogeochemistry* **2008**, *89*, 121–137.

78. Vicca, S.; Luyssaert, S.; Penuelas, J.; Campioli, M.; Chapin, F.S.; Ciais, P.; Heinemeyer, A.; Högberg, P.; Kutsch, W.L.; Law, B.E. Fertile forests produce biomass more efficiently. *Ecol. Lett.* **2012**, *15*, 520–526.

79. Wu, F.; Bao, W.; Li, F.; Wu, N. Effects of drought stress and n supply on the growth, biomass partitioning and water-use efficiency of sophora davidii seedlings. *Environ. Exp. Bot.* **2008**, *63*, 248–255.

80. Ibrom, A.; Oltchev, A.; June, T.; Kreilein, H.; Rakkibu, G.; Ross, T.; Panferov, O.; Gravenhorst, G. Variation in photosynthetic light-use efficiency in a mountainous tropical rain forest in Indonesia. *Tree Physiol.* **2008**, *28*, 499–508.

81. Schwalm, C.R.; Black, T.A.; Amiro, B.D.; Arain, M.A.; Barr, A.G.; Bourque, C.P.A.; Dunn, A.L.; Flanagan, L.B.; Giasson, M.-A.; Lafleur, P.M. Photosynthetic light use efficiency of three biomes across an east–west continental-scale transect in Canada. *Agric. For. Meteorol.* **2006**, *140*, 269–286.

82. Binkley, D.; Stape, J.L.; Bauerle, W.L.; Ryan, M.G. Explaining growth of individual trees: Light interception and efficiency of light use by eucalyptus at four sites in Brazil. *For. Ecol. Manag.* **2010**, *259*, 1704–1713.

Nitrogen Transfer to Forage Crops from a Caragana Shelterbelt

Gazali Issah, Anthony A. Kimaro, John Kort and J. Diane Knight

Abstract: Caragana shelterbelts are a common feature of farms in the Northern Great Plains of North America. We investigated if nitrogen (N) from this leguminous shrub contributed to the N nutrition of triticale and oat forage crops growing adjacent to the shelterbelt row. Nitrogen transfer was measured using ^{15}N isotope dilution at distances of 2 m, 4 m, 6 m, 15 m and 20 m from the shelterbelt. At 2 m caragana negatively impacted the growth of triticale and oat. At 4 m from the shelterbelt productivity was maximum for both forage crops and corresponded to the highest amount of N originating from caragana. The amount of N transferred from caragana decreased linearly with distance away from the shelterbelt, but even at 20 m from the shelterbelt row measureable amounts of N originating from caragana were detectable in the forage biomass. At 4 m from the shelterbelt approximately 40% of the N in both oat and triticale was from caragana, and at 20 m from the shelterbelt approximately 20% of the N in oat and 8% of the N in triticale was from caragana.

Reprinted from *Forests*. Cite as: Issah, G.; Kimaro, A.A.; Kort, J.; Knight, J.D. Nitrogen Transfer to Forage Crops from a Caragana Shelterbelt. *Forests* **2015**, *6*, 1922-1932.

1. Introduction

Field shelterbelts are a common feature on farms in the Northern Great Plains of North America. The first shelterbelts in Canada were planted in 1903 as windbreaks to protect soils, crops and farmyards [1]. In the USA, as part of the 1935 Soil Conservation Act, trees were planted in large numbers as a means of soil erosion control in response to the devastating drought, windstorms and resulting soil erosion that occurred during the Great Depression [2]. Shelterbelts consist of a variety of coniferous and deciduous trees, with specific species adapted to the variety of soils and environments encountered across the region. Planting of shelterbelts had a number of positive impacts including protecting soil from wind erosion [3,4], positively influencing field hydrology through trapping snow [5] and protecting crops from wind damage [3], among other beneficial impacts [6]. Shelterbelts do not only serve as physical barriers to soil and snow movement; they are living barriers that separate cropped fields and as such interact with the adjacent field crops in either a competitive or complementary fashion [7]. Trees compete with adjacent field crops for above-ground and below-ground resources. In the semi-arid climates of the Northern Great Plains, soil moisture is frequently the limiting resource [7,8]. Furthermore, crop yields near shelterbelts can be reduced due to allelopathy by the tree roots [9], nutrient leaching during snowmelt [10], shading [11] and changes to microclimate [12]. Nonetheless, field shelterbelts that typically occupy 5% to 6% of the cropland can produce economic returns to producers based entirely on the increased yields in sheltered areas [13]. In temperate climates, yield increments for a number of field crops affected by shelterbelts ranged from 6% to 44% [6].

Caragana (*Caragana arborescens* Lam.) is a common shrub found in shelterbelts all across the Canadian prairies. The species is one of the hardiest, small deciduous shrubs planted on the northern Great Plains [12]. It readily adapts to sandy, alkaline soils in open (unshaded) environments [14]. Because it is a legume, caragana has the capacity to form root nodules and fix atmospheric nitrogen (N) [14,15] and enhance N availability [16,17]. Biological N_2 fixation accounted for up to 80% of N in caragana leaves in a soil-less system [15] and up to 65% in field soil [17,18]. An intercropping study conducted in Saskatchewan, Canada provides indirect evidence for transfer of N from caragana to willow (*Salix miyabeana*). Caragana enhanced the growth of willow at two sites, but was outcompeted by willow at a third site [17]. At the first two sites, willow growth increased as a function of the proportion of caragana in the mixed stand suggesting that increased N availability was driving the relationship. Nitrogen transfer was not measured directly.

The objective of this study was to quantify the transfer of N from a caragana shelterbelt to adjacent triticale (X *Triticosecale* Wittmack) and oat (*Avena sativa* L.) crops grown for forage in successive years. The influence of distance from the shelterbelt was investigated to determine the extent of the influence of the shelterbelt on N nutrition and forage quality of the adjacent field crops.

2. Experimental Section

2.1. Experimental Setup

The experiment was established adjacent to an existing caragana shelterbelt at the Agroforestry Development Centre, Indian Head, SK, Canada (50°33' N 103°39' W) (Figure 1). The shelterbelt was planted on the nursery site in 1933 in a north-south orientation. Caragana was approximately 5 m tall at the time of the experiment. The shelterbelt is a single row shelterbelt, characteristic of most Canadian shelterbelt plantings [6]. The soil on site is an Orthic Oxbow with a loamy, morainal and hummocky landform, with a slope of 2% to 5% [19]. Because the site was originally a nursery site, historically the shelterbelt received periodic irrigation, fertilization and herbicide and pesticide applications uncharacteristic of shelterbelts on farmland. However, no applications of fertilizers, herbicides nor pesticides have been made for the last twenty-five years.

The shelterbelt is adjacent to an annually cropped field located on the eastern side of the shelterbelt. In the spring of 2011, prior to establishing the experiment, soils were sampled from the site in 15 cm increments to a 45 cm depth. Four soil cores were extracted at five distances from the shelterbelt row, at 2 m, 4 m, 6 m, 15 m and 20 m for physiochemical characterization (Table 1). All analyses were performed using standard protocols [20]. Electrical conductivity (EC) and pH were measured on 1:2 soil: water extracts. In 2011 the field was planted by staff at the Agroforestry Development Centre with spring triticale (X *Triticosecale* Wittmack *cv.* AC Ultima) at a seeding rate of 90 kg ha^{-1} and in 2012 the field was planted with Pinnacle oat (*Avena sativa* L.) at a seeding rate of 90 kg ha^{-1}. Total precipitation during the 2011 growing season (May to September) was 290 mm and average temperature was 15.3 °C. In the 2012 growing season (May to September) the

total precipitation was 285 mm and temperature 15.1 °C. The long-term (1971 to 2000) averages
for the area were 255 mm precipitation and temperature of 15.9 °C.

Table 1. Physiochemical properties of Orthic Oxbow soil at distances perpendicular to
a caragana shelterbelt.

Distance (m)	pH	EC (mS cm⁻¹)	P	NO₃⁻	NH₄⁺	S	OC	K⁺	Mg²⁺	Ca²⁺
				mg kg⁻¹			g kg⁻¹	cmol (+) kg⁻¹		
				0–15 cm						
2	8.04 †	0.26	22.04	8.27	3.72	5.66	7.55	0.70	1.86	9.83
4	8.05	0.17	26.21	2.40	3.09	5.15	7.05	0.62	1.66	9.29
6	8.06	0.19	21.34	1.52	2.85	4.47	7.00	0.52	1.98	9.50
15	8.01	0.22	27.23	2.23	3.36	7.65	7.20	0.63	2.21	7.75
20	7.93	0.26	31.94	2.99	3.77	9.00	7.45	0.65	2.44	7.27
				15–30 cm						
2	8.21	0.20	6.40	1.50	4.27	4.18	4.50	0.36	2.28	9.07
4	8.31	0.16	6.31	1.02	3.74	4.25	5.10	034	2.06	9.32
6	8.37	0.15	11.48	1.25	3.85	4.29	4.95	0.33	2.11	8.86
15	8.17	0.20	9.50	1.17	3.32	6.55	5.90	0.34	2.35	7.79
20	8.26	0.23	8.51	0.87	3.07	11.16	5.15	0.29	2.91	8.21
				30–45 cm						
2	6.40	0.13	5.36	0.22	3.33	3.37	2.45	0.23	2.58	6.13
4	8.49	0.18	9.77	0.29	3.94	4.18	3.00	0.31	3.10	8.34
6	8.45	0.20	6.09	0.67	3.66	4.11	3.55	0.24	2.70	8.27
15	8.44	0.21	5.64	0.49	3.30	7.80	3.45	0.23	2.75	7.90
20	8.56	0.29	5.16	0.46	4.25	nd ‡	3.80	0.25	3.34	7.79

† Mean values ($n = 4$); nd ‡ = no data.

In each of the two years of the study four 15 m × 20 m plots were established perpendicular to
the shelterbelt on the eastern side of the shelterbelt (Figure 1). Within each plot, 0.6 m × 0.6 m
subplots were established at distances 2 m, 4 m, 6 m, 15 m and 20 m from the centre of the
shelterbelt. Above-ground biomass from the triticale and oat sub-plots was hand-harvested
approximately 5 cm above the soil surface using scissors, 9 and 10 weeks after planting,
respectively. The time of sampling corresponded to the mid to late milk stage of development
(BBCH = 76) for each crop, after maximum N uptake occurred. In cereal crops maximum N uptake
typically occurs around 60 to 65 days after emergence [21]. Four additional reference samples were
collected from outside the main plots approximately 50 m from the shelterbelt. Plants from these
reference plots were assumed to be far enough from the caragana so as not to be influenced by the
caragana and not to have accessed N fixed by the caragana. Leaves from caragana were collected
from four randomly selected locations within the canopy of the shelterbelt in the experimental plot.
Delta ¹⁵N values from the reference crops and the caragana were used in calculating N transfer
from the caragana to the adjacent crops (Equation 1). Plant samples were oven-dried at 60 °C to
stable weight and ground using a Wiley mill (Thomas Scientific, Swedesboro, NJ, USA), and
re-ground to fine powder using a ball grinder (8000D Mixer/Mill, SPEX Sample Prep® LLC.,
Metuchen, NJ, USA). Subsamples (ca. 1 mg) were weighed using a micro-balance (Sartorius
Microbalance, CPA2P, Bradford, MA, USA) and encapsulated in 8 mm × 5 mm tin capsules for

analysis of N concentration and $\delta^{15}N$ on a Costech ECS4010 elemental analyzer (Costech Analytical Technologies Inc., Valencia, CA, USA) coupled to a Delta V Advantage mass spectrometer (Thermo Scientific, Bremen, Germany) at the University of Saskatchewan (Saskatoon, SK, Canada). The $\delta^{15}N$ of the samples was used to calculate the percentage of N derived from the atmosphere (% Ndfa) that was transferred from the caragana to the triticale and oats (% $N_{transfer}$) [21]:

$$\%N_{transfer} = \left(\frac{\delta^{15}N_{crop} - \delta^{15}N_{tree/crop}}{\delta^{15}N_{crop} - \delta^{15}N_{tree}} \right) \times 100\% \tag{1}$$

Where $\delta^{15}N_{crop}$ is the value for the reference triticale or oat not influenced by the caragana and accessing only soil available N; $\delta^{15}N_{tree/crop}$ is triticale or oat influenced by the caragana and assumed to access N from soil available N and caragana fixed N; and $\delta^{15}N_{tree}$ is the value for the caragana. Amount of N transferred from caragana to triticale or oat was calculated as the product of $\%N_{transfer}$ and amount of N in the crop.

Statistical analyses were performed on $\%N_{transfer}$ and amount of N transferred using SAS version 9.2 [22]. Simple linear regression was performed on $\%N_{transfer}$, the amount of N transferred, total N in the shoots of the forage crops and shoot biomass, as a function of distance from the shelterbelt using PROC REG procedure in SAS. Regression analysis was performed with data from the 2 m distance included and excluded. Only regressions where the 2 m distance measurements were excluded were significant ($P < 0.05$) and are reported.

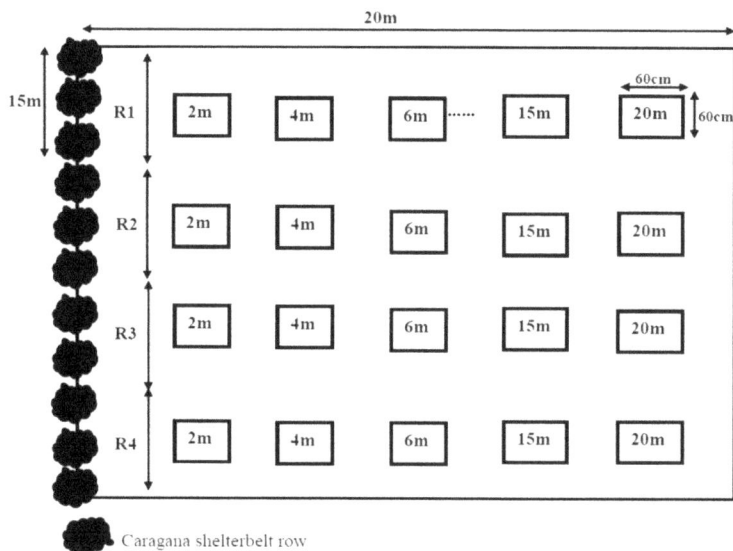

Figure 1. Diagram (not drawn to scale) of experimental plot layout showing replicates and sampling subplots at 2 m, 4 m, 6 m, 15 m and 20 m from the caragana shelterbelt (60 cm × 60 cm) within each 15 m swath.

3. Results and Discussion

Substantial N transfer from caragana to triticale and oat occurred in the two years of the study (Figure 2A and 2D). Except at the 2 m distance, both percentage of N and the amount of N (g N m^{-2}) originating from caragana in the triticale and oat biomass decreased linearly with increasing distance from the shelterbelt (Figure 2A and 2D). Both triticale and oat productivity were severely limited 2 m from the shelterbelt with both crops producing about 50% of the biomass produced 4 m from the shelterbelt (Figure 2C and 2F). Productivity of both crops was maximum at 4 m (Figure 2C and 2F) corresponding to the same distance at which the maximum amount of N was transferred (Figure 2A and 2D). The amount of N transferred to triticale at 2 m was about 58% of the transfer at 4 m and for oat only about 3.5% of the transfer at 4 m.

Low yields near shelterbelts have long been reported e.g., [11,23]. At the 2 m distance from the shelterbelt competition for water, light and/or nutrients between the caragana and the crop plants were probably responsible for the low yields of the crops. In temperate climates, competition for water is generally more limiting than competition for light [8] and has been reported to decrease yields of crops near tree rows [7]. Some early research focused on estimating the zone of influence of shelterbelts in terms of the negative impact on yield. Read [23] estimated yield reductions occurred to 1 h, where h is the height of the shelterbelt. Stoeckeler [11] suggested a range of 0.5 h to 1.5 h and Lyles et al. [24] reported reduced yields at a distance of 2 h. In our study the negative influence on forage yield of the approximately 5 m-tall shelterbelt was narrow at less than 4 m from the shelterbelt for both crops (i.e., <1 h). Considering that caragana has a very vigorous and aggressive growth habit and can be invasive in some conditions [14,25] this narrow zone of negative influence was somewhat surprising.

The benefits from the N$_2$-fixing capabilities of the plant might balance the negative impacts from the invasive nature of the shrub. Indeed the vigorous root growth of caragana might be responsible for the significant N transfer observed up to about 15 m in both crops. Considering tree root density decreases with increasing distance from the shelterbelt row, this approximately 15 to 20 m zone of influence in this study is assumed to be indicative of the zone where direct transfer of N takes place through root to root contact or mycorrhizal associations, probably augmented by indirect transfer through mineralization of below-ground roots and sloughed cells, nodules, and other organic compounds excreted from roots, and possibly leaf litter fall.

Studies using ^{15}N labeling have demonstrated net N transfer from legume tree saplings to grass grown together in containers [26] and have provided evidence of N transfer via mycorrhiza between N$_2$-fixing and non-fixing tree species [27–29]. Natural abundance methodologies have provided similar estimates of N transfer between trees [30–32]. The natural abundance method has an advantage over ^{15}N labeling studies in that N is not added to the soil or plant, enabling N transfer to be estimated without disturbing the system. However the method relies on very small differences in natural abundance values between the legume and non-legume plants and in the soil inorganic N pool. Any soil processes like nitrification, denitrification and ammonium volatilization that discriminate against ^{15}N can bias the estimates by altering the δ^{15}N signature of the inorganic N pool accessed by the plants. Furthermore, isotope ratio mass spectrometry (IRMS) analyses require a very small amount of an organic sample (typically 1 mg–10 mg) representative of the entire sample. It is essential that the organic material sampled is finely ground and homogeneously mixed. Because of the small size of the subsample for IRMS even small deviations in weighing samples can have a relatively large effect, particularly on the percent N values. Nonetheless natural abundance studies have been used successfully to estimate N transfer [30–32]. In our study, differences in δ^{15}N between the caragana and triticale and oat are sufficiently large and consistent to provide a confident estimate of N transfer (Table 2).

Table 2. Delta ^{15}N values of leaves (caragana) or shoots (triticale and oats) where triticale and oats are grown adjacent to a mature caragana shelterbelt row (distance 0) and sampled at perpendicular distances away from the shelterbelt.

	Caragana		Triticale		Oats	
Distance (m)	δ^{15}N (per mil)	*Std Dev*	δ^{15}N (per mil)	*Std Dev*	δ^{15}N (per mil)	*Std Dev*
0	4.55	0.44	-	-	-	-
2	-	-	7.03	0.65	8.175	1.20
4	-	-	7.38	0.44	5.845	0.75
6	-	-	8.22	0.78	6.268	0.18
15	-	-	8.99	1.91	6.285	0.19
20	-	-	10.83	1.19	7.935	1.18
P-value			0.0028			0.0047

The negative impact of the shelterbelt on forage crop yield 2 m from the shelterbelt was also observed for the amount of N transferred (Figure 2A and 2D). Despite lower yields and lower amounts of N transferred at 2 m compared to 4 m, in triticale approximately 60% of the N in the shoots was transferred from caragana at 2 m, whereas in oat only about 5% of N in the shoots was transferred from caragana. It is clear that the two forage crops interacted differently in the immediate vicinity of the shelterbelt trees. The concentration of N in the oat shoot biomass was higher at 2 m (2.3 ± 0.23 mg N g^{-1}) compared to the further distances (1.5 ± 0.13 mg g^{-1} averaged over the 4 to 20 m distance) indicating a concentration of N because of low dry matter production but similarly indicating that amounts of N present were sufficient to support the biomass produced at this distance. In contrast, in triticale N concentration was constant across the sampling transect (2.0 ± 0.13 mg N g^{-1}). Initial soil sampling in 2011 indicated that amounts of nitrate in the top 15 cm were higher 2 m from the shelterbelt than elsewhere in the field (Table 1). Soils were not sampled in 2012, but it appears that the levels of available N near the shelterbelt were sustained at high enough levels to fully supply the N requirement of oat at this distance. Amounts of soil N are typically higher near the tree line than in cropped areas in agroforestry systems [33–35].

Even though oat did not appear to rely on direct transfer from caragana through root to root contact or mycorrhizal connections, the lack of any apparent N transfer from caragana to the oat crop at the 2 m distance was unexpected. Oats at the 2 m distance should still be accessing N mineralized from caragana litter fall. In addition to the ^{15}N signature of organic matter inputs, changes in ^{15}N natural abundance of the available soil N can also occur through soil processes like nitrification, denitrification and NH$_3$ volatilization. Discrimination against ^{15}N occurs in all processes causing a decrease in $\delta^{15}N$ associated with nitrification or an increase in $\delta^{15}N$ associated with volatilization and denitrification [36]. Soil conditions near the shelterbelts should be conducive for denitrification in that snow melt in this snow accumulation zone could potentially provide the necessary anaerobic conditions, and mineralization of leaf litter the necessary nitrate and soil organic carbon substrates (Table 1). The high $\delta^{15}N$ signature of the oat growing at the 2 m distance could be a result of a high $\delta^{15}N$ signature in the soil N because of denitrification and/or volatilization. Alternatively, snowmelt and or/rainfall may have leached nitrate out of the rooting zone of the oat crop [10].

Despite N transfer to oat being negligible close to the caragana shelterbelt, the positive impact of the caragana on N transfer was sustained over larger distances for oat than triticale (Figure 2D and 2A). Whereas the amount of N transferred to triticale at 20 m was only 9% of N transferred at 4 m, the amount transferred to oat at 20 m was 22% of that transferred at 4 m. At 20 m approximately 20% of the N in oat was from caragana, whereas in triticale, N from caragana constituted only about 8% of the shoot N. The total amount of N in the shoot tissue of triticale (*i.e.*, transferred N + soil derived N) was about 1.3 times the amount in oat indicating a stronger sink for N (Figure 2B and 2E). It appears that the stronger demand for N probably led to the supply of available N being used up faster in triticale than oat.

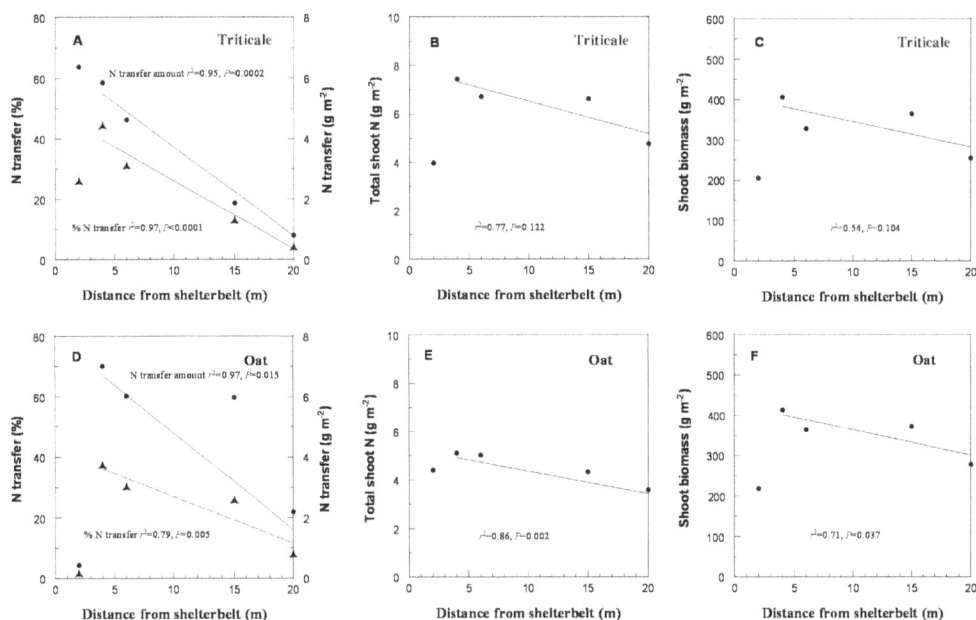

Figure 2. Nitrogen transfer characteristics and above-ground productivity in triticale (A–C) and oat (D–F) as a function of perpendicular distance away from a caragana shelterbelt. (**A,D**) are percentage of N in the forage crops transferred (dots) from caragana and amount of N transferred (triangles); (**B,E**) are total N in the above-ground biomass and (**C,F**) are above-ground biomass.

There are few studies in the literature examining the distance to which N transfer occurs particularly in mature shelterbelts. Daudin and Sierra [37] measured transfer between *Gliricidia sepium* and *Dichanhium aristatum* (Poir) in a 16-year old tropical agroforestry system over a 5 m distance. They measured a very similar pattern of decreasing N transfer to the grass with distance away from the tree row, and a decreasing trend in total N uptake from the soil. In general % N derived from the tree was relatively stable at around 50% over 3 m to 4 m, with percentages decreasing slightly at 5 m [37].

Isaac *et al.* [38] reported that N_2-fixation in *Acacia senegal* decreased with the age of the tree. In contrast, caragana appears to sustain relatively high N_2-fixation in mature trees at levels that sustain significant transfer to the adjacent crop.

4. Conclusions

Significant amounts of N were transferred from a mature caragana shelterbelt to adjacent oat and triticale forage crops. Close to the shelterbelt (<4 m) caragana negatively impacted the productivity of the adjacent forage crop. However at 4 m from the shelterbelt row, forage biomass and N content were at their maximum. Using the ^{15}N dilution protocol significant amounts of N originating from the caragana were detected in the shoot biomass of triticale and oat. Nitrogen transferred from caragana represented approximately 40% of the total N in the shoot of both triticale and oat 4 m from the shelterbelt. Both the percentage of N transferred and the amount of N transferred decreased linearly with distance away from the shelterbelt. Measureable amounts of N originating from caragana were detectable in the triticale and oat biomass 20 m from the shelterbelt row.

Acknowledgments

Funding from Agriculture and Agri-Food Canada (ADC) under the Research Affiliate Program (RAP) and University of Saskatchewan is appreciated. Authors also acknowledge the support from staff and the administration of the Agroforestry Development Centre, Indian Head, SK, Canada, during fieldwork. Our thanks to the Stable Isotope Laboratory, University of Saskatchewan for the ^{15}N analysis.

Author Contributions

This manuscript reports on the results of one chapter of Gazali Issah's MSc research supervised by Anthony A. Kimaro and Knight. J. Kort (retired) was the scientist at the AAFC-Agroforestry Development Centre. All authors provided input into the design and implementation of the experiment. The manuscript was written by Gazali Issah and J. Diane Knight and edited by the other authors. All authors contributed significantly to the research.

Conflicts of Interest

The authors declare no conflicts of interest.

References

1. One Hundred Years of Prairie Forestry. Available online: http://www.ourspace.uregina.ca/handle/10294/129 (accessed on 25 March 2015).

2. Arha, K.; Josling, T.; Sumner, D.A.; Thompson, B.H. National Forum on U.S. Agricultural Policy and the 2007 Farm Bill: Conserving the Ecological Integrity and Economic Resilience of American Working Lands; Woods Institute for the Environment, Stanford University: California, CA, USA, 2007; pp. 1–257. Available online: http://media.law.stanford.edu/publications/archive/pdf/farmbill_book. pdf (accessed on 23 March 2015).

3. Public Fundraising Regulatory Association. *A Study of Tree Planting on the Canadian Prairies*; Government of Canada-PFRA: Regina, SK, Canada, 1976; pp. 1–184.

4. Sutton, V. An Evaluation of Landowner's Attitudes toward Field Shelterbelts in Manitoba: A Case Study in the Lyleton Area. Master's Thesis, University of Manitoba, Winnipeg, Manitoba, Canada, 1983.

5. Scholten, H. Snow distribution on crop fields. *Agric. Ecosyst. Environ.* **1988**, *22–23*, 363–380.

6. Kort, J. Benefits of windbreaks to field and forage crops. *Agric. Ecosyst. Environ.* **1988**, *22–23*, 165–190.

7. Jose, S.; Gillespie. A.R.; Seifert, J.R.; Biehle, D.J. Defining competition vectors in a temperate alley cropping system in the Midwestern USA: 2. Competition for water. *Agroforest. Syst.* **2000**, *48*, 41–59.

8. Gillespie, A.R.; Jose, S.; Mengel, D.B.; Hoover, W.L.; Pope, P.E.; Seifert, J.R. Defining competition vectors in a temperate alley cropping system in the Midwestern USA: 1. Production physiology. *Agroforest. Syst.* **2000**, *48*, 25–40.

9. Zolotukhin, A.I. Allelopathic effect of shrubs used in steppe afforestation. *Ekologiya* **1980**, *11*, 13–17.

10. George, E.J. *Effect of Tree Windbreaks and Slat Barriers on Wind Velocity and Crop Yields*; USDA Agricultural Research Service: Washington, DC, USA, 1971; p. 23.

11. Stoeckeler, J.H. The design of shelterbelts in relation to crop yield improvement. *World Crop.* **1965**, *17*, 3–8.

12. George E.J. Tree and shrub species for the northern Great Plains. *Cir. USDA.* **1953**, *912*, 46.

13. Brandle, J.R.; Johnson, B.B.; Dearmont, D.D. Windbreak economics: The case of winter wheat production in Eastern Nebraska. *J. Soil Water Conserv.* **1984**, *39*, 339–343.

14. Dietz, D.R.; Slabaugh, P.E.; Bonner, F.T. *Caragana arborescens* Lam.: Siberian peashrub. In *The Woody Plant Seed Manual*; Franklin T., Karrfalt, R.P., Eds.; Department of Agriculture, Forest Service: Washington, DC, USA, 2008; Volume 727, pp. 321–323.

15. Moukoumi, J.; Hynes, R.K.; Dumonceaux, T.J.; Town, J.; Bélanger, N. Characterization and genus identification of rhizobial symbionts from *Caragana arborescens* in western Canada. *Can. J. Micro.* **2013**, *59*, 399–406.

16. Su, Y.Z.; Zhao, H.L. Soil properties and plant species in an age sequence of *Caragana microphylla* plantations in the Horqin Sandy Land, north China. *Ecol. Eng.* **2003**, *20*, 223–235.

17. Moukoumi, J.; Farrell, R.E.; van Rees, K.J.C.; Hynes, R.K.; Bélanger, N. Growth and nitrogen dynamics of juvenile short rotation intensive cultures of pure and mixed *Salix miyabeana* and *Caragana arborescens*. *Bioenergy Res.* **2012**, *5*, 719–732.

18. Issah, G.; Kimaro, A.A.; Kort, J.; Knight, J.D. Quantifying biological nitrogen fixation of agroforestry shrub species using ^{15}N dilution techniques under greenhouse conditions. *Agroforest. Syst.* **2014**, *8*, 607–617.

19. Agriculture and Agri-Food Canada. The soils of Indian Head Rural Municipality. *Saskatchewan Soil Survey*, 156; Saskatchewan Institute of Pedology, University of Saskatchewan: Saskatoon, SK, Canada, 1986.

20. Carter, M.R.; Gregorich, E.G. Soil sampling and methods of analysis. *Canadian Society of Soil Science*, 2nd ed.; CRC Press: Boca Raton, FL, USA, 2008.

21. Malhi, S.S.; Johnston, A.M.; Schoenau, J.J.; Wang, Z.H.; Vera, C.L. Seasonal biomass accumulation and nutrient uptake of wheat, barley and oat on a Black Chernozem soil in Saskatchewan. *Can. J. Plant Sci.* **2006**, *86*, 1005–1014.

22. *SAS Institute*, version 9.2; SAS/STAT user's guide; SAS Institute Inc.: Cary, NC, USA, 2008.

23. Read, R.A. Tree windbreaks for the central Great Plains. *Agriculture Handbook 250*; USDA forest Service: Washington, DC, USA, 1964; p. 68.

24. Lyles, L.; Tatarko, J.; Dickerson, J.D. Windbreak effects on soil water and wheat yield. *Trans. ASAE* **1984**, *20*, 69–72.

25. Henderson, D.C.; Chapman, R. *Caragana arborescens* invasion in Elk Island National Park, Canada. *Nat. Areas J.* **1996**, *26*, 261–266.

26. Rao, A.V.; Giller, K.E. Nitrogen fixation and its transfer from Leucaena to grass using [15]N. *For. Ecol. Manag.* **1993**, *61*, 221–227.

27. Arnebrandt, K.; Ek, H.; Finlay, R.D.; Soderstrom, B. Nitrogen translocation between *Alnus lutinosa* (L.) Gaertn. seedlings inoculated with *Frankia* sp. and *Pinus contorta* Doug, ex Loud seedlings connected by a common ectomycorrhizal mycelium. *New Phytol.* **1993**, *124*, 231–242.

28. Ekbland, A.; Huss-Danell, K. Nitrogen fixation by *Alnus incana* and nitrogen transfer from *A. incana* to *Pinus sylvestris* influenced by macronutrients and ectomycorrhiza. *New Phytol.* **1995**, *131*, 453–459.

29. Roggy, J.C.; Moiroud, A.; Lensi, R.; Domenach, A.M. Estimating N transfer between N2-fixing ctinorhizal species and the non-N2-fixing *Prunus avium* under partially controlled conditions. *Biol. Fert. Soils* **2004**, *39*, 312–319.

30. Binkley, D.; Sollins, P.; McGill, W.B. Natural abundance of nitrogen-15 as a tool for tracing alder-fixed nitrogen. *Soil Sci. Soc. Am. J.* **1985**, *49*, 444–447.

31. Snoeck, D.; Zapata, F.; Domenach, A.M. Isotopic evidence of the transfer of nitrogen fixed by legumes to coffee trees. *Biotech. Agron. Soc. Environ.* **2000**, *4*, 95–100.

32. Sierra, J.; Nygren, P. Transfer of N fixed by a legume tree to the associated grass in a tropical silvopastoral system. *Soil. Biol. Biochem.* **2006**, *3*, 1893–1903.

33. Kremer, R.J.; Li, J. Developing weed-suppressive soils through improved soil quality management. *Soil Tillage Res.* **2003**, *72*, 193–202.

34. Udawatta, R.P.; Kremer, R.J.; Adamson, B.W.; Anderson, S.H. Variations in soil aggregate stability and enzyme activities in a temperate agroforestry practice. *Appl. Soil Ecol.* **2008**, *39*, 153–160.

35. Paudel, B.R.; Udawatta, R.P.; Kremer, R.J.; Anderson, S.H. Agroforestry and grass buffer effects on soil quality parameters for grazed pasture and row-crop systems. *Appl. Soil Ecol.* **2011**, *48*, 125–132.

36. Hogberg, P. [15]N natural abundance in soil-plant systems. *New Phytol.* **1997**, *137*, 179–203.

37. Daudin, D.; Sierra, J. Spatial and temporal variation of below-ground N transfer from a leguminous tree to an associated grass in an agroforestry system. *Agri. Ecosyst. Environ.* **2008**, 126, 275–280.

38. Isaac, M.E.; Harmand, J.M.; Lesueur, D.; Lelon, J. Tree age and soils phosphorus conditions influence N$_2$-fixation rates and soil N dynamics in natural populations of *Acacia senegal*. *For. Ecol. Manag.* **2013**, *261*, 582–588.

Biomass, Carbon and Nutrient Storage in a 30-Year-Old Chinese Cork Oak (*Quercus Variabilis*) Forest on the South Slope of the Qinling Mountains, China

Yang Cao and Yunming Chen

Abstract: Chinese cork oak (*Quercus variabilis*) forests are protected on a large-scale under the Natural Forest Protection (NFP) program in China to improve the ecological environment. However, information about carbon (C) storage to increase C sequestration and sustainable management is lacking. Biomass, C, nitrogen (N) and phosphorus (P) storage of trees, shrubs, herb, litter and soil (0–100 cm) were determined from destructive tree sampling and plot level investigation in approximately 30-year old Chinese cork oak forests on the south slope of the Qinling Mountains. There was no significant difference in tree components' biomass estimation, with the exception of roots, among the available allometric equations developed from this study site and other previous study sites. Leaves had the highest C, N and P concentrations among tree components and stems were the major compartments for tree biomass, C, N and P storage. In contrast to finding no difference in N concentrations along the whole soil profile, higher C and P concentrations were observed in the upper 0–10 cm of soil than in the deeper soil layers. The ecosystem C, N, and P storage was 163.76, 18.54 and 2.50 t ha^{-1}, respectively. Soil (0–100 cm) contained the largest amount of C, N and P storage, accounting for 61.76%, 92.78% and 99.72% of the total ecosystem, followed by 36.14%, 6.03% and 0.23% for trees, and 2.10%, 1.19% and 0.03% for shrubs, herbs and litter, respectively. The equations accurately estimate ecosystem biomass, and the knowledge of the distribution of C, N and P storage will contribute to increased C sequestration and sustainable management of Chinese cork oak forests under the NFP program.

Reprinted from *Forests*. Cite as: Cao, Y.; Chen, Y. Biomass, Carbon and Nutrient Storage in a 30-Year-Old Chinese Cork Oak (*Quercus Variabilis*) Forest on the South Slope of the Qinling Mountains, China. *Forests* **2015**, *6*, 1239-1255.

1. Introduction

Protecting existing forests and planting new forests through reforestation and afforestation are important measures to enhance carbon (C) sequestration capacity and potential in terrestrial ecosystems [1–4]. A growing number of studies have addressed C storage and sequestration in key national ecological restoration programs in China, such as the Sloping Land Conversion program, Three-North Sheltbelt Forest program, Changjiang (Yangtze) River Basin Forest Protection program, Beijing-Tianjin Sandstorm Source Control program, Natural Forest Protection program (NFP), and other programs [2,5–9].

The NFP program was implemented in 1998 with the goal of promoting natural forest resource protection and cultivation to improve the ecological environment, has been implemented since 1998 [6,10]. It covers about 6.0×10^7 ha of natural forest land, which accounts for 50% of the total natural forest area within 17 provincial-level administrative units in China [10]. Hu and Liu [11] used the volume-biomass method and National Forestry Statistics to calculate C storage of the NFP program from 1998–2002 and found that in total it sequestered 44.07 Tg, including 21.32 Tg from reforestation and afforestation, and 22.75 Tg from reduced the timber production. The results of Wei *et al.* [10] showed that the tree C pool under the NFP program in northeastern China increased from 1998 to 2008, by 6.3 Tg C, which was mainly sequestrated by natural forests (5.1 Tg C year^{-1}). It has been projected that biomass C storage from afforestation under the NFP program could potentially increase from 33.67 Tg in 2011 to 96.03 Tg in 2020 [6]. However, most of these previous studies were mainly conducted on forest biomass C storage at the national and regional scales with different estimation methods and different forest resource data. Moreover, there are few precise studies concerning direct plot investigations for various forest types, C storage estimates that include understory, forest floor, and soil, and the relationship between climatic factors and forest types on regional scales [12–16]. An age-related study on C storage in a black locust forest ecosystem on the Loess Plateau showed that tree C storage increased from 5 to 38 years, but significantly decreased from 38 to 56 years owing to high tree mortality. Moreover, storage in the shrub layer increased with stand age, but it was age-independent in the herb layer and litter. Storage in the topsoil (0–20 cm) increased at a constant rate with stand age, while it was age-independent in sub-top soil [7]. With an increased numbers of local studies, we can gain a more comprehensive understanding of the complex nature of ecosystem C storage in order to scale up to regional and global levels.

Quercus species are a keystone species in a wide range of habitats from Mediterranean semi-desert woodlands to subtropical rainforest in Europe, North America, and Southeast Asia [17]. For example, the cork oak is the second most important Portuguese forest species both in terms of the country forest area and in terms of forest industry product exports [18]. Many studies have explored the temporal and spatial distribution of biomass and nutrient accumulation in *Quercus* species for suitable management and conservation, especially in Spain and Portugal [17–26]. Guyette, Dey and Stambaugh [17] documented the temporal distribution in C storage of oak wood at floodplains in northern Missouri, USA. In this study, we chose Chinese cork oak (*Quercus variabilis*) as a model system to carefully evaluate biomass and nutrient pools in different ecosystem components. Chinese cork oak is one of the major *Quercus* species in warm-temperate and subtropical forests, ranging from 22°–42° N to 99°–122° E. There were only a few studies that developed allometric equations (Table 1; Equations (1)–(3)) and addressed tree biomass allocation of Chinese cork oak in different sites in China [27–29]. The Qinling Mountains in the Shanxi Province is one of the major distribution areas of Chinese cork oak forests. Chinese cork oak forests in the Qinling Mountains make up the largest forest vegetation carbon sink based on the Shanxi Province forest resource inventory data. However, these forests have been under serious threat owing to excessive overexploitation and inappropriate management for timber and charcoal production, cultivation of edible wild mushrooms, and dye products. Therefore, many studies have been conducted to provide some recommendations toward sustainable management of the Chinese cork oak forests in the Qinling Mountains under the NFP program [30–33]. However, there is still a lack of information about C stocks, especially below-ground, for Chinese cork oak forests in the Qinling Mountain, China.

The focus of this study was to develop suitable allometric equations to calculate various tree biomass components, and to quantify the distribution patterns and quantities of C, N, and P among the major tree components, shrubs, herbs, litter, and soil (0–100 cm) in Chinese cork oak forest along the south slope of the Qinling Mountains. Results of this study may provide a crucial complement to previous studies on the understanding of C storage and forest management under the NFP program.

2. Materials and Methods

2.1. Study Site and Sampling

This study was conducted in the Shan Yang Country, Southern Shanxi Province, China (33°09′–33°42′ N, 109°32′–110°29′ E; Figure 1). The Shan Yang Country is located on the south slope of the Qinling Mountains, China. The study area is situated in the transitional area between the subtropical zone and the warm temperate zone, with a mean annual temperature 13.1 °C, a mean annual rainfall of 709 mm, and a mean frost-free period of 207 days. Forest coverage is 62.6%, and more than 80% of the forest area is dominated by pure Chinese cork oak forest, which originated from seedlings, with a few *Sabina chinensis*, and *Pinus armandii* species. Shrub species at the site include *Pyrus betulifolia*, *Lespedeza bicolor*, and *Platycarya strobilacea*, and the main herb species are *Carextristachya*, *Imperata cylindrical* and *Ophiopogon japonicas*.

With the guidance of local forestry bureau staff, three widely distributed Chinese cork oak stands with similar site conditions, little or no human disturbance, and approximately 30 years old were selected under from the Shan Yang Country. The sites were all located near the middle of slopes and there was little difference among the sites with regard to aspect (North West), gradient (30–35°), and elevation (845–1068 m). The distance between each stand was about 5–8 km. Historically, all study stands naturally regenerated after the natural Chinese cork oaks were harvested. A 20 m × 20 m plot was constructed in the central area of each stand for sampling. Diameter at breast height (DBH) and height (H) were measured for all trees (DBH ≥ 5.0 cm) in each plot. In early August 2013, five Chinese cork oak trees within representative stand-specific DBH range were selected and harvested destructively in each stand following a previously published harvest method [19,27–29,34–36]. Trees were cut near the ground surface, after measurement of the total H of each tree, the tree stem was first cut open at 1.3 m, and then the top part of stem (from 1.3 m to the tips) was divided into 1-m-long sections. All branches and leaves of each stem section were clipped from the tree stems and branches, respectively and weighed. Stem discs (5 cm thickness) were collected from of each section, and the stem bark was separated from each disc to measure the weight of the fresh stem bark and the stem wood without bark. The whole root system was manually excavated, washed lightly to remove soil particles, and weighed.

Table 1. Collection of the available allometric equations used to calculate biomass of the various tree components for Chinese cork oak forests in China.

NO.	Site	Stand [a]	Components	Allometric equation	Correlation coefficient (R)	Sources
1.	Xiaolong Mountains, Gansu Province (33°30′–34°49′ N, 104°22′–106°43′ E)	20–84 6.0–18.6 4.3–13.1	Stem wood	$LnW_S = -3.7447 + 0.9679\ Ln\ (DBH^2H)$	0.9981	[28]
			Stem bark	$LnW_{BA} = -3.2565 + 0.7156\ Ln\ (DBH^2H)$	0.9852	
			Branches	$LnW_{BR} = -4.8449 + 1.0013\ Ln\ (DBH^2H)$	0.9917	
			Leaves	$LnW_L = -3.3569 + 0.6050\ Ln\ (DBH^2H)$	0.9611	
			Roots	$LnW_R = -2.9066 + 0.8144\ Ln\ (DBH^2H)$	0.9941	
2.	Baotianman Natural Reserve, Henan Province (33°25′–33°33′ N, 111°53′–112° E)	45 6.0–30.5 8.3–18.5	Stem wood	$LgW_S = -0.5440 + 0.6796\ Lg\ (DBH^2H)$	0.9969	[27]
			Stem bark	$LgW_{BA} = -0.8246 + 0.5896\ Lg\ (DBH^2H)$	0.9983	
			Branches	$LgW_{BR} = -2.5609 + 1.1092\ Lg\ (DBH^2H)$	0.9750	
			Leaves	$LgW_L = -2.0038 + 0.7460\ Lg\ (DBH^2H)$	0.9851	
			Roots	$LgW_R = -0.2645 + 0.5173\ Lg\ (DBH^2H)$	0.9986	
3.	Xishan Mountains, Beijing (39°34′ N, 116°28′ E)	26 3.1–10.3 6.0–8.0	Stems	$W_{S+BA} = 0.0508(DBH^2H)^{0.92}$	0.9889	[29]
			Branches	$W_{BR} = 0.0197\ (DBH^2H)^{0.8944}$	0.9412	
			Leaves	$W_L = 0.0029\ (DBH^2H)^{0.9125}$	0.9557	
			Roots	$W_R = 0.0458\ (DBH^2H)^{0.7484}$	0.9485	
4.	Shanyang Country, Shanxi Province (33°9′–34°42′ N, 109°32′–110°29′ E)	30 5.0–34.2 4.0–15.0	Stem wood	$W_S = 0.0335\ (DBH^2H)^{0.9579}$	0.9949	This study
			Stem bark	$W_{BA} = 0.0458\ DBH^{2.1044}$	0.9836	
			Branches	$W_{BR} = 0.0047\ DBH^{2.9836}$	0.9830	
			Leaves	$W_L = 0.0128\ DBH^{2.1013}$	0.9916	
			Roots	$W_R = 0.0831\ DBH^{2.1980}$	0.9634	

[a] referring to stand age (year), DBH (cm) and H (m), respectively.

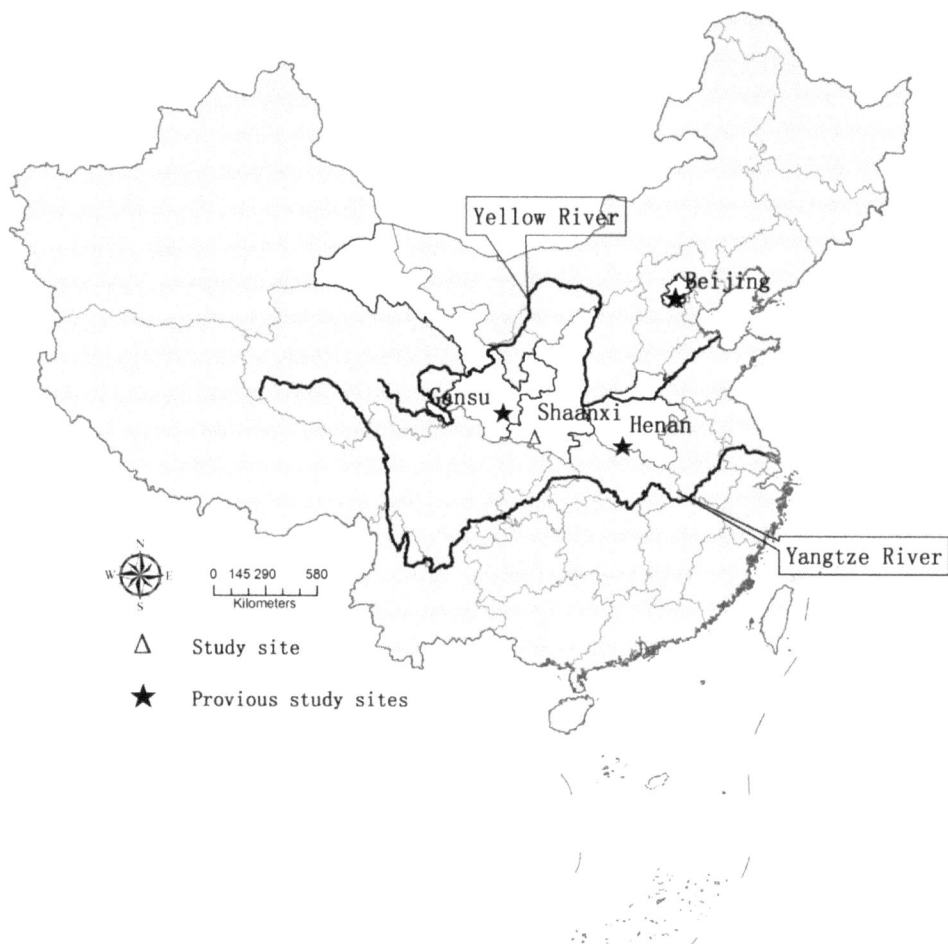

Figure 1. Site distribution of studies on Chinese cork oak allometric equations include Xiaolong Mountains, Gansu Province; the Xishan Mountains, Beijing; the Baotianman Natural Reserve, Henan Province; and the present study at Shanyang Country, Shanxi Province.

Shrub and herb biomass was determined using total harvesting destructive sampling techniques [34]. Sampling of the shrub layer and herb layer was conducted in five 2 m × 2 m subplots and 1 m × 1 m subplots, respectively. These subplots were randomly selected within each plot. Shrub plants were separated into leaves, stems and roots, and herbs were separated into aboveground and belowground components. Litter was sampled by collecting the entire organic material within five 1 m × 1 m subplots randomly chosen in each plot [34].

Subsamples of tree components (stem wood, stem bark, branches, foliage, and roots), shrub (leaves, stems and roots), herb (above- and belowground) and litter were sealed in plastic bags, and then oven dried at 80 °C in the laboratory to constant weight to obtain wet-to-dry mass conversion factors. The dried samples were ground and used to determine plant C and N concentrations by an elemental analyzer (Carlo Erba 1106, Milan, Italy). Total P was determined by the $HClO_4$-H_2SO_4 colorimetric method [37].

In each plot, five soil cores (5 cm in diameter) were randomly collected at 0–10 cm, 10–20 cm, 20–30 cm, 30–50 cm and 50–100 cm layers. After removing the plant roots, fauna and debris by hand, the soil was air dried at room temperature around 20 °C, and then ground and passed through a 0.25 mm sieve for determination of soil chemical properties. C concentration was determined by the $K_2Cr_2O_7$-H_2SO_4 method [38], total N concentration by the Kjeldahl method [39] and total P by the $HClO_4$-H_2SO_4 colorimetric method [37].A soil profile (1 m × 1 m × 1 m) in each plot was dug for measuring soil bulk density (g cm^{-3}). After excluding recognizable soil surface litter, stainless cutting rings (5 cm in diameter) were used to sample five replicated 100 cm^3 of soil at each layer at same depth intervals of 0–10 cm, 10–20 cm, 20–30 cm, 30–50 cm and 50–100 cm layers. The ring soil samples were scraped out and roots manually removed. The soils were dried at 105 °C to constant weight to calculate bulk density (oven-dried soil sample/volume of the metal ring). All of the samples were analyzed at the central laboratory of the Institute of Soil and Water Conservation, Chinese Academy of Sciences and Ministry of Water Resources (Yangling, China).

2.2. Data Calculation and Analysis

Before establishing the allometric equation, scatter plots were used to visualize whether the relationship between the independent (dry biomass (kg) of each tree component) and dependent variables (DBH and/or H of each tree) was linear. A power function was selected as an appropriate model in this study, with the following equations:

$$W = aDBH^b \tag{1}$$

$$W = a(DBH^2 \times H)^b \tag{2}$$

where W is the dry biomass (kg) of each tree component; a and b are allometric parameters; least squares linear regression was used to estimate the value of a and b. Ordinary least squares regression was performed to determine the coefficient of determination (R).

The developed allometric equations from this study (Table 1; Equations (4)) and the available allometric equations (Table 1; Equations (1)–(3)) developed previously in other study sites (Figure 1) were used to compare the differences in tree biomass (stem wood, stem bark, branches, leaves, and roots) estimations. Then, the total ecosystem C, N and P storage in the present study site were calculated based on the combination of tree (stem wood, stem bark, branches, leaves, and roots), shrub (leaves, stem and roots), herbs (above- and belowground), litter layers, and soil (0–100 cm) pools. C, N, and P concentrations of leaves, branches, stem bark, stem wood, and roots were multiplied by each tree component biomass from the developed species-specific allometric equations from the present site (Table 1, Equation (4)) to partition C, N, and P stocks among the tree components, and then summed the stocks for each tree and site to calculate the stand level stocks. For shrubs, herbs and litter, the C, N, and P concentrations were multiplied by their component mass at the plot level to calculate the stand level C, N and P stocks. The stock of soil C, N, and P in the different layers of soil were calculated by multiplying the soil bulk density with soil depth and C, N, and P concentrations in each layer. The carbon storage results in each layer of soil (0–10 cm, 10–20 cm, 20–30 cm, 30–50 cm and 50–100 cm) were then summed to compute the total C, N, and P storage to a depth of 100 cm.

All statistical analyses were performed with SPSS (version 20.0, SPSS Inc., Chicago, IL, USA) and the accepted significance level was $\alpha = 0.05$. Allometric equations were developed with linear regression. Goodness of fit was based on the coefficient of determination (R) and the level of probability (p). All comparisons of biomass among the different allometric equations were performed using ANOVA, followed by multiple comparisons (LSD tests).

3. Results

Allometric Equations and Tree Biomass

The allometric equations developed in this study were all power functions (Table 1). Only significant parameters were included in the equations; therefore, DBH was the only significant parameter for bark, leaves, and roots components. DBH squared multiplied H was the significant parameter for stem wood. All equations had significant linear relationships between dependent and independent variables, and their correlation coefficients ranged from 0.9634 to 0.9949 (Table 1).

Although allometric equations of different forms are available from the present and previous studies, there was no significant difference in the biomass estimation for the various tree components in this study, with the exception of roots (Table 2). The root biomass calculated by the equation of Bao *et al.* [19] was significantly lower (17.59 t ha^{-1}) than the values generated by the other equations. Therefore, biomass from other equations was distributed similarly in the following manner: stem wood > roots > branches > stem bark > leaves. According to the tree biomass distribution pattern in the present sites, stem wood made the largest contribution to the total tree biomass, accounting for around 50%, while the proportion of leaves to total tree biomass was only around 3%, and the ratio of below- to aboveground was 0.33 (Table 2).

Table 2. Comparison of the biomass estimates of the various tree components by different allometric equations for a Chinese cork oak forest in the south slope of the Qinling Mountains, China.

Tree component	Allometric Equations (1) [28]		Allometric Equations (2) [27]		Allometric Equations (3) [29]		Allometric Equations (4) (This Study)	
	Biomass (t ha^{-1})	Proportion (%)	Biomass (t ha^{-1})	Proportion (%)	Biomass (t ha^{-1})	Proportion (%)	Biomass (t ha^{-1})	Proportion (%)
Stem wood	50.56 ± 4.97a	40.96	64.53 ± 5.16a	46.42			66.20 ± 6.46a	49.13
Stem bark	11.48 ± 0.94a	9.30	16.99 ± 1.28a	12.22			14.48 ± 1.12a	10.75
Stem	62.05 ± 5.89a	50.26	81.54 ± 6.44a	58.65	74.46 ± 7.06a	62.22	80.68 ± 7.54a	59.87
Branches	21.92 ± 2.21a	17.76	18.07 ± 1.98a	13.00	23.61 ± 2.19a	19.73	16.30 ± 1.64a	12.10
Leaves	4.45 ± 0.34a	3.60	3.74 ± 0.31a	2.70	4.01 ± 0.38a	3.35	4.01 ± 0.31a	2.98
Aboveground	88.41 ± 8.42a	71.62	103.35 ± 8.61a	74.34	102.08 ± 9.63a	85.30	100.99 ± 9.47a	74.95
Root	35.05 ± 3.07a	28.39	35.67 ± 2.59a	25.66	17.59 ± 1.47b	14.70	33.76 ± 2.68a	25.05
Total tree	123.45 ± 11.48a		139.02 ± 11.14a		119.67 ± 11.09a		134.75 ± 12.11a	

Data is reported as mean ± standard error ($n = 3$); within a line, values followed by the same lowercase letter indicate that the estimated biomass of the same tree component did not differ significantly among different allometric equations ($p < 0.05$).

Higher C, N, and P concentrations were observed in the shrub leaves than in shrub roots (Table 3). Similarly, the C and N concentrations in aboveground herbs were significantly higher than in belowground herbs (Table 3). The C:N and C:P mass ratios in shrub leaves (19.63 and 369.08, respectively) were significant lower than that in branches (60.91 and 742.61) and roots (63.64 and 717.74) of shrubs, whereas the N:P mass ratio in shrub leaves (18.92) was significantly higher than in branches (12.51) and roots (13.52) of shrubs. However, there was no significant difference in the C:N, C:P and N:P mass ratios between aboveground and belowground herbs. In contrast, no significant differences were found in the N concentrations in the different soil layer. The highest C and P concentrations were in the surface 0–10 cm soil layer, and the values significant decreased with soil depth (Table 3). The C:N, C:P and N:P ratios did not change significantly with soil depths.

C, N and P storage in the Chinese cork oak ecosystem were 163.76, 18.54 and 2.50 t ha^{-1}, respectively. The C storage distribution among different tree components was in the following order: stem wood > roots > branches > stem bark > leaves. Whereas, the order was stem wood > roots > leaves > stem bark > branches for the N storage distribution, and stem wood > roots > branches > leaves > stem bark for the P storage distribution (Table 3). C, N and P storage in tree accounted 36.14%, 6.03%, and 0.23% of the total ecosystem C, N, and P pools, respectively. The C, N, and P storage in the shrub, herb and litter pools accounted 2.10%, 1.19% and 0.03% of the total ecosystem C, N, and P pools, respectively. Soil (0–100 cm) contained the largest amount of C, N, and P storage, accounting for 61.76%, 92.78% and 99.72% of total ecosystem C, N, and P pools, respectively. C and P storage in the 0–50 cm soil depthswere significantly greater than that in deeper soil, accounting for 61.35% and 60.02% of the entire soil profile (0–100 cm); whereas, N storage between 0 and 50 cm was almost equal to that in deeper soil, accounting for 47.40% of the entire soil profile (0–100 cm).

Table 3. C, N and P concentrations and storage distribution of a Chinese cork oak forest in the south slope of the Qinling Mountains, China.

Layer	Component	C Concentration (g kg^{-1})	C Storage (t ha^{-1})	N Concentration (g kg^{-1})	N Storage (kg ha^{-1})	P Concentration (g kg^{-1})	P Storage (kg ha^{-1})
Tree	Stem wood	433.28 ± 9.43c	28.68 ± 2.80a	2.63 ± 0.24c	174.11 ± 16.98a	0.35 ± 0.06b	23.17 ± 2.26a
	Stem bark	454.60 ± 2.20b	6.58 ± 0.51c	5.05 ± 0.10b	73.11 ± 5.66b	0.29 ± 0.01b	4.20 ± 0.32c
	Branches	433.81 ± 1.36c	7.07 ± 0.72c	4.50 ± 0.02bc	73.36 ± 7.42b	0.62 ± 0.08b	10.11 ± 1.02b
	Leaves	483.07 ± 2.59a	1.94 ± 0.15c	19.31 ± 1.19a	77.48 ± 5.99b	1.26 ± 0.16a	5.06 ± 0.39c
	Root	441.96 ± 4.03bc	14.92 ± 1.18b	4.54 ± 0.06bc	153.25 ± 12.17a	0.39 ± 0.02b	13.17 ± 1.04b
	Subtotal		59.19 ± 5.31		551.31 ± 47.76		55.70 ± 4.99
Shrub	Leaves	459.12 ± 4.96a	0.09 ± 0.02b	24.08 ± 0.81a	4.60 ± 0.90a	1.30 ± 0.06a	0.25 ± 0.05a
	Branch	449.30 ± 5.89a	0.37 ± 0.07ab	8.20 ± 0.60b	6.80 ± 1.34a	0.69 ± 0.06b	0.57 ± 0.11ab
	Root	428.63 ± 5.70b	0.59 ± 0.17a	7.85 ± 0.93b	10.84 ± 3.06a	0.85 ± 0.11b	1.18 ± 0.33a
	Subtotal		1.06 ± 0.22		22.24 ± 4.36		2.00 ± 0.42
Herb	Aboveground	428.73 ± 6.44a	0.08 ± 0.01a	14.83 ± 0.85a	2.81 ± 0.45a	1.34 ± 0.08a	0.25 ± 0.04a
	Belowground	262.03 ± 13.08b	0.11 ± 0.03a	7.61 ± 0.40b	3.14 ± 0.91a	1.02 ± 0.15a	0.42 ± 0.12a
	Subtotal		0.19 ± 0.04		5.95 ± 1.14		0.67 ± 0.14
Litter		364.17 ± 10.69	2.18 ± 0.44	12.49 ± 0.21	74.88 ± 15.20	0.97 ± 0.07	5.82 ± 1.18
Soil	0–10 cm	16.63 ± 1.19a	18.05 ± 1.86b	1.69 ± 0.21a	1.83 ± 0.25 t ha^{-1}b	0.43 ± 0.11a	0.46 ± 0.11 t ha^{-1}b
	10–20cm	10.12 ± 1.41bc	11.57 ± 2.05b	1.47 ± 0.27a	1.68 ± 0.34 t ha^{-1}b	0.29 ± 0.06ab	0.33 ± 0.07 t ha^{-1}b
	20–30cm	10.31 ± 0.64b	11.54 ± 0.81b	1.42 ± 0.29a	1.64 ± 0.38 t ha^{-1}b	0.27 ± 0.02ab	0.30 ± 0.02 t ha^{-1}b
	30–50 cm	8.63 ± 1.34bc	20.89 ± 3.77b	1.35 ± 0.33a	3.33 ± 0.87 t ha^{-1}b	0.16 ± 0.03b	0.38 ± 0.06 t ha^{-1}b
	50–100cm	5.76 ± 0.53c	39.10 ± 6.44a	1.26 ± 0.37a	9.41 ± 3.19 t ha^{-1}a	0.16 ± 0.03b	0.96 ± 0.13 t ha^{-1}a
	Subtotal		101.14 ± 11.03		17.89 ± 4.92 t ha^{-1}		2.43 ± 0.20 t ha^{-1}
Total ecosystem		163.76 ± 11.25		18.54 ± 4.96 t ha^{-1}		2.50 ± 0.21 t ha^{-1}	

Data is reported as mean ±standard error ($n = 3$); within a column, values followed by the same lowercase letter indicate that they did not differ significantly within the same layer ($p < 0.05$).

4. Discussion

Allometric equations are crucial in order to accurately estimate forest biomass for C accounting. A number of previous studies demonstrated that power function allometric equations based on DBH or squared DBH multiplied by H can be used to estimate tree biomass [29,35]. Allometric equations based on DBH are recommended because the measurement of H is time-consuming and less accurate than DBH [19,34,40,41]. In this study, we observed that DBH was the only significant parameter for bark, leaf, and root components; whereas, the squared DBH multiplied by H as the significant parameter for stem wood. Different models (Table 1) were used to estimate biomass distribution for Chinese cork oak in China due to various biotic and abiotic environmental factors. However, there was no significant difference in tree components biomass estimation, with the exception of roots, among all the available allometric equations. In some cases the power function failed, and then transformed models were needed to develop significant allometric equations for different tree species, locations, and specific-components [7,25,27,28,42–47].

The biomass distribution among tree components was as follows: stem wood > roots > branches > stem bark > leaves. Our findings were similar to the previous reports for Chinese cork oak in the temperate region of China [27–29]. However, the different biomass distribution also observed for Chinese cork oak. For example, the biomass distribution of a 20-year-old Chinese cork oak forest was distributed as follows: stem wood > branches > roots > stem bark > leaves at; while that in 30- and 40-year-old in hilly region of Taihang Mountain, the distribution was stem wood > stem bark > branches > roots > leaves at [48]. The total Chinese cork oak tree biomass was 158.84 t ha^{-1} in the Baotianman Natural Reserve, Henan Province [27], 134.75 t ha^{-1} in the south slope of the Qinling Mountains of the present study, 79.80 t ha^{-1} in the Xiaolong Mountains, Gansu Province [28] and 53.64 t ha^{-1} in the Xishan Mountains, Beijing [29]. These obvious differences were mainly caused by age, tree density, and climate factors. Zhao *et al.* [48] reported that in a hilly region of Taihang Mountain, tree biomass significantly increased from 131.65 t ha^{-1} in a 20 year old stand to 202.96 and 291.15 t ha^{-1} in 30 and 40 years old stands, respectively.

The nutrient concentrations among plant components were significantly different. The concentrations of C, N and P in tree leaves were the highest, but the lowest values of C and N were observed in tree stems. In addition to C, N and P, the highest mobile nutrient concentrations, such as Mg, K, and Mn, were also detected in the leaves of oak forests in Spain [19]. Moreover, the N, P, and K concentrations were the highest in spring and then decreased throughout the vegetative period [19]. Similarly, the highest concentrations were in shrub leaves, but the lowest values were in shrub roots. The C, N and P concentrations in aboveground portion of herb were higher than in the belowground portion. The leaves N:P ratio, which is relatively easy to determine, has been widely used to indicate limitations in soil N (N:P < 14) and soil P (N:P > 16) [49,50]. The leaf N (19.31 g kg^{-1}), P (1.26 g kg^{-1}) concentrations, and N:P ratio (16.10) of Chinese cork oak forests in the present study were similar with the average values of leaves N (18.33 g kg^{-1}) and P (1.18 g kg^{-1}) concentrations and the N:P ratio (16.56) in China [51]. Moreover, the N:P ratio of shrub leaves was also higher than 16. Therefore, soil P is a limiting nutrient in this study area for the growth demands of Chinese cork oak forests. In contrast, the leaves N:P ratios of sharptooth oak, Chinese pine, and Armand pine indicated that there is soil N limitation in the Qinling Mountains [52]. However, the nutrient concentrations in live leaves decreased and the N:P ratio increased during the growing seasons were observed in a Mediterranean cork oak forest in southwestern Spain [26]. Moreover, the leaf N and P concentrations of Chinese cork oak forests also decreased during stand development. It was shown that there is an disconnect bewteen soil P supply and plant growth demand indicated by an increased N:P ratio from 13.84 in 20-year-old stand to 16.05 and 19.75 in 30- and 40-year-old stands, respectively, in a hilly region of the Taihang Mountains [48]. Therefore, quantifying soil nutrient limitation by only using the leaf N:P ratio presents challenges because of the limited amount of data and only a partial understanding of the processes involved [50,53]. Although significant differences in leaves N and P concentrations were detected across all *Quercus* species in China, leaf N:P ratio (13.96) was well constrained to a relatively stable range for *Quercus* species and was less influenced by environmental variables across China [51].

Abundant precipitation, full sunshine and nutrient-rich soil were considered to be suitable for supporting plant growth in the southern slope of the Qinling Mountains [52]. C, N, and P were enriched at 0–10 cm soil depth, while C and P decreased with an increase in soil depth. However, N showed a stable trend. C, N, and P concentrations of 0–10 cm soil depth in this study area (16.63, 1.69 and 0.43 g kg^{-1}, respectively) were higher than the mean values (12.28, 0.94 and 0.38 g kg^{-1}, respectively) of soils in China. As a consequence, the C:N, C:P and N:P (10.55, 51.51 and 4.59 in molar, respectively) ratios of 0–10 cm soil depth in this study area were smaller than those mean values (14.4, 136 and 9.3, respectively) of soils in China [54]. C, N, and P concentrations increased during stand development of Chinese cork oak forests in the hilly region of the Taihang Mountains, possibly due to a larger accumulation of organic matter in older stands [48]. Under Mediterranean climate conditions in Mainland Spain, the soil C storage of cork oak forests was favored by large organic matter inputs, high soil clay contents, a calcium-saturated soil matrix and reduced summer aridity [23].

The proportion of stem to total tree biomass was around 60% in this study. The highest percentage was 73% for four Mediterranean oak forests in Spain [19]. The proportion of stem to total tree biomass has been used to infer the light conditions, soil nutrient and age stage at study sites [35,36]. For example, the ratios in Chinese pine were 46.9%, 72.2%, 70.6% and 70.7% for young, middle-aged, immature, and mature stands, respectively [34]. Although the low percentage of nutrients accumulated in the leaves because leaf biomass represents only around 3% of the total, the amount of nutrients accumulated in leaves was of great importance because the nutrients were subject to internal annual cycles within the tree, and some of them return to the soil in the form of leaf litter [19]. At the same time, root biomass accounted for a large proportion of the total tree biomass. The ratio of below- to aboveground biomass was 0.33 for Chinese cork oak forests in this study.The ratio value was 0.28, 0.29 and 0.23 for the Baotianman Natural Reserve of Henan Province [27], the Xiaolong Mountainsof Gansu Province [28], and the Xishan Mountains of Beijing [29], respectively. In contrast to the result [55] that there is a relatively constant ratio of below- to aboveground biomass during stand development globally, Zhao *et al.* [48] found that the ratio decreased during stand development, from 0.23 in a 20-year-old stand to 0.16 and 0.13 in 30- and 40-year-old stands, respectively, in a hilly region of the Taihang Mountains. The decreased below- to aboveground biomass ratio may demonstrate different strategies in nutrient cycling and water uptake potential [1]. Our study agrees with previous studies that soil is the largest C and nutrient element storage component, followed by tree, understory, and litter [21]. Soil C, N, and P storage were 1.62-, 27.52- and 34.71-fold higher than vegetation C, N, and P storage. In addition, soil C storage was higher in the topsoil than in deeper soil owing to soil organic matter is the main source of C stored in topsoil. Moreover, soil organic matter is essential to ecosystem productivity and regeneration. Recently, many studies have focused on biomass, C, and nutrient storage during stand development [1,7,34–36,48,55]. These studies provided a comprehensive understanding of the importance of considering the succession development of forest ecosystem C pools and the interaction with other nutrient elements, especially when estimating C sink potential over a life cycle. Because human disturbances can have huge impacts on certain age stage forests, we only selected 30-year-old Chinese cork oak stands with little or no human disturbance to describe the basic characteristics of biomass, C and nutrient storage distribution. Therefore, future studies should focus on improving our understand of the effects of age and disturbance in biomass, C concentrations and stocks for devising optimum forest management strategies aimed at mitigating climate change.

5. Conclusions

In this study, we presented the basic C, N and P storage distribution patterns among trees, shrubs, herbs, litter, and soil (0–100 cm) in a 30-year-old Chinese cork oak forest in the south slope of the Qinling Mountains, China. The biomass of tree components can be better predicted from allometric equations using DBH as the independent variable. There was no significant difference in various tree components biomass estimations, with the exception of roots, between different Chinese cork oak allometric equations developed from different sites. Stems and roots were the main proportion of total tree biomass. The proportion of stem to total tree biomass was around 60% and the ratio of below- to aboveground biomass was 0.33. There were significant differences in C, N and P concentrations between plant components. Plant leaves had the highest C, N, and P concentrations than other plant components. The N:P ratio was similar to the national level and indicated soil P limitation. However, C, N and P concentrations of the 0–10 cm soil depth were higher than those mean values of soils in China. C, N, and P storage of this Chinese cork oak ecosystem were 163.76, 18.54 and 2.50 t ha^{-1}, respectively. Soil was the largest C and nutrient element storage component, followed by trees. C, N, and P stocks in shrubs, herbs, and litter contributed only little to ecosystem C, N and P stocks. This study demonstrated that large-scale Chinese cork oak forests in the Qinling Mountains play an important role in C sequestration under NFP program. Furthermore, we suggest that further research on human-caused, natural disturbance and other influence factors on the continuous accumulation of C in both the plants and soils are especially critical for developing recommendations on appropriate forest management practices under the NFP program.

Acknowledgments

This research was supported by the National Nature Science Foundation of China (No. 41201088 and 41371506), the Strategic Priority Research Program and West Light Foundation of The Chinese

Academy of Sciences (XDA05050203 and K318021304), the Doctoral Fund of Ministry of Education of China (20120204120014), Specialized Research Fund for the Doctoral Program of Higher Education (2014YB056) and Institute of Soil and Water Conservation Funding (A315021380).

Author Contributions

The paper was written by Yang Cao with a contribution by Yunming Chen. Yunming Chen conceptualized the research design and site selection, and submitted the article. Yang Cao conducted field data collection and data processing.

Conflicts of Interest

The authors declare no conflict of interest.

References

1. Laclau, P. Biomass and carbon sequestration of ponderosa pine plantations and native cypress forests in northwest Patagonia. *For. Ecol. Manag.* **2003**, *180*, 317–333.
2. Chen, X.G.; Zhang, X.Q.; Zhang, Y.P.; Wan, C.B. Carbon sequestration potential of the stands under the grain for green program in Yunnan province, China. *For. Ecol. Manag.* **2009**, *258*, 199–206.
3. Bonner, M.T.L.; Schmidt, S.; Shoo, L.P. A meta-analytical global comparison of aboveground biomass accumulation between tropical secondary forests and monoculture plantations. *For. Ecol. Manag.* **2013**, *291*, 73–86.
4. He, Y.J.; Qin, L.; Li, Z.Y.; Liang, X.Y.; Shao, M.X.; Tan, L. Carbon storage capacity of monoculture and mixed-species plantations in subtropical China. *For. Ecol. Manag.* **2013**, *295*, 193–198.
5. Zeng, X.H.; Zhang, W.J.; Liu, X.P.; Cao, J.S.; Shen, H.T.; Zhao, X.; Zhang, N.N.; Bai, Y.R.; Yi, M. Change of soil organic carbon after cropland afforestation in "Beijing-Tianjin sandstorm source control" program area in China. *Chin. Geogra. Sci.* **2014**, *24*, 461–470.
6. Zhou, W.M.; Lewis, B.J.; Wu, S.N.; Yu, D.P.; Zhou, L.; Wei, Y.W.; Dai, L.M. Biomass carbon storage and its sequestration potential of afforestation under natural forest protection program in China. *Chin. Geogra. Sci.* **2014**, *24*, 406–413.
7. Li, T.; Liu, G. Age-related changes in carbon accumulation and allocation in plants and soil of a black locust forest on the loess plateau. *Chin. Geogra. Sci.* **2014**, *24*, 414–422.
8. Deng, L.; Liu, G.B.; Shangguan, Z.P. Land-use conversion and changing soil carbon stocks in China's "grain-for-green" program: A synthesis. *Glob. Change Biol.* **2014**, *20*, 3544–3556.
9. Chang, R.; Fu, B.; Liu, G.; Liu, S. Soil carbon sequestration potential for "grain for green" project in loess plateau, China. *Environ. Manag.* **2011**, *48*, 1158–1172.
10. Wei, Y.W.; Yu, D.P.; Lewis, B.J.; Zhou, L.; Zhou, W.M.; Fang, X.M.; Zhao, W.; Wu, S.N.; Dai, L.M. Forest carbon storage and tree carbon pool dynamics under natural forest protection program in northeastern China. *Chin. Geogra. Sci.* **2014**, *24*, 397–405.

11. Hu, H.; Liu, G. Carbon sequestration of China's national natural forest protection project. *Acta Ecol. Sin.* **2006**, *26*, 291–296.

12. Fang, J.; Brown, S.; Tang, Y.; Nabuurs, G.-J.; Wang, X.; Shen, H. Overestimated biomass carbon pools of the northern mid- and high latitude forests. *Clim. Chang.* **2006**, *74*, 355–368.

13. Liu, G.; Fu, B.; Fang, J. Carbon dynamics of Chinese forests and its contribution to global carbon balance. *Acta Ecol. Sin.* **2000**, *20*, 733–740.

14. Zhao, M.; Zhou, G. Forest inventory data (FID)-based biomass models and their prospects. *Chin. J. Appl. Ecol.* **2004**, *15*, 1468–1472.

15. Wei, Y.W.; Li, M.H.; Chen, H.; Lewis, B.J.; Yu, D.P.; Zhou, L.; Zhou, W.M.; Fang, X.M.; Zhao, W.; Dai, L.M. Variation in carbon storage and its distribution by stand age and forest type in boreal and temperate forests in northeastern China. *PLoS ONE* **2013**, *8*, doi:10.1371/journal.pone.0072201.

16. Du, L.; Zhou, T.; Zou, Z.H.; Zhao, X.; Huang, K.C.; Wu, H. Mapping forest biomass using remote sensing and national forest inventory in China. *Forests* **2014**, *5*, 1267–1283.

17. Guyette, R.P.; Dey, D.C.; Stambaugh, M.C. The temporal distribution and carbon storage of large oak wood in streams and floodplain deposits. *Ecosystems* **2008**, *11*, 643–653.

18. Borges, J.G.; Oliveira, A.C.; Costa, M.A. A quantitative approach to cork oak forest management. *For. Ecol. Manag.* **1997**, *97*, 223–229.

19. Santa-Regina, I. Biomass estimation and nutrient pools in four *Quercus pyrenaica* in sierra de gata mountains, Salamanca, Spain. *For. Ecol. Manag.* **2000**, *132*, 127–141.

20. Gardiner, E.S.; Hodges, J.D. Growth and biomass distribution of cherrybark oak (*Quercus pagoda raf.*) seedlings as influenced by light availability. *For. Ecol. Manag.* **1998**, *108*, 127–134.

21. Vallet, P.; Meredieu, C.; Seynave, I.; Belouard, T.; Dhote, J.F. Species substitution for carbon storage: Sessile oak *versus* corsican pine in France as a case study. *For. Ecol. Manag.* **2009**, *257*, 1314–1323.

22. Lopez, B.C.; Sabate, S.; Gracia, C.A. Thinning effects on carbon allocation to fine roots in a *Quercus ilex* forest. *Tree Physiol.* **2003**, *23*, 1217–1224.

23. Gonzalez, I.G.; Corbi, J.M.G.; Cancio, A.F.; Ballesta, R.J.; Cascon, M.R.G. Soil carbon stocks and soil solution chemistry in quercus ilex stands in mainland Spain. *Eur. J. For. Res.* **2012**, *131*, 1653–1667.

24. Cerasoll, S.; Scartazza, A.; Brugnoli, E.; Chaves, M.M.; Pereira, J.S. Effects of partial defoliation on carbon and nitrogen partitioning and photosynthetic carbon uptake by two-year-old cork oak (*Quercus suber*) saplings. *Tree Physiol.* **2004**, *24*, 83–90.

25. Fonseca, T.J.F.; Parresol, B.R. A new model for cork weight estimation in northern Portugal with methodology for construction of confidence intervals. *For. Ecol. Manag.* **2001**, *152*, 131–139.

26. Andivia, E.; Fernandez, M.; Vazquez-Pique, J.; Gonzalez-Perez, A.; Tapias, R. Nutrients return from leaves and litterfall in a mediterranean cork oak (*Quercus suber* L.) forest in southwestern Spain. *Eur. J. For. Res.* **2010**, *129*, 5–12.

27. Liu, Y.; Wu, M.; Guo, Z.M.; Jiang, Y.; Liu, S. Biomass and net productivity of *Quercus variabilis* forest in Baotianman natural reserve. *Chin. J. Appl. Ecol.* **1998**, *9*, 569–574.

28. Cheng, T.; Ma, Q.; Feng, Z.; Luo, X. Research on forest biomass in Xiaolong Mountains, Gansu province. *J. Beijing For. Univ.* **2007**, *29*, 31–36.

29. Bao, X.; Chen, L.; Chen, Q.; Ren, J.; Hu, Y.; Li, Y. The biomass of planted oriental oak (*Quercus variabilis*) forest. *Acta Phytoecol. et Geobot. Sin.* **1984**, *8*, 313–320.

30. Zhang, W.; Lu, Z. A study on the biological and ecological property and geographical distribution of *Quercus variabilis* population. *Acta Bot. Boreali-Occidential Sin.* **2001**, *22*, 1093–1101.

31. Ran, R.; Zhang, W.; Zhou, J.; He, J. Effects of thinning intensities on population regeneration of natural *Quercus variabilis* forest on the south slope of Qinling Mountains. *Chin. J. Appl. Ecol.* **2014**, *25*, 695–701.

32. Ran, R.; Zhang, W.; Zhou, J.; He, J. Effects of thinning intensity on the seed bank and seedling growth of *Quercus variabilis* on the south slope of Qinling mountains, northwest China. *Chin. J. Appl. Ecol.* **2013**, *24*, 1494–1500.

33. Yi, Q.Z.; Zhang, W.; Tang, D. Effects of reserved sprout number per stump on sprout development and biomass accumulation of *Quercus variabilis*. *Sci. Silvae Sin.* **2013**, *7*, 34–39.

34. Zhao, J.L.; Kang, F.F.; Wang, L.X.; Yu, X.W.; Zhao, W.H.; Song, X.S.; Zhang, Y.L.; Chen, F.; Sun, Y.; He, T.F.; *et al.* Patterns of biomass and carbon distribution across a chronosequence of Chinese pine (*Pinus tabulaeformis*) forests. *PLoS ONE* **2014**, *9*, doi:10.1371/journal.pone.0094966.

35. Li, H.; Li, C.; Zha, T.; Liu, J.; Jia, X.; Wang, X.; Chen, W.; He, G. Patterns of biomass allocation in an age-sequence of secondary *Pinus bungeana* forests in China. *For. Chron.* **2014**, *90*, 169–176.

36. Noh, N.J.; Son, Y.; Kim, R.H.; Seo, K.W.; Koo, J.W.; Park, I.H.; Lee, Y.J.; Lee, K.H.; Son, Y.M. Biomass accumulations and the distribution of nitrogen and phosphorus within three *Quercus acutissima* stands in central Korea. *J. Plant Biol.* **2007**, *50*, 461–466.

37. Parkinson, J.; Allen, S. A wet oxidation procedure suitable for the determination of nitrogen and mineral nutrients in biological material. *Commun. Soil Sci. Plant Anal.* **1975**, *6*, 1–11.

38. Nelson, D.W.; Sommers, L.E.; Sparks, D.; Page, A.; Helmke, P.; Loeppert, R.; Soltanpour, P.; Tabatabai, M.; Johnston, C.; Sumner, M. Total carbon, organic carbon, and organic matter. *Methods of Soil Analysis. Part 3—Chemical Methods,* SSSA Book Series 5.3; Soil Science Society of America, American Society of Agronomy: Madison, WI, UAS, 1996; pp. 961–1010.

39. Bremner, J.M. Total Nitrogen. In *Methods of Soil Analysis. Part 2. Chemical and Microbiological Properties, Agronomy Monograph 9.2*; SSSA: Madison, WI, USA, 1965; pp. 1149–1178.

40. Peichl, M.; Arain, M.A. Allometry and partitioning of above- and belowground tree biomass in an age-sequence of white pine forests. *For. Ecol. Manag.* **2007**, *253*, 68–80.

41. Gower, S.T.; Kucharik, C.J.; Norman, J.M. Direct and indirect estimation of leaf area index, fAPAR, and net primary production of terrestrial ecosystems. *Remote Sens. Environ.* **1999**, *70*, 29–51.

42. Basuki, T.M.; van Laake, P.E.; Skidmore, A.K.; Hussin, Y.A. Allometric equations for estimating the above-ground biomass in tropical lowland dipterocarp forests. *For. Ecol. Manag.* **2009**, *257*, 1684–1694.

43. Makungwa, S.D.; Chittock, A.; Skole, D.L.; Kanyama-Phiri, G.Y.; Woodhouse, I.H. Allometry for biomass estimation in *Jatropha* trees planted as boundary hedge in farmers' fields. *Forests* **2013**, *4*, 218–233.

44. Sitoe, A.A.; Mandlate, L.J.C.; Guedes, B.S. Biomass and carbon stocks of Sofala bay mangrove forests. *Forests* **2014**, *5*, 1967–1981.

45. Ritchie, M.W.; Zhang, J.W.; Hamilton, T.A. Aboveground tree biomass for *Pinus ponderosa* in northeastern California. *Forests* **2013**, *4*, 179–196.

46. Beets, P.N.; Kimberley, M.O.; Oliver, G.R.; Pearce, S.H.; Graham, J.D.; Brandon, A. Allometric equations for estimating carbon stocks in natural forest in New Zealand. *Forests* **2012**, *3*, 818–839.

47. Mate, R.; Johansson, T.; Sitoe, A. Biomass equations for tropical forest tree species in Mozambique. *Forests* **2014**, *5*, 535–556.

48. Zhao, Y.; Wang, P.; Fan, W.; Zhu, Y. Nutrient cycling in *Quercus varlabilis* plantations of different ages classes in hilly region of Taihang mountain. *Sci. Soil Water Conserv.* **2009**, *7*, 66–71.

49. Gusewell, S. N:P ratios in terrestrial plants: Variation and functional significance. *New Phytol.* **2004**, *164*, 243–266.

50. Townsend, A.R.; Cleveland, C.C.; Asner, G.P.; Bustamante, M.M.C. Controls over foliar N:P ratios in tropical rain forests. *Ecol.* **2007**, *88*, 107–118.

51. Wu, T.G.; Dong, Y.; Yu, M.K.; Wang, G.G.; Zeng, D.H. Leaf nitrogen and phosphorus stoichiometry of *Quercus* species across China. *For. Ecol. Manag.* **2012**, *284*, 116–123.

52. Liu, G.; Zhao, S.; Tu, X. Distributional characteristics on biomass and nutrient elements of pine-oak forest belt in Mt. Qinling. *Sci. Silvae Sin.* **2001**, *37*, 28–36.

53. Goll, D.S.; Brovkin, V.; Parida, B.R.; Reick, C.H.; Kattge, J.; Reich, P.B.; van Bodegom, P.M.; Niinemets, U. Nutrient limitation reduces land carbon uptake in simulations with a model of combined carbon, nitrogen and phosphorus cycling. *Biogeosciences* **2012**, *9*, 3547–3569.

54. Tian, H.Q.; Chen, G.S.; Zhang, C.; Melillo, J.M.; Hall, C.A.S. Pattern and variation of C:N:P ratios in China's soils: A synthesis of observational data. *Biogeochemistry* **2010**, *98*, 139–151.

55. Yang, Y.; Luo, Y. Carbon:Nitrogen stoichiometry in forest ecosystems during stand development. *Glob. Ecol. Biogeogr.* **2011**, *20*, 354–361.

Residual Long-Term Effects of Forest Fertilization on Tree Growth and Nitrogen Turnover in Boreal Forest

Fredrik From, Joachim Strengbom and Annika Nordin

Abstract: The growth enhancing effects of forest fertilizer is considered to level off within 10 years of the application, and be restricted to one forest stand rotation. However, fertilizer induced changes in plant community composition has been shown to occur in the following stand rotation. To clarify whether effects of forest fertilization have residual long-term effects, extending into the next rotation, we compared tree growth, needle N concentrations and the availability of mobile soil N in young (10 years) *Pinus sylvestris* L. and *Picea abies* (L.) H. Karst. stands. The sites were fertilized with 150 kg·N·ha^{-1} once or twice during the previous stand rotation, or unfertilized. Two fertilization events increased tree height by 24% compared to the controls. Needle N concentrations of the trees on previously fertilized sites were 15% higher than those of the controls. Soil N mineralization rates and the amounts of mobile soil NH_4-N and NO_3-N were higher on sites that were fertilized twice than on control sites. Our study demonstrates that operational forest fertilization can cause residual long-term effects on stand N dynamics, with subsequent effects on tree growth that may be more long-lasting than previously believed, *i.e.*, extending beyond one stand rotation.

Reprinted from *Forests*. Cite as: From, F.; Strengbom, J.; Nordin, A. Residual Long-Term Effects of Forest Fertilization on Tree Growth and Nitrogen Turnover in Boreal Forest. *Forests* **2015**, *6*, 1145-1156.

1. Introduction

Nitrogen (N) availability limits forest growth in most boreal and temperate forests [1,2], and fertilization with N is one of the most cost-effective silvicultural methods to increase boreal forest yield [3]. In fact, timber harvest profitability can increase by nearly 15% if forest stands are fertilized about 10 years before final felling [4]. Fertilization has been practiced commercially since the middle of the previous century in many parts of the boreal and temperate forest regions, for example, in Sweden [5,6], USA [7], Canada [8] and Finland [9]. During the 20th century approximately 10% of the productive forest land in Sweden, *i.e.*, circa (ca) 2 million hectares (ha), was fertilized at least once [3,6,10].

According to previous studies, forest fertilization with N additions corresponding to 150 kg·N·ha^{-1} application^{-1}, for up to three times during a forest rotation period [11], will only affect the subjected forest stand for up to 10 years after the last fertilizer application [3,9,12–16]. This conclusion is mainly based on studies of growth effects within the tree rotation period fertilized. Also a recent study, particularly targeting carry-over effects of fertilization between forest stand rotations, showed that growth of five year old pine seedlings was not affected by fertilization of the site three to nine years before harvest of the previous stand rotation [17], *i.e.*, supporting the view that there are no, or only very restricted long-term effects of fertilization. In contrast, a study from North America showed that the height growth of young Douglas fir (*Pseudotsuga menziesii* [Mirb.] Franco) trees were greater on plots that had been fertilized in the previous stand rotation period than on plots never fertilized [18].

Fertilization effects on species composition of the forest floor vegetation within one forest rotation have been extensively documented, e.g., summarized in Nohrstedt [3] and Saarsalmi and Malkonen [9], but only a few studies have addressed carry-over effects between forest stand rotations. In a study summarizing carry-over fertilization effects on forest floor vegetation, Hedwall, *et al.* [19] reported a higher abundance of graminoids and a lower abundance of dwarf-shrubs on experimental plots fertilized in the previous stand rotation period than on control plots never fertilized. Strengbom and Nordin [20] found carry-over effects between stand rotations on the composition and abundance of forest floor vegetation of operational forest fertilization, *i.e.*, with 150 kg·N·ha^{-1} at one or two occasions during the previous stand rotation.

When fertilizing forests, a large part of the N added is retained in the soil layer [21,22]. Strengbom and Nordin [23] suggested that the disturbance from harvest and site scarification leads to increased mineralization rates of immobilized soil N, explaining why effects of previous (more than 25 years ago) N fertilization was more evident in the early phase of the subsequent stand rotation than within the same stand rotation that were fertilized.

In this study, we revisited young forest stands with documented effects on species composition of the forest floor vegetation from fertilization of the previous stand rotation. We hypothesized that fertilization of the previous forest stand would cause (1) increased tree growth in the following rotation period; and (2) increased soil N mineralization rates, both in comparison to control stands that never had been fertilized. Thus, the aim of the study was to determine fertilization carry-over effects between tree rotations on tree height and diameter growth and on soil N turnover, *i.e.*, mineralization rates and amounts of mobile soil N.

2. Materials and Methods

2.1. Study Area

We studied forest sites each sized between 4.7 and 22.4 ha situated within an 8500 ha forest land area owned by the company Sveaskog. The forest land area is in the middle boreal zone [24] in central Sweden (63°00′ N, 16°40′ E). The studied forest sites, either had, or had not, been subjected to N fertilization before harvest and regeneration (Table 1). Control (C) sites ($n = 7$) were never fertilized; N1 sites ($n = 7$) were fertilized with 150 kg·N·ha^{-1} once (in 1985), while N2 sites ($n = 7$) were repeatedly fertilized with 150 kg·N·ha^{-1} (in 1977 and 1985), *i.e.*, our study included seven stands of each treatment. Nitrogen was added as granules of ammonium nitrate (NH$_4$NO$_3$) spread by tractor or aircraft.

Forest inventory records describing the forest sites before clear-cut kept by the land owner were used when selecting the sites. In particular care was taken to select sites with similar productivity (site) indices (Table 1). The other criteria for including sites in our study were temperature sum (growing degree days, GDD), *i.e.*, the annual temperature sum of all daily mean temperatures above 5 °C [25]; slope, *i.e.*, the forest floor incline, on a scale from 1 (<10% inclination) to 5 (>50%); current stand age; tree species; and soil conditions. The sites were spread out in the 8500 ha forest land area and spatially separated from each other (>1 km). The annual precipitation in the area is between 600 to 700 mm per year and the atmospheric background N deposition ranges from 2.5 to 3.5 kg ha^{-1} year^{-1} [26].

Table 1. Number of sites per treatment, elevation (above sea level, a.s.l.), temperature sum (GDD), forest floor slope, the site index (SI), the average numbers of trees ha^{-1} and the average sample tree ages (2010). The sites were fertilized once (N1) with 150 kg·N·ha^{-1} 25 years before the present study, twice (N2) 33 and 25 years before the present study or never fertilized (C).

Sites (n)	Control	N1	N2	F-value	p-value
	7	7	7		
Elevation a.s.l. (m)	334 ± 21	358 ± 20	334 ± 17	0.54	0.59
Temperature sum (GDD)	937 ± 6	900 ± 22	923 ± 11	1.59	0.23
Slope (1 to 5)	2.00 ± 0.22	1.86 ± 0.26	1.72 ± 0.18	0.41	0.67
Site index (H$_{100}$, m)	20.6 ± 0.2	19.7 ± 0.3	20.4 ± 0.3	3.00	0.08
Trees (ha^{-1})	2721 ± 378	2982 ± 416	2532 ± 206	0.43	0.65
Tree age (years)	10.4 ± 0.2	10.4 ± 0.2	10.3 ± 0.2	0.80	0.47

All stands were on mesic sites with moraine soil, geotechnical nomenclature coarse-grained till [27], with Udic moisture regime [28]. Before clear-felling all sites had a mixed tree species composition with *P. sylvestris* and *P. abies* as dominant trees and *Betula* spp. as subordinate trees. The relative abundance of the two dominant species prior to clear-felling ranged from 25% to 75% with no difference between fertilized and unfertilized stands (Mann–Whitney U-test: U = 83.5, p = 0.348) [20]. The forest field layer was dominated by dwarf shrubs such as *Vaccinium myrtillus* L. and *V. vitis-idaea* L., *i.e.*, the stands were classified as spruce forest of bilberry type [29]. All sites were harvested by clear-felling between 1997 and 2000 and were thereafter treated with soil scarification and regenerated by planting *P. sylvestris* and *P. abies* seedlings at a density of 2200 to 2300 stems ha^{-1}. Due natural regeneration, all forest stands at the time of our study had a mix of both coniferous species and a sub population of *Betula pendula* Roth, *Betula pubescence* Ehrh and *Populus tremula* L.

2.2. Data Collection

Data on tree growth and needle N concentration were collected in August 2010 from four 100 m^2 (r = 5.64 m) circular plots per forest site. The plots were distributed along transects that were approximately stand-centered, described in Strengbom and Nordin [20], to minimize the influence of surrounding stands. Tree height (cm), diameter (mm) at breast height (DBH, at 1.3 m above ground) and species were noted for all trees taller than 1.3 m using cross-callipering and a five meter long, telescopic measuring stick (Teleskopmeter, Skogma, Hammerdal, Sweden). Across all plots, 38% of the trees were *P. abies*, 29% were *P. sylvestris* and 33% *Betula* spp. Less than 1% were *P. tremula*. From each site, 16 undamaged sample-trees (the four tallest from each circular plot) were selected. *P. abies* (56%) and *P. sylvestris* (44%) were exclusively used as sample-trees. Sites were classified as dominated by *P. abies* or *P. sylvestris* (*P. abies* = four C, three N1 and four N2;

P. sylvestris = three C, four N1 and three N2). There was no statistically significant difference in site species classification between treatments (Kruskal-Wallis one-way analysis of variance test = 0.364, $p = 0.834$). Sample-tree annual shoot growth was measured from first visible node and upwards toward full tree height. Tree age was determined from the number of nodes plus one per sample tree.

Current year needles were collected for C and N analyses in early august 2010. Needles were cut from the sample tree's top branches facing the centre of the circle. All needle samples were dried in 70 °C for 24 h and grinded to a fine powder using a bead mill (five minutes per sample), the powder was stored in clear glass vials. In late September, small amounts (ca 3 mg per sample tree) of the ground needle samples were sealed in small tin foil cups, one per sample tree, and sent to a lab (Umeå Plant Science Centre, UPSC) for C and N content analyses using an AutoAnalyzer 3 (SEAL Analytical, OmniProcess AB, Solna, Sweden). Before statistical analyses replicate samples from the same site were pooled.

The buried bag technique was used to investigate humus N mineralization. At each site seven samples from the organic mor-layer were collected between 1st and 3rd June in 2009 with a cylindrical (diameter 10 cm) soil corer. After gently removing large roots and living green material, half of the sample from each mor-layer core was placed in plastic bags in a cooler and transported to the lab for analyses. The second half was placed in plastic bags and buried in the mor-layer *in situ*, and were retrieved between the 28th and 30th September the same year, stored in a cooler and transported to the lab for analyses. The mor-layer samples were extracted in 1 m KCL and analyzed for NH_4 and NO_3 using a FIAstar 5012 Analyzer (Tecator, Höganäs, Sweden). Mineralization was calculated through differences in NH_4-N and NO_3-N content ($mg \cdot g^{-1}$ dry weight [DW] soil) between sample times, *i.e.*, early June and late September. All contaminated samples (*i.e.*, damaged bags) were excluded before estimating mineralization. In total 17 samples were excluded from analyses (2 from C sites, 9 from N1 sites and 6 from N2 sites). Before statistical analyses replicate samples of NH_4-N and NO_3-N from the same site were pooled to obtain one value of mineralization per site.

Resin ion-exchange capsules (PST-1, Universal Bioavailability Environment/Soil Test, MT, USA) were used to estimate the amount of soil mobile NH_4-N and NO_3-N (mg per capsule). Six capsules were buried just beneath the mor-layer at each site in early June in 2009 and retrieved in late September in the same year. Following retrieval the capsules were placed in plastic bags in a cooler and transported to the lab for analyses. The ion-exchange resins were brushed off to get them as clean as possible before analyses. The capsules were placed in 50 mL Falcon tubes and 7 mL of 1 m KCL was pipetted into the tubes that thereafter were oscillated for 30 min in room temperature, thereafter the 7 mL was poured into a second collecting falcon tube. This process was repeated three times ($3 \times 7 = 21$ mL), and the tubes were stored in a cooler between oscillations. Soil mobile NH_4-N and NO_3-N were analyzed using a FIAstar 5012 Analyzer (Tecator, Höganäs, Sweden). Before statistical analyses replicate samples of NH_4-N and NO_3-N from the same site were pooled to obtain one value of soil mobile N per site.

2.3. Experimental Design and Statistical Analyses

For all statistical analyses the experiment is regarded as a completely randomized design. To test for differences in annual shoot growth among treatments (C, N1 and N2) we used repeated measures ANOVA, with annual growth (cumulative height growth during 10 years) as response and treatment and time as dependent variables. The degrees of freedom were corrected with the Greenhouse-Geisser elipson as the assumption of sphericity was not met, tested by the Mauchly's test of sphericity [30]. Sphericity is not met when the variances of the differences between all combinations of associated groups are not equal; correcting the degrees of freedom with the Greenhouse-Geisser epsilon produces a more accurate, upward adjusted p-value in the subsequent analysis. After adjusting the degrees of freedom, Tukey's post hoc HSD method was applied to detect significant ($\alpha = 0.05$) differences between treatments when the analysis of variance showed significant main N treatment effects. For the repeated measures ANOVA we used the software package IBM SPSS (v. 20).

Differences in tree height (in 2010), trees per ha, sample tree age, needle N concentration (%-DW), sample tree DBH, total soil mobile N (NH_4-N + NO_3-N, mg per ion capsule) and total mineralization rates (ammonification plus nitrification) between the treatments were analyzed with one-way ANOVAs followed by Tukey's post hoc test for pairwise comparison when the analysis of variance showed significant ($\alpha = 0.05$) main N treatment effects. Response variables were checked for normal distribution and homoscedasticity, and when these requirements were not met (for mineralization rate, amount of mobile soil N and needle N), data was transformed with the natural logarithm (LN). For the one-way ANOVAs we used the software package Minitab (v. 16). All data showed in figures and tables are original, untransformed mean values ± 1 standard error (SE).

3. Results

On the sites that were fertilized twice (N2) during the preceding stand rotation (fertilizer applied 25 and 33 years before our study), the annual shoot height growth over time was greater ($F_{2.857,25.715} = 5.09$, $p = 0.018$; Table 2) than those of trees on N1 sites and on control sites never fertilized (Figure 1). Annual shoot height growth over time on N1 plots was, however, not significantly greater than that on control plots ($F_{2.857,25.715} = 5.09$, $p = 0.255$). In 2010 the 10 year old trees on N2 sites were on average 24% higher than trees on control sites ($F_{2,18} = 4.50$, $p = 0.026$, Table 3). Average tree height on N1 sites was not different from that on control sites ($F_{2,18} = 4.50$, $p = 0.214$). Tree diameter (DBH) was not affected by the previous fertilization treatments ($F_{2,18} = 0.75$, $p = 0.486$, Table 3).

Table 2. Results from repeated measures ANOVA testing the effects of fertilization of the previous stand rotation on annual tree shoot height growth of the current stand rotation.

Within Subjects	Degrees of Freedom	F-value	p-value
Year	1.43	948.88	<0.001
Year · Treatment	2.86	3.33	0.037
Error (year)	25.72	1217.1	
Between subjects			
Treatment	2	5.09	0.018
Error	18		

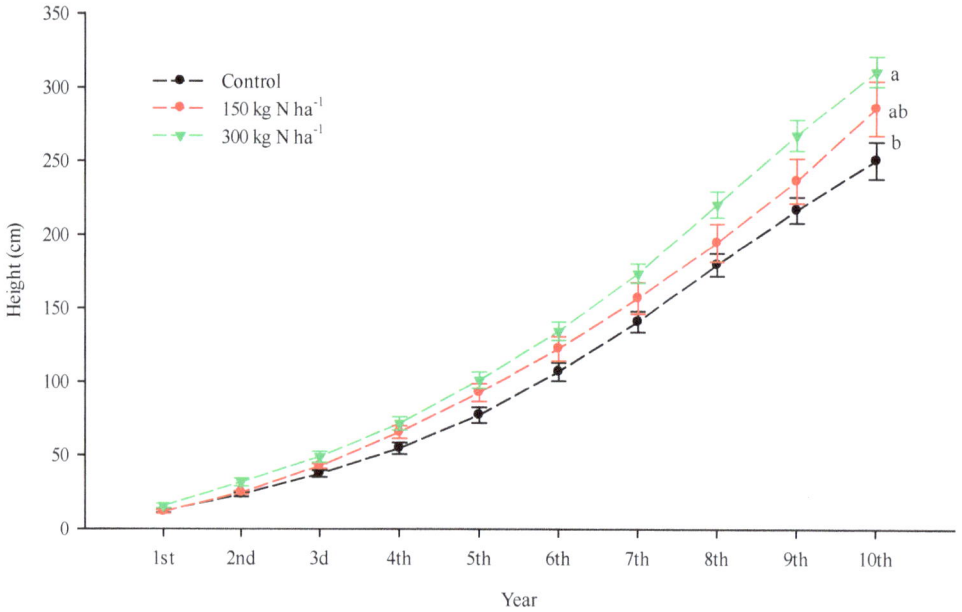

Figure 1. The mean (± 1 SE) annual tree height on sites with no fertilization (C), on sites fertilized once with 150 kg·N·ha^{-1} during the previous stand rotation (N1) and on sites fertilized twice with 150 kg·N·ha^{-1} during the previous stand rotation (N2). Different lower case letters (a or b) indicate a significant difference ($p < 0.05$) between sites analyzed with a repeated measures ANOVA followed by a Tukey's post hoc HSD test.

Trees on control sites had lower needle N concentration than trees on both N1 and N2 sites ($F_{2,17} = 5.04$, $p = 0.018$; Table 3). On N1 and N2 sites the needle N concentration was approximately the same, about 16% and 14% higher on N2 and N1 sites respectively compared to the control. There were no differences in needle carbon concentration (data not shown) between sites with the different N treatments.

Although the variation among sites with the same treatment was large, soil N mineralization rates and soil mobile N differed between the treatments (Table 3). Soil N mineralization rates were nearly four times higher on N2 sites than on control sites ($F_{2,18} = 5.22$, $p = 0.016$), whereas there were no differences in N mineralization rates between the control and the N1 sites. The total amount of soil mobile N that were captured on ion exchange resins placed in the soil was 74% higher on N2 sites than on control sites ($F_{2,18} = 3.71$, $p = 0.045$; Table 3).

Table 3. The mean (± 1 SE) tree needle N concentration (%-DW), diameter at breast height (DBH, mm), tree height in 2010, the total mineralization rates (NH_4-N and NO_3-N mg g^{-1} DW soil) and the total soil mobile N (NH_4-N and NO_3-N) concentration. The 10 year old trees grew on sites never fertilized (C), fertilized once with 150 kg·N·ha^{-1} during the previous stand rotation (N1) and fertilized twice with 150 kg·N·ha^{-1} during the previous stand rotation (N2). Different lower case letters (a or b) besides means in each column indicate significant post hoc differences between treatments (Tukey's HSD test).

Treatment	Needle N Concentration (%-DW)	DBH (mm, 2010)	Tree Height (cm, 2010)	Total Mineralization (N mg g^{-1} DW soil)	Total Soil Mobile N (mg capsule^{-1})
C	1.19 ± 0.05 a	31 ± 5 a	251 ± 12 a	0.083 ± 0.012 a	0.057 ± 0.023 a
N1	1.38 ± 0.05 b	34 ± 5 a	286 ± 13 ab	0.090 ± 0.021 a	0.072 ± 0.022 ab
N2	1.35 ± 0.04 b	39 ± 4 a	311 ± 17 b	0.293 ± 0.109 b	0.099 ± 0.020 b
F-value	5.04	0.75	4.50	5.22	3.71
P-value	0.018	0.486	0.026	0.016	0.045

4. Discussion

According to common forestry practice the growth enhancing effect of N fertilization is expected to diminish within 10 years following treatment [3,9,31]. However, 25 years following the latest fertilization event and 10 years after harvest, scarification and tree planting, we detected positive effects of the previous N fertilization on tree growth in the subsequent young forest stands (Figure 1, Table 2). Our study revealed a fertilizer induced increment in tree growth and needle N concentration. A positive relationship between tree growth and needle N concentration is reported in previous studies, e.g., Bauer, *et al.* [32] and Iivonen, *et al.* [33], and needle N concentration is in turn indicative of the plant available N pool [34]. The higher needle N concentration, in combination with the higher soil N mineralization rates and higher amount of soil mobile N on previously fertilized sites, indicate that N dynamics in the previously fertilized sites remained altered despite that more than 25 years had passed since the last fertilization event.

When fertilizing forests, a large proportion of the N added is normally retained in the soil layer [21,22,35], and disturbances such as clear-felling and soil scarification have been shown to increase soil N mineralization rates and the amount of available soil nutrients [36–38], generally due to increased soil microbial activity. Previously added N retained in the soil layer may thus be accessible to above ground vegetation as N mineralization rates increases after clear-felling and soil scarification. Previous studies of the same sites as used in the present study revealed fertilizer induced shifts in species composition of the forest floor vegetation [20,23]. In combination with the results from the present study this clearly demonstrates that carry-over effects from previous fertilization events influence several components of the ecosystem, resulting in a long-lasting effect on site productivity.

Our results are in contrast to several studies of long-term effects of forest fertilization. For example, Pettersson and Högbom [39] studied old fertilization experiments in boreal forests where

100 to 240 kg of N had been applied one to four times, 14 to 28 years before their study. They found that the previous N addition did not increase *P. abies* or *P. sylvestris* growth in the long term. The majority of the previous studies have, however, addressed the fertilizer effects on site productivity in the forest stand subjected to fertilization, *i.e.*, within the same stand rotation. There are only a few studies targeting carry-over effects of N fertilization across stand rotation periods. In accordance with our results Högbom *et al.* [40,41] found that sites fertilized during the preceding stand rotation had an increased amount of inorganic soil N as well as increased soil N mineralization rates in comparison to unfertilized sites. Worth noting is that this study used N doses largely exceeding the doses used in our study (up to 1800 kg·N·ha^{-1}). Generally much higher amounts of N fertilizer than used in operational forestry practice, appears to be a common feature of most previous studies that have addressed carry-over effects between stand rotations, e.g., Smolander, Priha, Paavolainen, Steer and Malkonen [14], Pettersson and Högbom [39] and Olsson and Kellner [42]. Hence, an important contribution from our study is that N doses used in operational forestry practice can also result in a significant positive carry-over effect on tree growth. Moreover, the only other study we are aware of that targets carry-over effects across stand rotation periods from forest fertilization conducted according to the standard practice, report results that contrast ours, *i.e.*, no positive effect on *P. sylvestris* seedlings growth during the initial five year period following plantation [17]. Although there might be several potential explanations to the contrasting results, our study indicates that it may take more than five years to distinguish a consistent effect of previous fertilization on tree height growth. In accordance with our results, a North American study showed that *Pseudotsuga menziesii* height growth in the first couple of years after seedling establishment was unaffected by fertilization of the preceding forest stand, but that the positive response from past fertilization gradually increased over the years [18]. In their study, a significant difference with approximately 15% greater mean height in fertilized stands than in unfertilized stands became evident when the trees were seven to nine years old. In light of current and previous studies it could be argued that growth enhancement is cumulative, *i.e.*, the growth that the fertilization treatments trigger is built up gradually over time, to eventually entail significant differences in tree height between treatments.

Our study system is situated in a forest land area managed according to common operational practices. Since we lack information on initial site productivity conditions and it is not possible to find detailed information on the criteria used to select sites to be fertilized, we cannot exclude the possibility that initial differences may have influenced our final results, *i.e.*, that N2 sites were more fertile than C sites also before fertilization. However, all available information suggests that the sites selected for fertilization were not biased towards more productive sites. For example, the similar site productivity indices, expressed as estimated average tree height at tree age 100 years (20.6 on C sites and 20.4 on N2 sites) [43,44], and the overall similarity in site characteristics between the sites (see Table 1) supports the information provided by the forest companies employees that the sites selected for fertilization in 1977 and 1985 were, from a productivity point of view, randomly distributed over the forest land area. In addition, if results as those obtained in our study, with N1 sites consistently showing intermediate responses compared to the unfertilized controls and N2 sites (Table 3), were to be generated by pre-fertilization differences in productivity, sites selected for being fertilized would always have had to been more productive than unfertilized sites, and N2 sites would always have had to been more productive than N1 sites. Besides that it appears unlikely that sites by coincidence ended up being selected in such a way, we cannot see any logic reasons to why sites intentionally should be selected in such a biased way. Thus, we are confident that our results indicating residual effects of past fertilization are valid and not generated by initial differences in site productivity.

5. Conclusions

Our study highlighted that forest fertilization according to the standard practice may cause residual long-term effects on tree growth, needle N concentration, soil mineralization rates and amounts of mobile soil N in the stand rotation following the one fertilized. The findings presented herein emphasize that the effects from operational N fertilization on tree growth and biogeochemical ecosystem features can be of a more long-lasting character than previously thought. Whether this increase in site productivity will persist or in time level off is unclear, and our findings advocate that long-term effects from forest fertilization need to be studied further. Further research is also needed to elucidate the mechanisms on how fertilizer N is immobilized following fertilization, and how it is subsequently released following the disturbance caused by tree harvest and soil scarification to thereafter enhance growth of the young forest stand.

Acknowledgments

This work was financially supported by Future Forests and grants from the Swedish Research Council for Environment, Agricultural Sciences and Spatial Planning (FORMAS) to A. Nordin. We would like to thank Ann Sehlstedt and Sonja Wahlberg for assisting with the field work. We would also like to thank the two anonymous reviewers for their valuable suggestions.

Author Contributions

Annika Nordin and Joachim Strengbom conceived and designed the experiment. Fredrik From carried out the field work, data analysis and prepared the manuscript. Annika Nordin and Joachim Strengbom contributed to the data analysis and prepared the manuscript.

Conflicts of Interest

The authors declare no conflict of interest.

References

1. Tamm, C.O. Nitrogen in terrestrial ecosystems: Questions of productivity, vegetational changes, and ecosystem stability. In *Ecological Studies Analysis and Synthesis*; Springer-Verlag Berlin Heidelberg: Berlin, Germany, 1991; p. 115.
2. Vitousek, P.M.; Howarth, R.W. Nitrogen limitation on land and in the sea: How can it occur? *Biogeochemistry* **1991**, *13*, 87–115.

166

3. Nohrstedt, H.O. Response of coniferous forest ecosystems on mineral soils to nutrient additions: A review of Swedish experiences. *Scand. J. For. Res.* **2001**, *16*, 555–573.

4. Simonsen, R.; Rosvall, O.; Gong, P.C.; Wibe, S. Profitability of measures to increase forest growth. *For. Policy Econ.* **2010**, *12*, 473–482.

5. Kardell, L. Skogsgödslingen I backspegeln. In *Debatten om Storskogsbrukets Kvävegödsling I Sverige ca 1960–2009*; Future Forests Working Report; Future Forests: Umeå, Sweden, 2010.

6. Lindkvist, A.; Kardell, O.; Nordlund, C. Intensive forestry as progress or decay? An analysis of the debate about forest fertilization in Sweden, 1960–2010. *Forests* **2011**, *2*, 112–146.

7. Albaugh, T.J.; Allen, H.L.; Fox, T.R. Historical patterns of forest fertilization in the southeastern United States from 1969 to 2004. *South. J. Appl. For.* **2007**, *31*, 129–137.

8. Brockley, R.P. Fertilization of lodgepole pine in western Canada. In Proceedings of the Enhanced Forest Management, Fertilization & Economics Conference, Edmonton, AB, Canada, 1–2 March 2001; pp 44–55.

9. Saarsalmi, A.; Malkonen, E. Forest fertilization research in Finland: A literature review. *Scand. J. For. Res.* **2001**, *16*, 514–535.

10. Näslund, B.Å.; Stendahl, J.; Samuelsson, H.; Karlsson, L.; Kock-Hansson, G.; Svensson, H.; Engvall, C. Kvävegödsling på skogsmark. *Underlag för Skogsstyrelsen Föreskrifter och Allmänna Råd om Kvävegödsling*; Näslund, B.Å., Ed.; Swedish Forest Agency: Jönköping, Sweden, 2013; p. 48. Available online: http://www.regelradet.se/wp-content/files_mf/13702638072013_178_Rapport_kvavegodsling.pdf (accessed on 7 April 2015).

11. Enander, G.; Samuelsson, H. Skogsstyrelsens Allmänna Råd Till Ledning för Hänsyn Enligt 30 § Skogsvårdslagen (1979:429) vid Användning av Kvävegödselmedel på Skogsmark; Skogssstyrelsen. SKSFS **2007**, *3*, 5.

12. Pettersson, F. *New Predictive Functions Reveal Same Growth Response to N Fertilization as 30 Years Ago*; SkogsForsk (Forestry Research Institute of Sweden): Uppsala, Sweden, 1994; p. 6.

13. Priha, O.; Smolander, A. Nitrification, denitrification and microbial biomass N in soil from 2 N-fertilized and limed norway spruce forests. *Soil Biol. Biochem.* **1995**, *27*, 305–310.

14. Smolander, A.; Priha, O.; Paavolainen, L.; Steer, J.; Malkonen, E. Nitrogen and carbon transformations before and after clear-cutting in repeatedly N-fertilized and limed forest soil. *Soil Biol. Biochem.* **1998**, *30*, 477–490.

15. Högbom, L.; Jacobsson, S. Kväve 2002—En konsekvensbeskrivning av skogsmarksgödsling i sverige. *Skogforsk Redogörelse* **2002**, *6*. Available online: http://shop.skogsstyrelsen.se/shop/9098/art64/4645964-efc63c-1547.pdf (accessed on 7 April 2015).

16. Peterson, C.E.; Gessel, S.P. *Forest Fertilization in the Pacific Northwest: Results of the Regional Forest Nutrition Research Project*; General Technical Report; Pacific Northwest Forest and Range Experiment Station, USDA Forest Service: Washington, DC, USA, 1983; pp. 365–369.

17. Johansson, K.; Ring, E.; Hogbom, L. Effects of pre-harvest fertilization and subsequent soil scarification on the growth of planted Pinus Sylvestris seedlings and ground vegetation after clear-felling. *Silva Fennica* **2013**, *47*. doi:10.14214/sf.1016.

18. Footen, P.W.; Harrison, R.B.; Strahm, B.D. Long-term effects of nitrogen fertilization on the productivity of subsequent stands of douglas-fir in the pacific northwest. *For. Ecol. Manage.* **2009**, *258*, 2194–2198.

19. Hedwall, P.O.; Nordin, A.; Strengbom, J.; Brunet, J.; Olsson, B. Does background nitrogen deposition affect the response of boreal vegetation to fertilization? *Oecologia* **2013**, *173*, 615–624.

20. Strengbom, J.; Nordin, A. Commercial forest fertilization causes long-term residual effects in ground vegetation of boreal forests. *For. Ecol. Manag.* **2008**, *256*, 2175–2181.

21. Melin, J.; Nommik, H. Fertilizer nitrogen distribution in a *Pinus slvestris/Picea abies* ecosystem, central Sweden. *Scand. J. For. Res.* **1988**, *3*, 3–15.

22. Nohrstedt, H.O. Effects of repeated nitrogen fertilization with different doses on soil properties in a *Pinus-sylvestris* stand. *Scand. J. For. Res.* **1990**, *5*, 3–16.

23. Strengbom, J.; Nordin, A. Physical disturbance determines effects from nitrogen addition on ground vegetation in boreal coniferous forests. *J. Veg. Sci.* **2012**, *23*, 361–371.

24. Ahti, T.; Hämet-Ahti, L.; Jalas, J. Vegetation zones and their sections in northwestern Europe. *Ann. Bot. Fenn.* **1968**, *5*, 169–211.

25. Womach, J.; Becker, G.S.; Blodgett, J.; Buck, G.; Canada, C.; Chite, R.; Cody, B.; Copeland, C.; Corn, L.; Cowan, T.; *et al. Agriculture: A Glossary of Terms, Programs, and Laws*, 2005th ed.; BiblioGov: Columbus, OH, USA, 2005; p. 282.

26. Phil-Karlsson, G.; Akselsson, C.; Ferm, M.; Hellsten, S.; Hultberg, H.; Karlsson, P.E. *Totaldeposition av Kväve Till Skog*; Svensk Miljöinstitutet AB: Stockholm, Sweden, 2011.

27. Driessen, P.; Deckers, J.; Spaargaren, O.; Nachtergaele, F. Lecture notes on the major soils of the world. Diagnostic horizons, properties and materials. In *World Reference Base for Soil Resources (WRB)*; Driessen, P.M., Deckers, J., Spaargaren, O., Eds.; Food and Agriculture Organization: Rome, Italy, 2000; pp. 121–124.

28. Chesworth, W. Moisture regimes. In *Encyclopedia of Soil Science*; Chesworth, W., Ed.; Springer: Dordrecht, Netherlands, 2008; p. 485.

29. Påhlsson, L. Vegetationstyper I norden. In *Nordiska Ministerrådet Copenhagen*; Nordic Council of Ministers: Copenhagen, Denmark, 1995; p. 145–146.

30. Mauchly, J.W. Significance test for sphericity of a normal *n*-variate distribution. *Ann. Math. Stat.* **1940**, *11*, 204–209.

31. Nason, G.; Myrold, D.D. *Nitrogen Fertilizers: Fates and Environmental Effects in Forests*; Institute of Forest Resources Contrib. 73. College of Forest Resources, University of Washington: Seattle, WA, USA, 1992; pp. 67–81. Available online: http://www.cfr.washington.edu/research.smc/rfnrp/2FFC_Chap6.pdf (accessed on 7 April 2015).

32. Bauer, G.; Schulze, E.D.; Mund, M. Nutrient contents and concentrations in relation to growth of *Picea abies* and *Fagus sylvatica* along a European transect. *Tree Physiol.* **1997**, *17*, 777–786.

33. Iivonen, S.; Kaakinen, S.; Jolkkonen, A.; Vapaavuori, E.; Linder, S. Influence of long-term nutrient optimization on biomass, carbon, and nitrogen acquisition and allocation in Norway spruce. *Can. J. For. Res.* **2006**, *36*, 1563–1571.

34. Binkley, D.; Reid, P. Long-term increase of nitrogen availability from fertilization of douglas-fir. *Can. J. For. Res.* **1985**, *15*, 723–724.

35. Melin, J.; Nommik, H.; Lohm, U.; Flowerellis, J. Fertilizer nitrogen budget in a Scots pine ecosystem attained by using root-isolated plots and N^{15} tracer technique. *Plant Soil* **1983**, *74*, 249–263.

36. Lundmark-Thelin, A.; Johansson, M.B. Influence of mechanical site preparation on decomposition and nutrient dynamics of Norway spruce (*Picea abies* (L.) Karst) needle litter and slash needles. *For. Ecol. Manag.* **1997**, *96*, 101–110.

37. Vitousek, P.M.; Matson, P.A. Disturbance, nitrogen availability, and nitrogen losses in an intensively managed loblolly pine plantation. *Ecology* **1985**, *66*, 1360–1376.

38. Rosén, K.; Aronson, J.A.; Eriksson, H.M. Effects of clear-cutting on streamwater quality in forest catchments in central Sweden. *For. Ecol. Manag.* **1996**, *83*, 237–244.

39. Pettersson, F.; Högbom, L. Long-term growth effects following forest nitrogen fertilization in *Pinus sylvestris* and *Picea abies* stands in Sweden. *Scand. J. For. Res.* **2004**, *19*, 339–347.

40. Hogbom, L.; Nohrstedt, H.O.; Lundstrom, H.; Nordlund, S. Soil conditions and regeneration after clear felling of a *Pinus sylvestris* l. Stand in a nitrogen experiment, central Sweden. *Plant Soil* **2001**, *233*, 241–250.

41. Högbom, L.; Nohrstedt, H.Ö.; Lundström, H.; Nordlund, S. *Kvävegödsling Kan ge Varaktiga Effekter i Marken (Nitrogen Fertilization Can Have Long Term Effects on Soils)*; Skogforsk: Uppsala, Sweden, 2000; p. 4. (In Swedish)

42. Olsson, B.A.; Kellner, O. Long-term effects of nitrogen fertilization on ground vegetation in coniferous forests. *For. Ecol. Manag.* **2006**, *237*, 458–470.

43. Hägglund, B. *Forecasting Growth And Yield In Established Forests. An Outline And Analysis Of The Outcome Of A Subprogram Within The Hugin Project*; Swedish University of Agricultural Sciences: Umeå, Sweden, 1981; p. 145.

44. Hägglund, B. *Samband Mellan Ståndortsindex h100 Och Bonitet För Tall Och Gran i Sverige*; Swedish University of Agricultural Sciences: Umeå, Sweden, 1981; p. 90.

Mid-Rotation Silviculture Timing Influences Nitrogen Mineralization of Loblolly Pine Plantations in the Mid-South USA

Michael A. Blazier, D. Andrew Scott and Ryan Coleman

Abstract: Intensively managed loblolly pine (*Pinus taeda* L.) plantations often develop nutrient deficiencies near mid-rotation. Common silvicultural treatments for improving stand nutrition at this stage include thinning, fertilization, and vegetation control. It is important to better understand the influence of timing fertilization and vegetation control in relation to thinning as part of improving the efficiency of these practices. The objective of this study was to determine the effects of fertilization and vegetation control conducted within a year prior to thinning and within a year after thinning on soil N supply in mid-rotation loblolly pine plantations on a gradient of soil textures. Net N mineralization (N_{min}) and exchangeable N were measured monthly. Fertilization increased annual N_{min} at all sites irrespective of timing relative to thinning, with the increase more pronounced when combined with vegetation control. This finding suggests some management flexibility in the timing of mid-rotation fertilization relative to thinning for increasing soil N supply. However, the site with the highest total soil N and the lowest C:N ratio was more prone to NO_3-N increases after fertilization conducted pre- and post-thinning. At all sites, fertilization with vegetation control promoted increases in NO_3-N when done after thinning, which may indicate that this practice increased soil N supply to levels that exceeded stand N demand.

Reprinted from *Forests*. Cite as: Blazier, M.A.; Scott, D.A.; Coleman, R. Mid-Rotation Silviculture Timing Influences Nitrogen Mineralization of Loblolly Pine Plantations in the Mid-South USA. *Forests* **2015**, *6*, 1061-1082.

1. Introduction

As intensively managed loblolly pine plantations in the southeastern United States approach mid-rotation, stand nutrient demand increases and soil nutrient supply decreases [1,2]. To synchronize nutrient supply with plant demand, increase leaf area, and improve productivity, plantations are often fertilized, typically after thinning [3–5]. Nitrogen is typically the limiting nutrient in these plantations at this stage because it is needed in relatively high quantities for photosynthesis, carbohydrate assimilation, and foliage growth [6]. Herbicides are also sometimes applied at this stage to control competing vegetation and improve site resource availability for the crop tree [7]. Further improvement of loblolly pine production efficiency via mid-rotation silviculture will likely require more than simply increasing resource inputs; silvicultural practices that better manage the interactions among crop trees, non-crop vegetation, soil properties, and applied nutrients will likely be necessary [8]. Altering the timing of fertilization and vegetation control treatments relative to mid-rotation thinning is one potential method of influencing resource use efficiency. Fertilizing mid-rotation stands prior to thinning may promote higher stand-level fertilizer uptake, which could in turn reduce risks of N leaching on well-drained soils and denitrification on poorly drained soils. With pre-thinning fertilization, nutrient levels of fertilized trees remaining after thinning may be elevated, which could enable trees to more rapidly exploit increased growing space. However, increased intraspecific competition in stands at canopy closure can temper response to fertilization, with loblolly pine stands generally exhibiting declining potential response to fertilization as they approach basal areas greater than 35 m²·ha⁻¹ [8].

A vital component of understanding loblolly pine plantation response to timing of fertilization relative to thinning is observing its effects on soil N supply. Soil N supply has been defined as net N mineralization [9]. Fertilization of mid-rotation loblolly pine plantations can increase N mineralization [2,10]. Non-crop vegetation suppression can likewise increase N mineralization in mid-rotation loblolly pine plantations, particularly when combined with fertilization [2]. Nitrogen mineralization rates may be relatively high in the year of thinning because stand conditions could be similar to those of complete harvesting, with increased radiation to the forest floor and decreased inputs of soluble organic C because of tree harvesting. Such conditions have been shown to result in relatively high N mineralization in the initial years of loblolly pine plantations following clearcuts [11].

Microbial biomass and activity are closely associated with soil N mineralization, and they are affected by fertilization and vegetation control in forests [12]. As such, it is appropriate to explore these variables in tandem with soil N mineralization trends. Fertilization can lead to a "priming effect" in which microbial N immobilization potential is reduced, leading to increases in N mineralization that lead to increases in soil N availability beyond that applied in fertilizer [9]. Microbial biomass and activity increased [13], decreased [14], or had no response [15] to forest fertilization in previous studies. Combining non-crop vegetation control with fertilization reduced microbial biomass and activity to a greater extent than to fertilization alone in a juvenile loblolly pine plantation, which was attributed to reduction in labile C sources for microbes due to vegetation control [16].

The influence of altering the timing of fertilization and vegetation control relative to thinning at mid-rotation loblolly pine plantations on soil N dynamics and its associated microbial parameters is relatively unstudied. Accordingly, the objectives of this study were to determine the effects of fertilization and vegetation control conducted within a year prior to thinning and within a year after thinning on soil N mineralization, exchangeable N, soil microbial biomass C, and dehydrogenase activity in mid-rotation loblolly pine plantations on a gradient of soil textures in the mid-South USA. Study sites were selected along a gradient of soil textures because soil texture influences nutrient cycling; mineral particle size distribution can affect organic matter retention, N mineralization, microbial biomass, and other soil properties [17].

2. Materials and Methods

2.1. Study Sites

In September and October 2003, three study sites were established in mid-rotation loblolly pine plantations in north central and southwest Louisiana. The sites were selected for their similarities in age and management history and the range of soil drainage classes they provided (Table 1). All sites had chop and burn site preparation; the Oakdale site had been fertilized with 50 kg·N·ha^{-1} and 56 kg·P·ha^{-1} as diammonium phosphate in 1997.

Table 1. Location, stand age and basal area in 2003, and U.S. Department of Agriculture (USDA) Natural Resource Conservation Service soil type classification of study sites established in loblolly pine plantations in north central and southwest Louisiana, USA.

Nearest Town	Geographic Coordinates	Stand Age (Years)	Stand Basal Area (m²·ha⁻¹)	Soil Type
Dodson	32.0727° N, 92.6697° W	13	23.5	Bowie (moderately well-drained, fine-loamy, siliceous, thermic Plinthic Paleudult)
Lucky	32.3029° N, 92.9240° W	12	19.9	Betis (excessively-drained, sandy siliceous, thermic Lamellic Paleudult)
Oakdale	30.8164° N, 92.6914° W	15	20.5	Caddo (poorly-drained, fine-silty, siliceous, thermic Typic Glossaqualf)

Understory and mid-story of the sites at the initiation of the study were dominated by hardwood tree species at the Dodson and Lucky sites and shrub species at the Oakdale site [7]. Red maple (*Acer rubrum* L.) and sweetgum (*Liquidambar styraciflua* L.) were frequently observed species at the Dodson and Lucky sites. Southern red oak (*Quercus falcata* Michx.) was present in high numbers at the Lucky site. The most common understory and mid-story species at the Oakdale site were yaupon holly (*Ilex vomitoria* Sol. Ex Aiton), wax myrtle (*Morella cerifera* L. (Small)), and Chinese tallow tree (*Triadica sebifera* L.).

2.2. Study Design and Treatments

Study design was a randomized complete block with soil texture differences to a 40-cm depth used as a blocking factor at each site. Experimental units were 0.1-ha plots, and all treatment combinations were replicated three times at each study site. Plots were separated by at least 13 m to insure independence of treatments. Measurement zones were established within the central 0.08 ha of each plot. Blocks, which contained one replicate of each treatment combination and buffer spaces between plots, were 1 ha in size.

Treatments conducted in this study were: (1) no fertilization or herbicide (CONTROL), (2) fertilization (FERT), and (3) fertilization and vegetation suppression (FERTVS). The FERT and FERTVS treatments were conducted prior to thinning or after thinning to separate sets of plots. The CONTROL treatment was applied to plots observed in pre-thin and post-thin condition, *i.e.*, the same CONTROL plots were observed pre- and post-thinning. The pre-thin FERT and FERTVS plots were sampled in only the year prior to thinning, and the post-thin FERT and FERTVS plots were observed in only the year after thinning. There were three replications for each treatment. With this treatment structure there were three CONTROL plots, three pre-thin FERT plots, three post-thin FERT plots, three pre-thin FERTVS plots, and three post-thin FERTVS plots.

Fertilization was conducted pre-thin in April 2004 and post-thin in April 2005. A mixture of urea and diammonium phosphate supplying 135 kg·N and 13 kg·P·ha^{-1} was applied using shoulder-mounted spreaders. Fertilizer was applied at each site within one week of predicted rainfall in order to minimize variability in the dissolution of granules [18].

Non-crop vegetation suppression treatments (herbicides) were applied as part of the FERTVS treatments by backpack sprayers in March 2004 (pre-thin) and March 2005 (post-thin). A 20% solution of triclopyr mixed with diesel as a penetrant was applied as a basal bark spray to suppress hardwood saplings and shrubs. A 5% solution of glyphosate mixed with water was used to suppress herbaceous vegetation and hardwood sprouts. These herbicide mixtures were re-applied as necessary for the remainder of each year in order to prevent regrowth of competing vegetation. Understory biomass was reduced by 95.4, 51.4, and 71.4% at the Dodson, Lucky, and Oakdale sites, respectively, in the FERTVS treatment relative to the CONTROL and FERT treatments [7].

Thinning was conducted operationally using feller-bunchers and skidders in October 2004 for the Dodson and Oakdale sites and in November 2004 for the Lucky site. Thinning was conducted as a fifth-row removal thinning from below at each site, with crown biomass returned to down rows as conventionally conducted in operational thinning. Crown biomass was separated from boles by gate de-limber at the logging deck, and skidders were used to transport and deposit the crown biomass along down rows as the skidders returned into the stand to continue tree felling. A target residual density of 494 trees·ha^{-1} was created for each plot by selecting and marking residual trees after down rows were cut in the plantations. Residual trees were selected on the basis of dominant or codominant height and DBH, good form, and spacing. Spacing between trees within rows was approximately 6 m, although this spacing varied depending on availability of trees of appropriate height, DBH, and form within 6 m of each tree. After marking residual trees between down rows, the areas between down rows were thinned by feller-bunchers. Achieved average residual densities at the Dodson, Lucky, and Oakdale sites were 421, 420, and 466 trees·ha^{-1}, respectively. Post-thinning basal areas were respectively 11.2, 8.5, and 8.9 m^2·ha^{-1} for the Dodson, Lucky, and Oakdale sites.

2.3. Nitrogen Mineralization and Exchangeable N

Net N mineralization (N$_{min}$) was measured using the *in situ* buried bag method [19] as described for a sweetgum (*Liquidambar styraciflua* L.) plantation by Scott *et al.* [9]. Soil samples for N$_{min}$ were collected to a depth of 20 cm using punch augers at two randomly located subsample points within the central 0.08 ha of each plot. Sampling occurred every four to six weeks, with the six-week incubation durations occurring in fall and winter months from January 2004 through November 2005. For each sample period, two composite samples were taken at each sample point. Each composite sample was comprised of three punch auger samples; organic matter was removed from the soil surface immediately prior to sampling. One composite sample at each sample point was used for *in situ* incubation, and the other was used for pre-incubation analysis. Each composited sample for incubation was used to fill sterile 14 cm × 23 cm sampling bags with round-wire closure. The bags had a 0.08 mm thickness to allow gas exchange while retaining moisture. After bags were filled with soil, any surface organic matter was moved aside and a small hole was dug to a depth of 20 cm in the center of the three sampling points used to create the composite sample. The incubation bag was then placed in the hole, and any surface organic matter was replaced. Samples were incubated until the subsequent sample period. All pre- and post-incubation samples were stored in a cooler for transport to the laboratory, and the samples were then air-dried to constant weight and sieved to pass a 2 mm mesh for preparation of extraction of NH_4^+-N and NO_3^--N to be used for determination of N_{min}. Soil NH_4^+-N and NO_3^--N were extracted using a 2 M solution of KCl at a 10:1 ratio of solution to soil [20]. Concentrations of NH_4^+-N and NO_3^--N were quantified by the microdiffusion-conductivity method [21] using a Bran Luebbe AutoAnalyzer 3 (SPX Equipment, Delavan, WI, USA).

Net N mineralization was calculated as the difference in mineral N (mg·N·kg^{-1} soil) following incubation and prior to incubation (NH_4-N + NO_3-N)$_{t+1}$ − (NH_4-N + NO_3-N)$_t$ [22]. The N_{min} rates for each incubation period were adjusted to a one-month (30-day) basis by dividing net N_{min} by the number of days in the incubation period and then multiplying by 30. Annual N_{min} was estimated by summing N_{min} measurements for all incubation periods for each plot, dividing by the total days of incubation for the year, and then multiplying by 365. Bulk density samples were collected to a 20-cm depth in all plots with a 5-cm diameter soil core sampler in spring 2004, and these measurements were used to convert annualized N_{min} to a kg·ha^{-1} basis.

The NH_4-N and NO_3-N concentrations of the pre-incubation bags were used to determine changes in exchangeable N within and between the years of this study. In addition to NH_4-N and NO_3-N concentrations, total exchangeable N was assessed as the sum of NH_4-N and NO_3-N. The ratio of NO_3-N to total exchangeable N was calculated for the pre-incubation bag samples as well.

2.4. Soil Microbial Biomass C, Dehydrogenase Activity

Soil samples used for assays of soil microbial biomass C (C_{mic}) and microbial activity were collected to a 20 cm depth with punch augers. Sampling for these assays was conducted in tandem with some dates at which N_{min} samples were collected using sampling and compositing protocol similar to that described above for N_{min}. In 2004, soil samples for which C_{mic} and microbial activity assays were conducted were collected in April, May, July, and September. In 2005, samples for which C_{mic} was assayed were collected monthly from March through September. Microbial activity assays were also conducted for the May, June, and September 2005 soil samples. All samples were refrigerated at 5 °C until analyses, which was initiated within 14 days of sampling.

The chloroform fumigation-incubation method was used to determine C_{mic} [23,24]. A 10-day pre-incubation was carried out prior to fumigation of soil samples in order to allow the influence of soil disturbance to subside and living fragments of roots to die [25,26]. Microbial biomass C was determined by fumigating soil samples with alcohol-free $CHCl_3$ vapor for 24 h, incubating soil samples at 25 °C for 10 days, collecting respired CO_2 in 2 M NaOH, measuring CO_2-C by titration of the NaOH with 0.1 N HCl, and using a proportionality constant of 0.45 to convert CO_2-C to C_{mic} after subtracting the CO_2-C produced by a non-fumigated control [23,24]. All results are expressed on an oven-dry soil basis (105 °C, 24 h).

Microbial activity was estimated by determining dehydrogenase activity [27,28]. Dehydrogenase, which is only active in viable living cells, serves as an indicator of total microbial metabolic activity [29,30]. To quantify dehydrogenase, triphenyltetrazolium chloride (TTC) was used as an artificial electron acceptor. Dehydrogenase reduces TTC to red-colored triphenyl formazan (TPF) that can be extracted with methanol and quantified colorimetrically [29]. Results are expressed as μg TPF kg^{-1} soil on an oven-dry soil basis (105 °C, 24 h).

2.5. Soil Labile C

Labile C was measured using the sequential fumigation-incubation procedure [31] for soil samples collected in September 2005 as described above for C$_{mic}$. The September sample was selected for this assay because above- and below-ground plant biomass is near its seasonal peak at that time. Soil samples were fumigated with alcohol-free CHCl$_3$ vapor for 24 h, and respired CO$_2$ was measured as described above for the chloroform fumigation-incubation method for C$_{mic}$ measurement for a series of eight 10-day fumigation-incubation cycles. The method is based on the assumption that soil labile C is decayed during these cycles according to the negative exponential equation as widely reported in plant decomposition studies, so labile C was estimated from CO$_2$ measurements using a first-order kinetics model [31].

2.6. Soil Carbon and Nitrogen

In April 2005, five (for the Lucky site) and six (for the Dodson and Oakdale sites) months after thinning, soil samples were collected from all plots to characterize C and N concentrations of surface soil. Ten samples were randomly sampled in the central 0.08 ha of each plot to a 20 cm depth using punch augers; the ten samples were composited for each plot. Organic matter was removed from the soil surface prior to sampling. Carbon and nitrogen concentrations were measured by combustion using a LECO CNS 2000 analyzer (Leco, Inc., St. Joseph, MI, USA).

2.7. Soil Temperature and Volumetric Water Content

Soil temperature and volumetric water content (VWC) measurements in the top 20 cm of soil were collected concurrently with all soil sampling for soil N and microbial assays using a K-type thermocouple thermometer (Hanna Instruments, Inc., Woonsocket, RI, USA) and a TDR 200 time-domain reflectometry probe (Spectrum Technologies, Inc., Paxinos, PA, USA), respectively. Soil temperature and water content was measured within 10 cm of all samples collected for soil N and microbial assays.

2.8. Environmental Conditions

Precipitation and temperature trends for the regions in which all study sites were located were obtained from the Historical Climate Trends Tool of the Southern Climate Impacts Planning Program (SCIPP) [32]. This database facilitates comparison of average monthly precipitation and temperature trends relative to long-term (1971–2000) average trends for each month. The Lucky and Dodson sites were in the North Central Louisiana SCIPP climate region, and the Oakdale site was in the Southwest Louisiana SCIPP climate region.

2.9. Statistical Analysis

All treatment effects were analyzed for variance (ANOVA) at α = 0.05 using the MIXED procedure of the SAS System (SAS Institute, Inc., Cary, NC, USA). When the null model likelihood ratio test revealed heterogeneous variances in a dataset, the GROUP option of MIXED was utilized to perform ANOVA using different variances for all treatment combinations. When an ANOVA indicated significant treatment effects, treatment means were calculated and separated by the DIFF and SLICE options of the LSMEANS procedure. The DIFF option provided multiple comparisons of treatment means by invoking *t*-tests to determine significant differences between all possible treatment combinations. The SLICE option provided *t*-tests of treatment means in which the effect of one treatment is evaluated at each level of another treatment. The SLICE option was used to investigate treatment main effects when significant two-way interactions were observed.

Models used in analyses of parameters measured in this study differed depending on sampling frequency. Analyses of N_{min}, NH_4-N, NO_3-N, total exchangeable N, ratio of NO_3-N to total exchangeable N, soil volumetric water content, and soil temperature were each performed with a model that included these parameters as a dependent variable and block, treatment, sampling date, and the interaction of treatment and date as independent variables. All independent variables were considered fixed effects. The model was a repeated measures model with an autoregressive correlation structure. These analyses were performed by site and year because of differences in sampling dates for each site and year. There were two numerator degrees of freedom for block and treatment associated with the model for all sites and both years. Due to sampling date differences for each site and year, there were differences in date and treatment × date numerator degrees of freedom associated with the model. For the Dodson and Oakdale sites in 2004 there were seven numerator degrees of freedom and 14 numerator degrees of freedom for date and treatment × date, respectively. For the Lucky site in 2004, the model was associated with six numerator degrees of freedom for date and 12 numerator degrees of freedom for treatment × date. For the Lucky and Oakdale sites in 2005, the model was associated with eight numerator degrees of freedom for date and 16 numerator degrees of freedom for treatment × date. The model had nine and 18 numerator degrees of freedom, respectively, for date and treatment × date for the Dodson site in 2005.

Microbial biomass C and dehydrogenase activity were sampled at the same dates for each site within both years of the study, but there were differences in the number of sampling dates between 2004 and 2005. As such, analyses of these parameters were performed by year. The model used in analyses included C_{mic} and dehydrogenase as dependent variables and block, treatment, site, sampling date, and all possible interactions of treatment, site, and sampling date as independent variables. All independent variables were fixed effects. The model was a repeated measures model with an autoregressive correlation structure. For C_{mic} and dehydrogenase in both years, numerator degrees of freedom associated with the model were two, two, two, and four for block, site, treatment, and site × treatment, respectively. For C_{mic} and dehydrogenase activity in 2004, numerator degrees of freedom associated with the model were six, six, and 12 for site × sampling date, treatment × sampling date, and site × sampling date × treatment, respectively. There were eight, eight, and 16 numerator degrees of freedom associated with site × sampling date, treatment ×

sampling date, and site × sampling date × treatment, respectively, for the model for C_{mic} in 2005. For dehydrogenase activity in 2005 there were four, four, and eight numerator degrees of freedom for site × month, treatment × sampling date, and site × sampling date × treatment, respectively, associated with the model.

Annualized N_{min} was measured over the same sample period for each site in both years of the study. As such, the model used in its analysis included annualized N_{min} as a dependent variable and block, site, treatment, year, and all possible interactions between site, treatment, and year as independent variables. All independent variables were fixed effects, and the model was a repeated measures model with an autoregressive correlation structure. Block, site, treatment, site × year, and treatment × year were each associated with two numerator degrees of freedom in analysis. Site × treatment and site × treatment × year were each associated with four numerator degrees of freedom, and year was associated with one numerator degree of freedom in analysis.

Labile C, soil C concentrations, soil N concentrations, and soil C:N ratios were each measured a single time at each site in this study. The model used in their analyses therefore included block, site, treatment, and site × treatment as independent variables, which were all fixed effects. For labile C (which was sampled only in CONTROL and post-thin FERT and FERTVS plots), block, site, and treatment each had two numerator degrees of freedom in analysis, and site × treatment had four numerator degrees of freedom. For C, N, and C:N ratios (which was sampled post-thinning in all plots) block and site each had two numerator degrees of freedom in analysis, treatment had four numerator degrees of freedom, and site × treatment had eight numerator degrees of freedom.

To provide greater insight into how thinning affected N_{min}, exchangeable N, C_{mic}, dehydrogenase activity, soil temperature, and soil moisture, cross-year comparisons were performed using data from CONTROL plots. CONTROL plot data from common sample dates for 2004 and 2005 of all sites were pooled. Analyses were performed using a repeated measures model with an autoregressive correlation structure that included block, site, year, and the interaction of site and year as fixed effects. Block, site, and site × year each had two numerator degrees of freedom in analysis, and year was associated with one numerator degree of freedom.

Correlations among N_{min}, volumetric soil moisture, and soil temperature were quantified for each site using the PROC CORR procedure of the SAS System, which produced Pearson correlation coefficients and the probabilities associated with these statistics. For each site, correlations were also determined between N_{min}, C_{mic}, and dehydrogenase activity for months in which all variables were measured.

3. Results

3.1. Environmental and Edaphic Conditions

Temperature trends for the regions of Louisiana in which the study sites were located were similar to 30-year trends in both years of the study. In both regions, 2004 precipitation exceeded the 30-year average by approximately 25% whereas precipitation for 2005 was 40 to 50% below the 30-year average. Deviations from the long-term trend were greatest in mid-summer and fall each year, with

2004 precipitation higher than the 30-year average and 2005 precipitation lower than the 30-year average during these months.

Soil temperature was similar among treatments at all sites prior to thinning in 2004. After thinning, the CONTROL treatment soils were about 2–3 °C warmer than soils in the FERT and FERTVS treatments throughout most of the summer and early autumn at the Dodson ($P = 0.001$) and Lucky sites ($P = 0.002$) (Figure 1). Soil temperatures were similar ($P = 0.88$) among all treatments at Oakdale. In the cross-year comparison of CONTROL plot data, across all sites soil temperature was significantly ($P = 0.03$) warmer (2–3 °C) in the CONTROL plots post-thinning than pre-thinning.

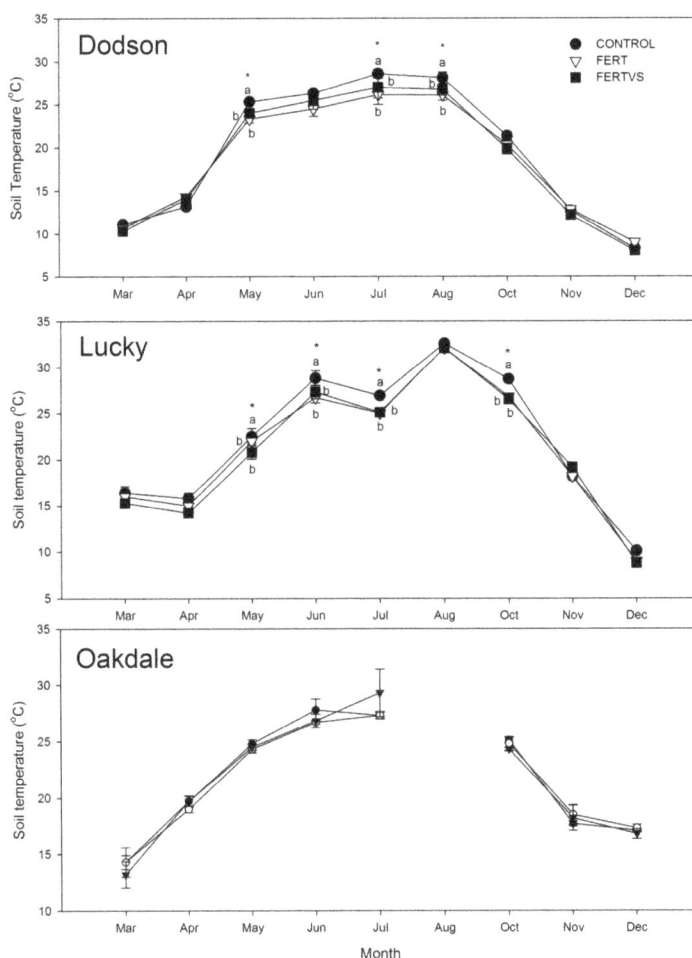

Figure 1. Monthly soil temperature in 2005 in thinned mid-rotation plantations (identified by the name of the nearest city) in north central (Dodson, Lucky) and southwest (Oakdale) Louisiana, USA, in response to no fertilization or herbicide (CONTROL), fertilization (FERT), and fertilization and vegetation suppression (FERTVS) treatments. The plantations were thinned in October and November 2004. For months headed by an asterisk, symbols noted by different letters differ at $P < 0.05$.

Fertilized soils were generally drier at the Dodson and Lucky sites but unaffected at the Oakdale site. In 2004, there was a significant ($P = 0.01$) treatment effect at the Lucky site, at which VWC of the CONTROL treatment (mean 11.5% v/v) was 40% wetter than that of the FERT (mean 6.4% v/v) and FERTVS (mean 7.7% v/v) treatments. Similarly, the CONTROL soils at Dodson were wetter than the FERT and FERTVS treatments in 2004, but only significantly in March and December (*i.e.*, there was a significant ($P = 0.04$) treatment × date interaction). At the Dodson site in 2005 there was a significant ($P = 0.03$) treatment effect; the CONTROL (mean 16.9% v/v) and FERTVS (mean 17.2% v/v) treatments had greater VWC than the FERT (mean 13.5% v/v) treatment. On average, soils were the wettest at Oakdale (25.1% v/v), moderate at Dodson (18.9% v/v), and driest at Lucky (12.0% v/v).

3.2. Nitrogen Mineralization and Exchangeable N

Monthly N_{min} differed among treatments at one or more sites in both years of the study (Figure 2). At Dodson in 2004, N_{min} in the FERTVS soils was more than twice that of the CONTROL and FERT soils ($P = 0.04$). No differences in monthly N_{min} were observed at the Lucky ($P = 0.71$) and Oakdale ($P = 0.57$) sites in 2004. In 2005, the FERT and FERTVS treatment soils had nearly double the average monthly N_{min} than the CONTROL treatment soils at the Dodson ($P = 0.02$) and Oakdale ($P = 0.02$) sites. No differences ($P = 0.66$) in monthly N_{min} among treatments were observed at the Lucky site in 2005.

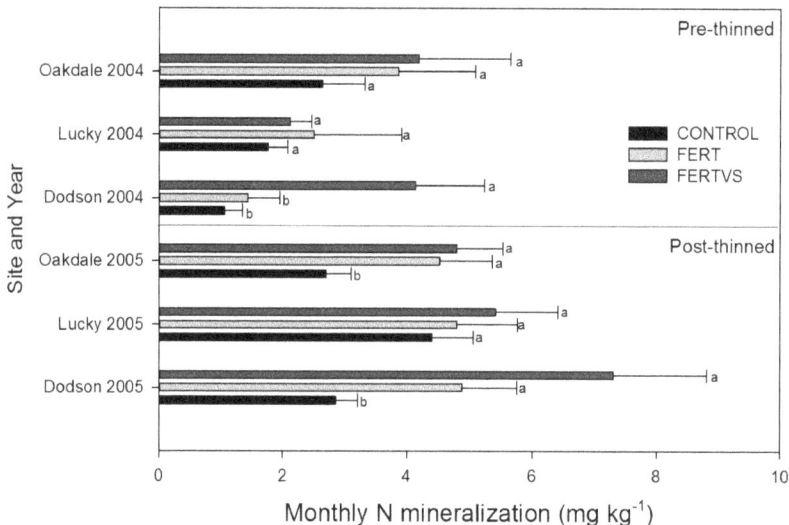

Figure 2. Average monthly N mineralization of non-thinned and thinned mid-rotation loblolly pine plantations in north central (Dodson, Lucky) and southwest Louisiana (Oakdale), USA, in response to no fertilization or herbicide (CONTROL), fertilization (FERT), and fertilization and vegetation suppression (FERTVS) treatments. Plantations were thinned in October and November 2004. For each site and year, columns headed by a different letter differ at $P < 0.05$.

Across all sites and years of the study, FERT (mean 113.5 kg·ha^{-1}·year^{-1}) and FERTVS (mean 146.3 kg·ha^{-1}·year^{-1}) treatments had greater annualized N_{min} than the CONTROL (mean 81.1 kg·ha^{-1}·year^{-1}) treatment; annualized N_{min} of the FERTVS treatment was also greater than that of the FERT treatment ($P = 0.004$). Annualized N_{min} was similar ($P = 0.98$) among sites. Across all sites, annual N_{min} significantly differed ($P < 0.0001$) between years of the study, with annual N_{min} nearly twice as high after thinning (mean 142.3 kg·ha^{-1}·year^{-1}) as prior to thinning (mean 85.0 kg·ha^{-1}·year^{-1}).

Soil NO_3-N concentrations were different among treatments at one site or more in both years of the study. Concentrations of NO_3-N differed at only the Oakdale site in 2004, for which there was a significant ($P = 0.004$) treatment × date interaction (Figure 3). Soil NO_3-N concentrations were greater for FERTVS treatment than the other treatments in May, June, and August. The FERT treatment had greater NO_3-N than that of the CONTROL treatment in May and June. A significant treatment × date effect ($P = 0.01$) was also observed for the Oakdale site for 2005 (Figure 4). In 2005, NO_3-N of the FERTVS treatment exceeded that of the CONTROL and FERT treatments in all months observed after April. The FERT treatment had greater NO_3-N concentrations than the CONTROL treatment in all months observed after April. Significant treatment effects were found for the Dodson ($P = 0.0003$) and Lucky ($P = 0.004$) sites in analyses of NO_3-N for 2005 (Table 2). At the Dodson site the FERTVS treatment had the highest NO_3-N concentrations, and the FERT treatment had greater NO_3-N concentrations than the CONTROL treatment. The FERT and FERTVS treatments had greater NO_3-N concentrations than the CONTROL treatment at the Lucky site.

Ratios of NO_3-N to total exchangeable N were different among treatments in both years at the Oakdale site and at the Lucky site in 2005. There was a significant treatment effect for the Oakdale site in 2004, with the FERT (mean 34.0%) and FERTVS (mean 35.0%) treatments had a higher ($P = 0.04$) NO_3:exchangeable N ratio than the CONTROL treatment (mean 29.7%). In 2005, there was a significant treatment × date effect ($P = 0.01$) for the Oakdale site (Figure 4). In all months observed after May, the FERTVS treatment had a higher ratio than the CONTROL. The FERTVS treatment also had a higher ratio than the FERT treatment in all months observed between June and November except October. The FERT treatment had a greater NO_3:exchangeable N ratio than the CONTROL treatment in June, July, September, and November. At the Lucky site there was a significant treatment effect ($P = 0.02$), with the FERTVS treatment having a higher ratio than the FERT and CONTROL treatments (Table 2).

In 2004, there were significant treatment × date interactions for the Oakdale ($P < 0.0001$) and Lucky ($P < 0.0001$) sites in analyses of NH_4-N (Figures 3 and 5). At Oakdale, the FERT and FERTVS treatments had greater NH_4-N concentrations than the CONTROL treatment in May and June (Figure 3). At the Lucky site differences were observed among treatments in only April, in which the FERT treatment had NH_4-N concentrations that exceeded those of the CONTROL and FERTVS treatments (Figure 5). Soil NH_4-N concentrations of the FERTVS treatment were significantly greater than those of the CONTROL treatment in that month as well.

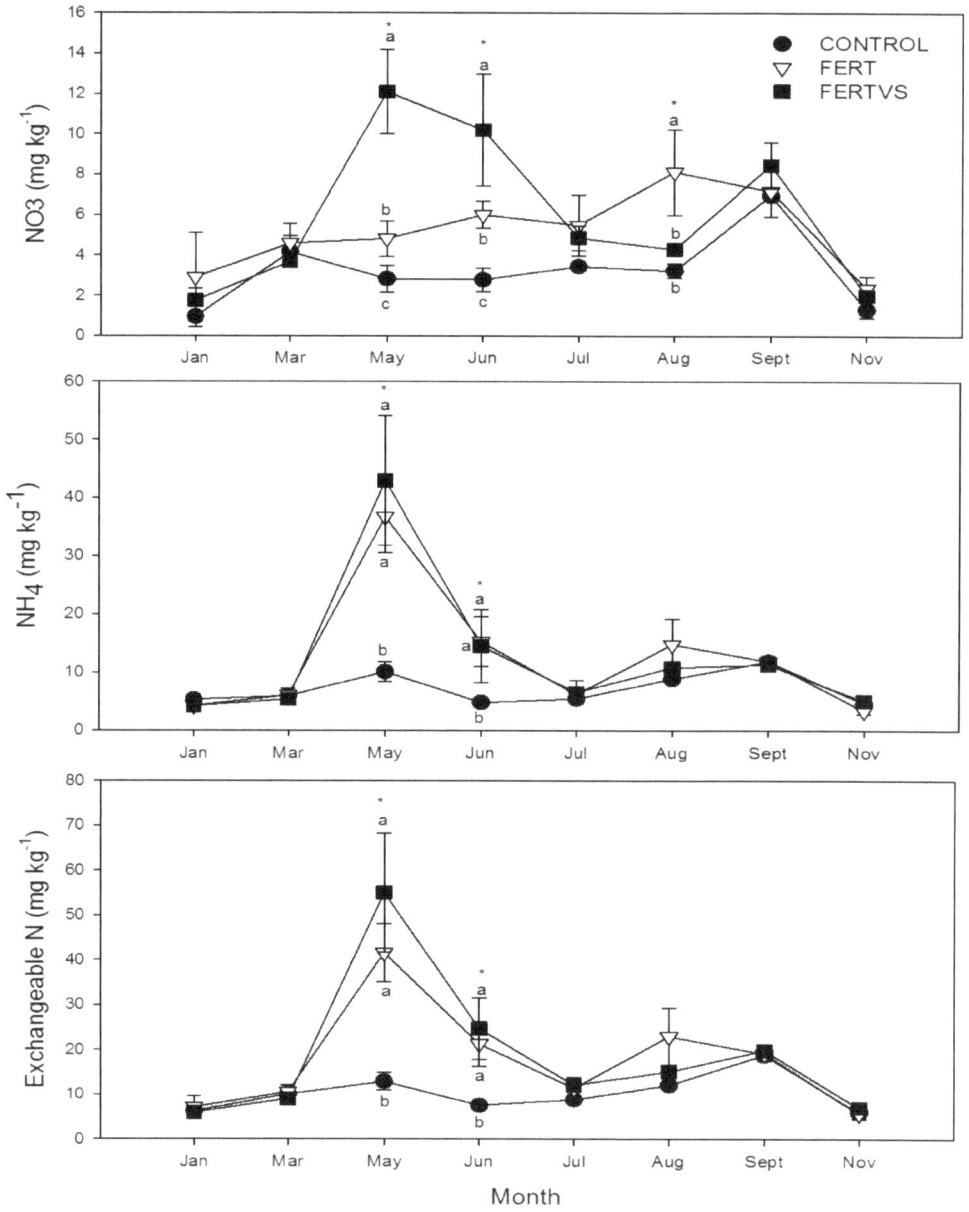

Figure 3. Monthly NO₃-N, NH₄-N, and total exchangeable N concentrations of a non-thinned mid-rotation loblolly pine plantation in 2004 at the Oakdale site in southwest Louisiana, USA, in response to no fertilization or herbicide (CONTROL), fertilization (FERT), and fertilization and vegetation suppression (FERTVS) treatments. For months headed by an asterisk, symbols noted by different letters differ at $P < 0.05$.

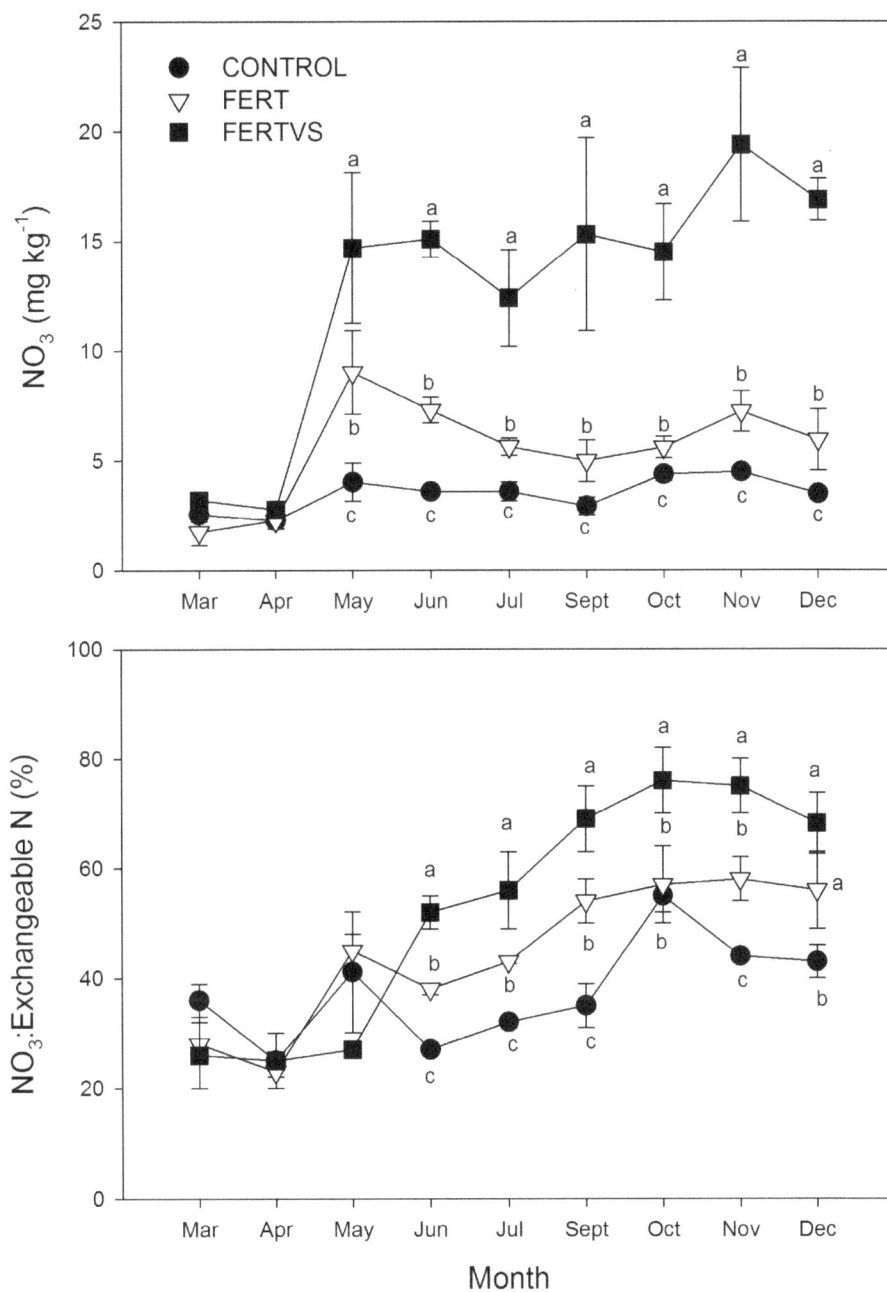

Figure 4. Monthly NO_3-N concentrations and ratios of NO_3 to total exchangeable N of a thinned mid-rotation loblolly pine plantation in 2005 at the Oakdale site in southwest Louisiana, USA, in response to no fertilization or herbicide (CONTROL), fertilization (FERT), and fertilization and vegetation suppression (FERTVS) treatments. For months headed by an asterisk, symbols noted by different letters differ at $P < 0.05$.

Table 2. Soil exchangeable N of thinned mid-rotation loblolly pine plantations in 2004 and 2005 in north central (Dodson, Lucky) and southwest Louisiana (Oakdale), USA in response to no fertilization or herbicide (CONTROL), fertilization (FERT), and fertilization and vegetation suppression (FERTVS) treatments. Means within a row followed by a different letter differ at $P < 0.05$; rows in which no letters are provided had a significant interaction between treatment and another parameter in analyses of variance. Values in parentheses are standard errors.

Exchangeable N	Year	Site	Treatment		
			CONTROL	FERT	FERTVS
NO$_3$-N (mg·kg^{-1})	2004	Dodson	3.4 (0.6) *a*	3.7 (0.7) *a*	3.9 (0.8) *a*
		Lucky	2.5 (0.3) *a*	2.9 (0.5) *a*	2.5 (0.3) *a*
		Oakdale	3.2 (0.4)	5.2 (0.6)	5.7 (0.8)
	2005	Dodson	3.3 (0.2) *c*	4.4 (0.4) *b*	8.4 (0.8) *a*
		Lucky	2.9 (0.1) *b*	4.2 (0.4) *a*	5.6 (0.6) *a*
		Oakdale	10.0 (0.6)	12.5 (1.0)	22.2 (1.8)
NH$_4$-N (mg·kg^{-1})	2004	Dodson	5.4 (0.6) *a*	7.3 (1.3) *a*	7.1 (1.3) *a*
		Lucky	6.5 (1.5)	16.5 (6.9)	11.0 (3.0)
		Oakdale	7.3 (0.7)	11.9 (2.2)	11.1 (2.5)
	2005	Dodson	7.6 (1.0) *b*	11.5 (2.5) *ab*	12.7 (1.5) *a*
		Lucky	5.6 (0.6)	8.3 (1.3)	8.5 (1.4)
		Oakdale	6.6 (0.6) *c*	7.0 (0.8) *b*	9.5 (1.1) *a*
Total exchangeable N (mg·kg^{-1})	2004	Dodson	8.8 (1.0) *a*	10.9 (1.7) *a*	11.0 (1.8) *a*
		Lucky	9.0 (1.7)	19.4 (7.0)	13.4 (3.1)
		Oakdale	10.5 (1.0)	17.1 (2.4)	16.8 (3.1)
	2005	Dodson	10.9 (1.0) *b*	15.9 (2.7) *ab*	21.0 (1.8) *a*
		Lucky	8.6 (0.6)	12.5 (1.4)	14.1 (1.7)
		Oakdale	10.0 (0.7) *c*	12.5 (1.0) *b*	22.2 (1.8) *a*
NO$_3$: total exchangeable N (%)	2004	Dodson	37.9 (2.9) *a*	35.6 (2.9) *a*	35.7 (2.8) *a*
		Lucky	33.9 (2.2) *a*	28.4 (3.9) *a*	27.3 (3.0) *a*
		Oakdale	28.8 (2.3) *a*	34.9 (2.8) *a*	36.3 (2.9) *a*
	2005	Dodson	35.5 (2.6)	36.3 (2.6)	42.5 (3.2)
		Lucky	37.0 (2.4) *b*	37.5 (2.1) *b*	43.3 (2.4) *a*
		Oakdale	35.8 (1.9)	44.2 (2.9)	54.8 (3.8)

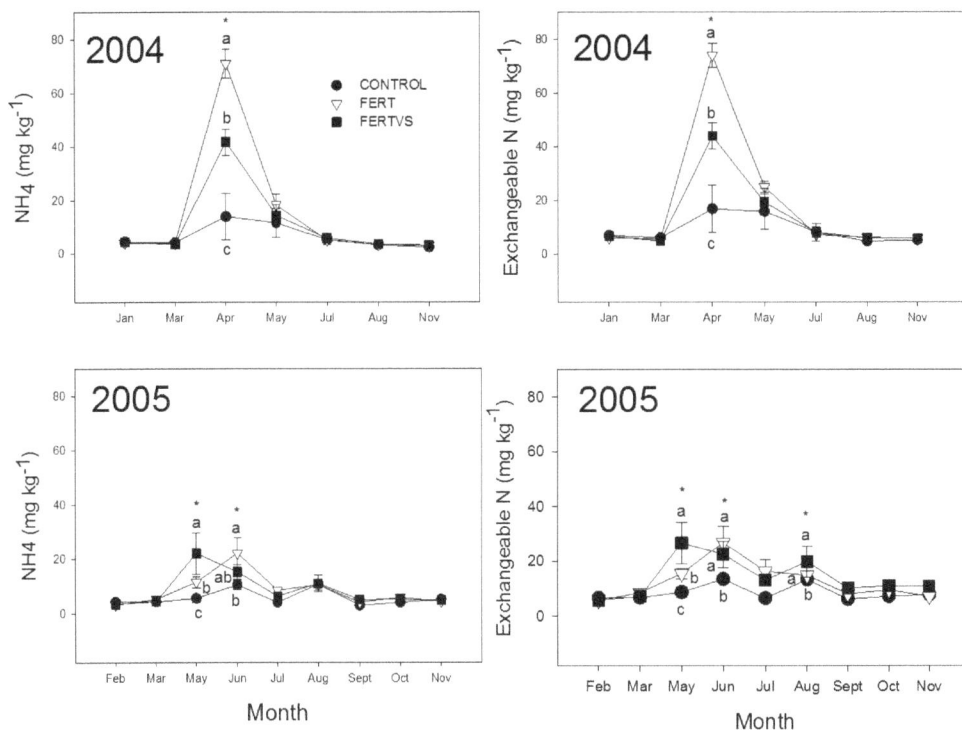

Figure 5. Monthly NH4-N and total exchangeable N concentrations of a mid-rotation loblolly pine plantation at the Lucky site in north central Louisiana, USA, that was thinned in November 2004 in response to no fertilization or herbicide (CONTROL), fertilization (FERT), and fertilization and vegetation suppression (FERTVS) treatments. For months headed by an asterisk, symbols noted by different letters differ at $P < 0.05$.

Soil NH4-N concentrations differed among treatments at all sites in 2005. The treatment effect was significant for the Dodson ($P = 0.003$) and Oakdale ($P = 0.03$) sites (Table 2), and the treatment × date interaction was significant ($P = 0.03$) for the Lucky site (Figure 5). At the Dodson site, the FERTVS treatment had greater NH4-N concentrations than the CONTROL treatment (Table 2). At the Oakdale site, NH4-N concentrations of the FERTVS treatment were higher than those of the other two treatments and NH4-N of the FERT treatment greater than that of the CONTROL treatment (Table 2). This treatment ranking was also observed for the Lucky site in May, and in June the FERTVS treatment had greater NH4-N than the CONTROL treatment (Figure 5).

Significant treatment × date interactions were observed in analyses of total exchangeable N for the Lucky ($P < 0.0001$) and Oakdale ($P < 0.0001$) sites for 2004 (Figures 3 and 5). At the Lucky site, treatments differed as described above for NH4-N of 2004 (Figure 5). Differences among treatments were similarly identical to those described above for NH4-N of 2004 at the Oakdale site (Figure 3).

For the Dodson ($P = 0.0002$) and Oakdale ($P < 0.0001$) sites, significant treatment effects were found in analyses of total exchangeable N for 2005 (Table 2). Total exchangeable N differences among treatments at both sites were identical to those described above for NH4-N. At the Lucky site there was a significant ($P = 0.03$) treatment × date interaction (Figure 5). In May the FERTVS treatment had greater total exchangeable N than the CONTROL and FERT treatments, and total exchangeable N of the FERT treatment exceeded that of the CONTROL treatment. In June and August the FERT and FERTVS treatments had greater total exchangeable N than the CONTROL treatment.

All exchangeable N variables differed between years in the CONTROL treatment (Table 3). Soil NO_3-N concentrations of 2005 were greater ($P = 0.002$) than those of 2004 across all sites. Sites differed ($P = 0.03$) in NO_3-N concentrations across years in the CONTROL treatment as well, with the Oakdale site (mean 5.6 mg·kg^{-1}) having greater NO_3-N concentrations than the Lucky site (mean 3.4 mg·kg^{-1}). Total exchangeable N in 2005 was greater ($P = 0.005$) than that of 2004 across all sites (Table 3). The Dodson (mean 14.1 mg·kg^{-1}) and Oakdale (mean 15.1 mg·kg^{-1}) sites had total exchangeable N concentrations across years greater ($P = 0.04$) than those of the Lucky site (mean 11.1 mg·kg^{-1}). The site × year interaction was significant in analyses of NH_4-N ($P = 0.0001$) and the NO_3: exchangeable N ratio ($P = 0.002$). At the Dodson site NH_4-N concentrations of 2005 were greater than those of 2004, but an opposite trend was observed for NH_4-N of the Oakdale site, with concentrations of 2004 exceeding those of 2005. The NO_3:exchangeable N ratio was different at only the Oakdale site between years in the CONTROL treatment. At that site, the ratio was greater for 2005 than 2004.

Table 3. Exchangeable N in 2004 and 2005 of mid-rotation loblolly pine plantations in north central (Dodson, Lucky) and southwest Louisiana (Oakdale) in CONTROL treatment of no fertilizer or herbicide. Plantations were thinned in October and November 2004. For each site, means within a row followed by a different letter differ at $P < 0.05$. All values except NO_3:exchangeable N ratio are in mg·kg^{-1}.

	Site					
	Dodson		**Lucky**		**Oakdale**	
	2004	**2005**	**2004**	**2005**	**2004**	**2005**
NO_3-N	3.9 b	5.3 a	2.5 b	4.2 a	4.1 b	7.2 a
NH_4-N	6.5 b	12.4 a	6.7 a	8.5 a	11.0 a	7.6 b
Total exchangeable N	10.6 b	17.5 a	9.3 b	12.8 a	15.0 b	15.1 a
NO_3:exchangeable N	0.40 a	0.36 a	0.33 a	0.37 a	0.34 b	0.47 a

3.3. Soil Microbial Biomass C, Dehydrogenase Activity, and Labile C

Overall, soil microbial biomass C and dehydrogenase activity were relatively unaffected by site, treatment, or year. In only April 2004 ($P = 0.0004$), the FERT (mean 443.5 mg·kg^{-1}) and FERTVS (mean 464.8 mg·kg^{-1}) treatments had greater C_{mic} than the CONTROL treatment (mean 373.1 mg·kg^{-1}). The FERT (mean 30.1 mg·kg^{-1}) and FERTVS (mean 37.8 mg·kg^{-1}) treatments had greater ($P = 0.003$) dehydrogenase activity than the CONTROL treatment (mean 13.6 mg·kg^{-1}) in only April 2004 as well. Monthly average C_{mic} ranged from 126.9 to 460.8 mg·kg^{-1} over the years of the study, and monthly average dehydrogenase activity ranged from 13.4 to 77.1 mg·kg^{-1}.

The site $\times$ treatment interaction was significant ($P = 0.02$) for the labile C analysis (Table 4). The CONTROL treatment had 6 and 10% greater labile C than the FERTVS treatment at the Dodson and Lucky sites, respectively. At the Oakdale site the FERTVS soils had about 6% greater labile C than the CONTROL and FERT soils.

Table 4. Labile C (mg·kg^{-1}) of thinned mid-rotation loblolly pine plantations during 2005 in north central (Dodson, Lucky) and southwest Louisiana (Oakdale), USA in response to no fertilization or herbicide (CONTROL), fertilization (FERT), and fertilization and vegetation suppression (FERTVS) treatments. Means within a row followed by a different letter differ at $P < 0.05$; values in parentheses are standard errors.

Site	CONTROL	FERT	FERTVS
Dodson	6.23 (0.07) *a*	6.18 (0.31) *ab*	5.88 (0.05) *b*
Lucky	6.33 (0.09) *a*	6.16 (0.14) *ab*	5.78 (0.13) *b*
Oakdale	5.97 (0.14) *b*	5.89 (0.08) *b*	6.30 (0.05) *a*

3.4. Soil C and N

No differences were observed among treatments in soil C, N, and C:N ratios, but differences ($P = 0.04, 0.0002, <0.0001$, respectively) were observed among sites for each of these parameters (Table 5). The Dodson site had higher C concentrations than the Oakdale site. The Oakdale site had greater N concentrations than both other sites, and the Dodson site had greater N concentrations than the Lucky site. The Lucky site had the highest C:N ratio of the sites, while the Oakdale site had the lowest C:N ratio.

Table 5. Soil C and N concentrations and C:N ratios of thinned mid-rotation loblolly pine plantations in 2005 in north central (Dodson, Lucky) and southwest Louisiana (Oakdale), USA. Means within a row differ at $P < 0.05$; values in parentheses are standard errors.

Soil Attribute	Site		
	Dodson	Lucky	Oakdale
C (mg·kg^{-1})	13,138 (676) *a*	12,153 (813) *ab*	10,554 (676) *b*
N (mg·kg^{-1})	557 (23) *b*	465 (29) *c*	658 (23) *a*
C:N ratio	23.5 (0.66) *b*	26.5 (0.79) *a*	16.1 (0.66) *c*

3.5. Correlations

Monthly N$_{min}$ was significantly positively correlated ($r = 0.18$, $P = 0.03$) with soil temperature. Soil N$_{min}$, C$_{mic}$, and dehydrogenase activity were uncorrelated. Microbial biomass C was significantly positively correlated ($r = 0.21$, $P < 0.0001$) with VWC and negatively correlated ($r = -0.21$, $P < 0.0001$) with soil temperature. Dehydrogenase activity was positively correlated with soil temperature ($r = 0.33$, $P < 0.0001$) and C$_{mic}$ ($r = 0.19$, $P = 0.002$).

4. Discussion

This study yielded N_{min} values within the range reported in previous studies of loblolly pine that used similar buried bag procedures. The 81 kg·ha^{-1} annual N_{min} mean of the CONTROL treatment in this study was similar to the 75 kg·ha^{-1} annual N_{min} reported for a 7-year-old loblolly pine plantation in Florida, USA on sandy loam soil [33]. The 81 to 146 kg·ha^{-1} annual N_{min} range of this study was somewhat higher than the 22 to 96 kg·ha^{-1} annual N_{min} range for a 14-year-old loblolly pine plantation in North Carolina, USA on a highly weathered, eroded clay loam receiving fertilizer and herbicide [2]. Across sites and years, annual N_{min} in this study was 6%, 8%, and 10% of total N measured in spring 2005 for the CONTROL, FERT, and FERTVS treatments, respectively. These proportions of annual N_{min} to total N are slightly higher than the 2%, 3%, and 6% for control, fertilizer, and fertilizer with herbicide treatments, respectively, conducted in a non-thinned 14-year-old loblolly pine plantation in North Carolina, USA [2]. However, proportions of annual N_{min} to total N of pre-thin CONTROL, FERT, and FERTVS treatments in this study were closer to those of the North Carolina loblolly pine plantation [2], respectively averaging 4%, 6%, and 8% across sites.

Annual N_{min} was increased by fertilization and to a greater extent by the combination of fertilization and vegetation control irrespective of timing relative to thinning and across a gradient of soil conditions. These results suggest that annual N supply can be increased by fertilization in the year of application in mid-rotation loblolly pine plantations prior to or after thinning. Increases in N_{min} after forest fertilization are well-demonstrated [2,9,10,12], but this study is the first to the authors' knowledge to show increases in N_{min} irrespective of timing relative to thinning. Similar to this study, combination treatments of fertilization and vegetation suppression increased annual N_{min} more than fertilization alone when conducted in an unthinned mid-rotation loblolly pine plantation with basal area comparable with that of the stands prior to thinning in this study [2].

The increased annual N_{min} in response to FERT and FERTVS treatments suggested that a priming effect occurred, but not all mechanisms conventionally defined for that effect occurred. The priming effect for N has been defined as extra release of soil-derived N in response to substances added to the soil [34]. The increased annual N_{min} of the FERT and FERTVS treatments are consistent with this definition. Soil microbes are often directly or indirectly responsible for priming effects, and increases in N_{min} after fertilization have been attributed to reduced microbial N immobilization when soil C:N ratios are below 30:1 [9,34]. All sites in this study had C:N ratios below 30:1, but the soil microbial parameters were relatively unaffected by treatments. The soil C_{mic} and dehydrogenase activity data was variable, so the procedures conducted may not have been sensitive enough to detect changes in soil microbial biomass and activity. There are also priming effect pathways that may function independent of soil microbes [34]. Labile C has been identified as an important parameter affecting the priming effect, particularly in response to forest management treatments such as vegetation control that alter organic matter inputs to soil. Such treatments can reduce labile C, which in turn reduces microbial N immobilization and increases N_{min} [35]. The greater annual N_{min} of the FERTVS treatment relative to the FERT and CONTROL treatments is consistent with increasing N_{min} in response to reducing vegetation C inputs. Labile C was lower for the FERTVS treatment relative to the CONTROL treatment at the Dodson and Lucky sites, but an opposite trend was observed at the Oakdale site. The contrasting trend at Oakdale may have been influenced by its under- and mid-story vegetation. Oakdale had sweetgum (*Liquidambar styraciflua* L.) greater than 8 cm in diameter as part of the mid-story, and it was resistant to the basal spray of triclopyr used in this study. Multiple applications of triclopyr were required to suppress sweetgum trees of that diameter. Oakdale also had abundant Chinese tallow tree that was resistant to triclopyr basal spray, requiring multiple applications. Chinese tallow tree sprouting also occurred once the mid-story was reduced, and the sprouts were relatively resistant to broadcast glyphosate and directed spray of triclopyr [7]. This herbicide resistance made suppression take several months, which may have affected labile C inputs.

As with fertilization and fertilization with vegetation control, thinning impacted annual N_{min} across all site conditions. Thinning nearly doubled annual N_{min} across all sites and treatments, although this comparison was made with non-thinned stands receiving higher than average annual precipitation and thinned stands that received lower than average annual precipitation. Studies of loblolly pine harvesting have demonstrated N_{min} increases after clearcutting, with N_{min} increases attributed to increased soil temperatures after harvest as well as mixing of organic matter with mineral soil during harvest [36,37]. Recently thinning was shown to increase N_{min} of *Cryptomeria japonica* forests, with N_{min} increases related to increases in soil temperature [38]. This study similarly observed that soil temperature was significantly increased by 2–3 °C after thinning across all sites and treatments. Soil temperature was also positively correlated with N_{min}.

Although annual N_{min} trends were similar across years and sites, monthly patterns in N_{min} were more affected by site conditions. Monthly N_{min} was relatively variable, which made differences among treatments, if present, difficult to detect. When fertilization was done prior to thinning, only at the Dodson site was a difference in average monthly N_{min} observed. At Dodson, a difference in monthly N_{min} was found only with the FERTVS treatment. Dodson had the highest understory biomass among all sites prior to thinning [7], which may have tempered the effect of the FERT treatment on monthly N_{min}. Average monthly N_{min} was increased by fertilization at both the Dodson and Oakdale sites. At each site average monthly N_{min} of the FERT and FERTVS treatments were similar; vegetation control may not have been as necessary in the year of thinning to foster increases in monthly N_{min}.

Monthly exchangeable N (NH_4-N, NO_3-N, total exchangeable N) trends were affected by timing of treatments and site conditions. The dominant form of exchangeable N was NH_4-N, so total exchangeable N trends were nearly identical to NH_4-N trends. Loblolly pine plantation soils are typically characterized by dominance of NH_4-N as exchangeable N [39]. Fertilization conducted prior to thinning had little effect on NH_4-N and total exchangeable N, with increases occurring only in the two months after fertilization at the two sites with lower soil N, Dodson and Lucky. When fertilization was done after thinning monthly NH_4-N and total exchangeable N concentrations increased after fertilization at all sites, but responses were different at each site. At the Lucky site, NH_4-N and total exchangeable N increases occurred only in the two months after fertilization. At Dodson, fertilization after thinning consistently increased monthly NH_4-N and total exchangeable N, but only when combined with vegetation suppression. Both fertilization treatments increased monthly NH_4-N at the Oakdale site, with greater increases occurring when combined with vegetation suppression. This gradient in NH₄-N and total exchangeable N increases across the sites was likely related to the gradient in total soil N and C:N ratios of the sites. Total soil N affects N transformations in soil, with transformations occurring more readily in soils with higher total soil N [40]. Nitrogen transformations in soil are also affected by C:N ratios, with transformations occurring more readily in soils with lower C:N ratios [40].

Monthly NO₃-N trends were affected by timing of fertilization and site conditions. Changes in monthly NO₃-N were relatively low when fertilization was conducted prior to thinning. Only at the Oakdale site were monthly NO₃-N concentrations elevated by fertilizer treatments, and the duration of response was the four-month period after application. As mentioned above, the higher soil N and lower C:N ratio of the Oakdale site likely made it more prone to changes in N. Soil NO₃-N concentrations similarly increased in response to fertilization and fertilization and vegetation control in an unthinned mid-rotation loblolly pine plantation in North Carolina, USA [39]. Average monthly NO₃-N was changed by fertilization at all sites when conducted post-thinning. Both fertilization treatments increased NO₃-N at the Lucky site, and the FERTVS treatment increased monthly NO₃-N more than the FERT treatment at the Dodson and Oakdale sites.

The increases in NO₃-N and the ratio of NO₃:total exchangeable N have implications for how the timing of fertilization matched site N demand. Increases in soil NO₃-N concentrations after fertilization can be indicative of low plant uptake of applied N [41]. Increases in the proportion of NO₃-N in the total exchangeable N pool in soil are also an index of N saturation [12]. The increases in NO₃-N and NO₃:exchangeable N ratios at the Oakdale site when fertilized before and after thinning suggest that soil N was sufficient for the site to meet stand N demand, particularly after thinning. Foliage nutrient testing determined that foliage N concentrations at the Oakdale site were above critical values [7]. Soil N of the Oakdale site was likely enhanced by the presence of wax myrtle, a N-fixing shrub, in its understory [7]. Wax myrtle added 1.9 kg $N \cdot ha^{-1} \cdot year^{-1}$ in a slash pine (*Pinus elliotti* Engelm. var. *elliotti*) plantation in Florida, USA [42]. The increases in soil NO₃-N and the NO₃:exchangeable N ratio at the Lucky site and the increases in soil NO₃-N at the Dodson in response to FERTVS treatment conducted post-thinning suggest that this treatment exceeded stand N demand. These results show that avoiding fertilization pre- or post-thinning at an N-sufficient site and applying herbicide in the year of thinning and fertilization at the range of site conditions in this study could minimize NO₃-N concentration increases, which can help in reducing N leaching loss potential and improving fertilizer use efficiency [33].

5. Conclusions

For the range of site conditions in this study, fertilizing at mid-rotation stimulated increases in annual soil N supply irrespective of timing relative to thinning and the effect was more pronounced when vegetation control was conducted in concert with fertilization. This finding suggests management flexibility in the timing of fertilization at mid-rotation to increase soil N supply. However, site conditions were important considerations for minimizing the potential for NO₃-N leaching and matching stand N demand. The site with the highest total soil N and the lowest C:N ratio was more prone to NO₃-N increases after fertilization whether done prior to or after thinning. At all sites, combining vegetation suppression with fertilization promoted increases in NO₃-N when done after thinning. As such, combining vegetation suppression with fertilization in the year of thinning may have increased soil N supply to levels that exceeded stand N demand.

Acknowledgments

The authors gratefully acknowledge funding from the USDA Agenda 2020 program and the USDA Forest Service, Southern Research Station, the cooperation of Weyerhaeuser NR, International Paper, and Boise, and all research technicians and students that assisted in field and laboratory to make this project possible.

Author Contributions

Michael Blazier led the writing, data analyses, and co-conceived the development of the project; Andrew Scott co-conceived the development of the project, collaborated on the analyses and interpretation of the data and writing; Ryan Coleman conducted much of the field and laboratory procedures and collaborated on the writing and data analyses.

Conflicts of Interest

The authors have no conflict of interest to declare.

References

1. Allen, H.L.; Dougherty, P.M.; Campbell, R.G. Manipulation of water and nutrients: Practice and opportunity in southern U.S. pine forests. *Forest Ecol. Manag.* **1990**, *30*, 437–453.
2. Gurlevik, N.; Kelting, D.L.; Allen, H.L. Nitrogen mineralization following vegetation control and fertilization in a 14-year-old loblolly pine plantation. *Soil Sci. Soc. Am. J.* **2004**, *68*, 272–281.
3. Dickens, E.D.; Moorhead, D.J.; McElvany, B. Pine plantation fertilization. *Better Crops* **2003**, *87*, 12–15.
4. Jokela, E.J. Nutrient management of southern pines. In *Slash Pine: Still Growing and Growing*; General Technical Report SRS-6; Dickens, E.D., Barnett, J.P., Hubbard, W.G., Jokela, E.J., Eds.; USDA Forest Service, Southern Research Station: Asheville, NC, USA, 2004; pp. 27–35.
5. Albaugh, T.J.; Allen, H.L.; Fox, T.R. Historical patterns of forest fertilization in the southeastern United States from 1969 to 2004. *South. J. Appl. For.* **2007**, *31*, 129–137.
6. Zerpa, J.L.; Fox, T.R. Controls of volatile ammonia losses from loblolly pine plantations fertilized with urea in the Southeast USA. *Soil Sci. Soc. Am. J.* **2011**, *75*, 257–266.
7. Blazier, M.A.; Scott, D.A. Nitrogen distribution within the soil-plant-microbial system in response to pre-thinning fertilization treatments in Louisiana. In Proceedings of the 13th Biennial Southern Silvicultural Research Conference, Memphis, TN, USA, 28 February–4 March 2005; Conner, K.F., Ed.; General Technical Report SRS-92; USDA Forest Service, Southern Research Station: Asheville, NC, USA, 2006; pp. 129–134.

8. Jokela, E.J.; Dougherty, P.M.; Martin, T.A. Production dynamics of intensively managed loblolly pine stands in the southern United States: A synthesis of seven long-term experiments. *For. Ecol. Manag.* **2004**, *192*, 117–130.

9. Scott, D.A.; Burger, J.A.; Kaczamarek, D.J.; Kane, M.B. Nitrogen supply and demand in short-rotation sweetgum plantations. *For. Ecol. Manag.* **2004**, *189*, 331–343.

10. Maimone, R.A.; Morris, L.A.; Fox, T.R. Soil nitrogen mineralization potential in a fertilized loblolly pine plantation. *Soil Sci. Soc. Am. J.* **1991**, *55*, 522–527.

11. Li, Q.; Allen, H.L.; Wilson, C.A. Nitrogen mineralization dynamics following the establishment of a loblolly pine plantation. *Can. J. For. Res.* **2003**, *33*, 364–374.

12. Fenn, M.E.; Poth, M.A.; Terry, J.D.; Blubaugh, T.J. Nitrogen mineralization and nitrification in a mixed-conifer forest in southern California: Controlling factors, fluxes, and nitrogen fertilization response at a high and low nitrogen deposition site. *Can. J. For. Res.* **2005**, *35*, 1464–1486.

13. Allen, H.L.; Schlesinger, W.H. Nutrient limitations to soil microbial biomass and activity in loblolly pine forests. *Soil Biol. Biochem.* **2004**, *36*, 581–589.

14. Soderstrom, B.; Baath, E.; Lundgren, B. Decrease in soil microbial activity and biomass owing to nitrogen amendments. *Can. J. Microbiol.* **1983**, *29*, 1500–1506.

15. Thirukkumaran, C.M.; Parkinson, D. Microbial respiration, biomass, metabolic quotient and litter decomposition in a lodgepole pine forest floor amended with nitrogen and phosphorus fertilizers. *For. Ecol. Manag.* **2002**, *159*, 187–201.

16. Blazier, M.A.; Hennessey, T.C.; Deng, S. Effects of fertilization and vegetation control on microbial biomass carbon and dehydrogenase activity in a juvenile loblolly pine plantation. *For. Sci.* **2005**, *51*, 449–459.

17. Bechtold, J.S.; Naiman, R.J. Soil texture and nitrogen mineralization potential across a riparian toposequence in a semi-arid savanna. *Soil Biol. Biochem.* **2006**, *38*, 1325–1333.

18. Kissel, D.E.; Cabrera, M.L.; Vaio, N.; Craig, J.R.; Rema, J.A.; Morris, L.A. Rainfall timing and ammonia loss from urea in a loblolly pine plantation. *Soil Sci. Soc. Am. J.* **2004**, *68*, 1744–1750.

19. Eno, C.F. Nitrate production in the field by incubating the soil in polyethylene bags. *Soil Sci. Soc. Am. Proc.* **1960**, *24*, 277–279.

20. Bremner, J.M. Inorganic forms of nitrogen. In *Methods of Soil Analysis, Part 2*; Black, C.A., Ed.; American Society of Agronomy: Madison, WI, USA, 1965; pp. 1179–1237.

21. Carlson, R.M. Automated separation and conductimetric determination of ammonia and dissolved carbon dioxide. *Anal. Chem.* **1978**, *50*, 1528–1531.

22. Hart, S.C.; Stark, J.M.; Davidson, E.A.; Firestone, M.K. Nitrogen mineralization, immobilization, and nitrification. In *Methods of Soil Analysis Part 2, Microbiological and Biochemical Properties*; Weaver, R.W., Angele, S., Bottomly, P., Eds.; Soil Science Society of America: Madison, WI, USA, 1994; pp. 985–1018.

23. Jenkinson, D.S.; Powlson, D.S. The effects of biocidal treatments on metabolism in soil-I. Fumigation with chloroform. *Soil Biol. Biochem.* **1976**, *8*, 167–177.

24. Jenkinson, D.S.; Powlson, D.S. The effects of biocidal treatments on metabolism in soil-V. A method for measuring soil biomass. *Soil Biol. Biochem.* **1976**, *8*, 209–213.

25. Sparling, G.P.; West, A.W.; Whale, K.N. Interference from plant roots in the estimation of soil microbial ATP, C, N, and P. *Soil Biol. Biochem.* **1985**, *17*, 275–278.

26. Jenkinson, D.S. Determination of microbial biomass carbon and nitrogen in soil. In *Advances in Nitrogen Cycling in Agricultural Ecosystems*; Wilson, J.R., Ed.; CAB International: Wallingford, UK, 1988; pp. 368–386.

27. Lenhard, G. The dehydrogenase activity in soil as a measure of the activity of soil microorganisms. *Z. Pflanzenernäh Düng Bodenkd* **1956**, *73*, 1–11.

28. Alef, K. Dehydrogenase activity. In *Methods in Applied Soil Microbiology and Biochemistry*; Alef, K., Nannipieri, P., Eds.; Academic Press: San Diego, CA, USA, 1995; pp. 228–231.

29. Tabatabai, M.A. Soil enzymes. In *Methods of Soil Analysis Part 2: Microbiological and Biochemical Properties*; Bigham, J.M., Ed.; Soil Science Society of America Inc.: Madison, WI, USA, 1994; pp. 775–833.

30. Camiña, F.; Trasar-Cepeda, C.; Gil-Sotres, F.; Leirós, C. Measurement of dehydrogenase activity in acid soils rich in organic matter. *Soil Biol. Biochem.* **1998**, *30*, 1005–1011.

31. Zou, X.M.; Ruan, H.H.; Fu, Y.; Yang, X.D.; Sha, L.Q. Estimating soil labile organic carbon and potential turnover rates using a sequential fumigation-incubation procedure. *Soil Biol. Biochem.* **2005**, *37*, 1923–1928.

32. Southern Climate Impacts Planning Program (SCIPP). Historical Climate Trends Tool. Available online: http://www.southernclimate.org/data.php (accessed on 30 December 2014).

33. Lee, K.; Jose, S. Nitrogen mineralization in short-rotation tree plantations along a soil nitrogen gradient. *Can. J. For. Res.* **2006**, *36*, 1236–1242.

34. Kuzyakov, Y.; Friedel, J.K.; Stahr, K. Review of mechanisms and quantification of priming effects. *Soil Biol. Biochem.* **2000**, *32*, 1485–1498.

35. Allen, H.L. *Thirteenth Annual Report*; North Carolina State University Forest Nutrition Cooperative: Raleigh, NC, USA, 2001; p. 8.

36. Vitousek, P.M.; Matson, P.A. Disturbance, nitrogen availability, and nitrogen losses in an intensively managed loblolly pine plantation. *Ecology* **1985**, *66*, 1360–1376.

37. Piatek, K.B.; Allen, H.L. Nitrogen mineralization in a pine plantation fifteen years after harvesting and site preparation. *Soil Sci. Soc. Am. J.* **1999**, *63*, 990–998.

38. Zhuang, S.; Wang, J.; Sun, X.; Wang, M. Effect of forest thinning on soil net nitrogen mineralization and nitrification in a *Crytomeria japonica* plantation in Taiwan. *J. For. Res.* **2014**, *25*, 571–578.

39. Gurlevik, N. Stand and Soil Responses of a Loblolly Pine Plantation to Midrotation Fertilization and Vegetation Control. Ph.D. Thesis, North Carolina State University, Raleigh, NC, USA, January 2002.

40. Knoepp, J.D.; Vose, J.M. Regulation of nitrogen mineralization and nitrification in Southern Appalachian ecosystems: Separating the relative importance of biotic *vs.* abiotic controls. *Pedobiologia* **2007**, *51*, 89–97.

41. Adeli, A.; Rowe, D.E.; Read, J.J. Effects of soil type on bermudagrass response to broiler litter application. *Agron. J.* **2006**, *98*, 148–155.
42. Permar, T.A.; Fisher, R.F. Nitrogen fixation and accretion by wax myrtle (*Myrica cerifera*) in slash pine (*Pinus elliotti*) plantations. *For. Ecol. Manag.* **1983**, *5*, 39–46.

Influence of Tree Spacing on Soil Nitrogen Mineralization and Availability in Hybrid Poplar Plantations

Yafei Yan, Shengzuo Fang, Ye Tian, Shiping Deng, Luozhong Tang and Dao Ngoc Chuong

Abstract: Nitrogen (N) availability and mineralization are key parameters and transformation processes that impact plant growth and forest productivity. We hypothesized that suitable plantation spacing can lead to enhanced soil N mineralization and nitrification, which in turn promote tree growth. Studies were conducted to evaluate seasonal patterns of soil inorganic N pools as well as rates of nitrification and N mineralization of three soil layers under four tree spacing treatments. Results showed tree spacing significantly affected annual net N mineralization, whereas inorganic N content in surface soils was significantly affected by tree spacing only during the growing season. The total annual cumulative net N mineralization ranged from 80.3–136.0 mg·kg^{-1} in the surface soils (0–20 cm), whereas the cumulative net N mineralization of 6 × 6 m and 4.5 × 8 m spacings was 65% and 24% higher than that of the 5 × 5 m, respectively. In general, tree spacing would affect N availability in soil by altering N mineralization rates, while high annual N mineralization was found in soils of low density plantations, with higher rates in square spacing than rectangular spacing. The obtained results suggest that suitable spacing could lead to enhanced N mineralization, but seasonal variation of soil N mineralization may not only be directly related to plantation productivity but also to understory vegetation productivity.

Reprinted from *Forests*. Cite as: Yan, Y.; Fang, S.; Tian, Y.; Deng, S.; Tang, L.; Chuong, D.N. Influence of Tree Spacing on Soil Nitrogen Mineralization and Availability in Hybrid Poplar Plantations. *Forests* **2015**, *6*, 636-649.

1. Introduction

Initial spacing is an important determinant of site utilization and subsequent selection and harvesting options [1]. Planting density and tree spacing not only strongly affect soil nutrient dynamics, aboveground biomass production, and crown characteristics [2–4] but also wood quality [5]. Previous researches suggested that height, diameter at breast height (DBH) and survival of planted trees were affected by density and spacing with DBH being the most sensitive and height being the least sensitive to variations in density [6,7]. Kang *et al.* [8] found that yield, wood quality, and pulp fiber properties of jack pine were improved through stand density regulation. However, recent forest management practices have been focusing mostly on issues surrounding biodiversity and nutrient status [9]. Changes of forest structure can lead to different forest environments and understory vegetation. Some studies suggested that a suitable spacing treatment of various plantations such as red pine and spruce was vital in optimizing nutrient dynamics such as the nitrogen (N) mineralization process in support of optimal plant growth [10]. Therefore, understanding impact of tree spacing treatment on soil nutrient dynamics is important in developing effective forest management practices for sustainable production.

Of nutrients essential for plant growth, N is frequently the most limiting in terrestrial ecosystems [11,12]. Nitrogen availability and mineralization are key parameters and transformation processes that impact plant growth and forest productivity. Mineralization of soil organic N has been shown to be fundamentally linked with productivity in most ecosystems [13–15]. With more than 90% of soil N in organic forms, N availability to plants and soil inorganic N contents, such as NO_3^- and NH_4^+ are largely governed by mineralization processes [16]. Some of the released NH_4^+ may be oxidized to NO_3^- through well recognized nitrification processes which may result in N loss from the ecosystem if plants cannot take up the formed NO_3^- timely. Therefore, both nitrification and N mineralization are key processes influencing N availability in soils, which are in turn being influenced by substrate quality and accessibility, as well as abiotic factors such as soil moisture, temperature, and pH [17,18]. Although N mineralization and nitrification processes and their governing factors have been well evaluated and documented, understanding their impact to forest ecosystem requires targeted effort that is tailored to meet the needs of specific terrestrial ecosystems under evaluation.

In recent years, there has been growing interest in poplar production in China, due to its ability to adapt to a wide range of growth environments, its responsiveness to tree breeding and silviculture, as well as its high timber yield. In the past decades, more than 7.0 million ha poplar plantations have been established in China that serve as wood resources [19]. Such practices have not only promoted economic development in the region, but have also facilitated ecosystem improvements. Plantation density and spacing affect the growing resources available to each tree and the size and form of logs available at harvest [20]. However, long-term effects of planting spacing and density management on soil nutrient dynamics remain unclear and deserve attention for the practices to be sustainable. It is generally recognized that improved synchrony between N supply and demand will increase N use efficiency and decrease offsite N transport from agricultural systems. In this study, we hypothesized that suitable stand structure (such as density and spacing) could lead to enhanced nitrification and N mineralization in soils, which would result in increased N availability and improved plant growth. The main objective of this study was to investigate the effects of planting density and tree spacing on soil N transformation and availability in the poplar plantations.

2. Materials and Methods

2.1. Site Description and Plantation Establishment

The experiment was conducted at Chenwei forest farm located in Jiangsu, China ($33°16'$ N, $118°21'$ E). This area is under a semi-humid region in mid-latitude warm zones with long-periods of sunshine duration and plentiful rainfall. The mean annual precipitation is 972.5 mm, occurring mostly from June to August. The mean annual sunshine duration is 2250–2350 h and mean annual temperature is 14.4 °C which varies from −7 °C in January to 28 °C in July. The soil is a Gley soil with clay loam texture derived from lacustrine sediments. The main understory plant species includes *Echinochloa crusgalli*, *Youngia japonica*, *Geranium wilfordii*, *Duchesnea indica*, and *Herba Cirsii*.

Poplar plantations were established in March 2007 by using one-year old seedlings of clone "Nanlin-95", a hybrid of clone I-69 (*Populus deltoides* Bartr. cv. "Lux") × clone I-45 (*P.* ×

euramericana (Dode) Guineir cv. I-45/51′) bred by Nanjing Forestry University, China. Four levels of spacing (6 × 6, 4.5 × 8, 5 × 5, and 3 × 8 m), including two rectangles and two squares, were evaluated. Spacing 4.5 × 8 and 6 × 6 m were of the same density (278 stems·ha^{-1}), while spacings of 3 × 8 and 5 × 5 m were also regarded as of the same density (400 stems·ha^{-1}). Three replicates of each treatment were randomly arranged in 12 plots which were established at the same topographic position with each plot about 1200–1800 m^2 (50 trees per plot).

Five years following the plantation establishment, three soil subsamples and each at three layers (0–5, 5–10, and 10–20 cm) were taken from each plot in May 2012. At each plot, three sampling points were randomly selected 2.5–4 m from the base of a tree trunk to minimize the direct physical and chemical influences of roots [21]. A total of 108 soil samples at a time were collected from the four treatments. The obtained basic properties of soils for each treatment are shown in Table 1.

Table 1. Basic properties of soils in the poplar plantation following five years of tree spacing treatment *.

Spacing (m)	Soil Layer (cm)	SOC (g·kg^{-1})	TN (g·kg^{-1})	SOC:TN	pH	Bulk Density (g·cm^{-1})
	0–5	31.4	1.1	16.4	7.1	1.10
6 × 6	5–10	21.2	0.6	22.3	7.2	1.19
	10–20	13.9	0.6	14.2	7.4	1.28
	0–5	26.1	0.9	16.1	7.0	1.13
4.5 × 8	5–10	16.4	0.5	19.5	7.0	1.21
	10–20	12.0	0.4	16.2	7.3	1.35
	0–5	23.9	0.9	15.6	7.1	1.19
5 × 5	5–10	18.1	0.5	22.8	7.3	1.24
	10–20	13.5	0.5	15.9	7.4	1.34
	0–5	22.9	1.1	12.7	7.0	1.22
3 × 8	5–10	14.2	0.6	13.9	7.1	1.36
	10–20	11.0	0.5	13.6	7.3	1.39

* SOC = Soil organic carbon and TN = Total nitrogen; pH was determined in soil:water = 1:2.5.

2.2. Soil N Mineralization and Sample Collection

Soil N mineralization was determined using the resin core method and through *in situ* soil incubation [22,23] based on the assumption that N mineralization rates are similar inside and outside a sample core. Briefly, soil cores were prepared by hammering sharp-edged PVC tubes into soil layers 0–5, 5–10, and 10–20 cm. The PVC tubes were internal diameter (ID) × height (H) = 5 × 5 cm for soil layers 0–5 and 5–10 cm, and ID × H = 5 × 10 cm for soil layer 10–20 cm. Resin layers (mixture of Dowex 50W-X8 cation and Dowex l-X8 anion exchange resins) were placed above and below the soil cores using separate PVC tubes (ID × H = 5 × 2 cm). The resin tube was covered with a 0.8 mm pore-diameter nylon mesh on top to allow gas, water and nutrients exchange, and mesh with a 0.3 mm pore-diameter nylon mesh on bottom to prevent resin loss during incubation. Nitrogen input from rainfall and litter decomposition could be absorbed by the above resin layer, while N leached from the soil core could be captured by the resin layer below.

Three sub-plots were established randomly within each plot for resin-core incubation. On 10 May 2012, following clipping and removing plants and litter, sharp-edged PVC tubes were driven into three soil layers (0–5, 5–10, and 10–20 cm) to sample soil cores and make resin-soil core combinations within each sub-plot. The resin-soil core combinations were then re-buried and incubated *in situ*. After 3 months, the resin-soil core combinations were retrieved for chemical analysis, and additional new resin-soil core combinations were prepared and incubated again for the next 3-months. The procedure was repeated at a 3-month interval (from 10 May to 10 August 2012 (A); from 10 August to 10 November 2012 (B); from 10 November 2012 to 10 February 2013 (C); and from 10 February to 10 May 2013 (D)) continuously over the following year. Additional soil cores were sampled in adjacent locations simultaneously at each time for the quantification of soil initial ammonium and nitrate-N contents. The collected soil and tube samples were transported to the laboratory and stored at 4 °C until analysis.

2.3. Soil Chemical Analysis and Data Calculation

Soil ammonium and nitrate-N levels were determined by adding 2 M KCl [23], followed by shaking for an hour on a reciprocal shaker and filtration with Whatman-Xinhua filter paper before quantifying NH_4^+-N and NO_3^--N using a Flow Injection Autoanalyzer (Bran+Luebbe, Hamburg, Germany). The resin bags were washed with distilled water and dried at room temperature (28–32 °C). Resin NH_4^+-N and NO_3^--N were removed by shaking the resin bags with 2 M KCl for 12 h.

Soil moisture content was determined by oven drying at 105 °C for 48 h. Soil pH was measured in a soil-water suspension (soil:water = 1:2.5) using an automatic titrator (SH/T0983, Changsha, China). Soil bulk density was determined using the core method [24] using three undisturbed soil cores per plot, collected randomly from each soil layer using cylindrical cores. Total N was determined by wet-digestion, following by quantification using a Flow Injection Auto-analyzer (Bran+Luebbe, Hamburg, Germany).

Soil nitrification and N mineralization were estimated based on N levels in the resin and in soils before and after incubation. Net N mineralization rates during each incubation period were calculated from differences of inorganic N concentrations between initial and incubated samples. Similarly, net N nitrification rates were calculated from differences of NO_3^--N concentrations between initial and incubated samples. Net N mineralization and nitrification rates during each incubation period were calculated as follows:

$$R_M = (T_{m1} - T_{m0}) \cdot t^{-1}, R_N = (T_{n1} - T_{n0}) \cdot t^{-1} \tag{1}$$

$$T_{m1} = S_{m1} + R_{m1}, T_{n1} = S_{n1} + R_{n1} \tag{2}$$

where R_M ($mg \cdot kg^{-1} \cdot day^{-1}$) is the net N mineralization rate, and R_N ($mg \cdot kg^{-1} \cdot day^{-1}$) is the net nitrification rate. T_{m1} and T_{m0} ($mg \cdot kg^{-1}$) represent total inorganic N (sum of NH_4^+-N and NO_3^--N) concentrations after and before incubation, respectively; T_{n1} and T_{n0} ($mg \cdot kg^{-1}$) represent total NO_3^--N concentrations after and before incubation; and t is the number of incubation days. S_{m1} and

R_{m1} (mg·kg^{-1}) are the total inorganic N (NH$_4^+$-N and NO$_3^-$-N) concentrations in the soil and resin samples; while S_{n1} and R_{n1} (mg·kg^{-1}) are the NO$_3^-$-N concentrations in the soil and resin samples.

Annual net N mineralization and nitrification were estimated by summing the results from four incubation periods in the year [25]. Nitrogen content in the experiment area (kg N·ha^{-1}) was estimated based on mean inorganic N concentrations in the top 20 cm surface soils and mean soil bulk density (kg·m^{-3}) over this soil layer.

2.4. Statistical Analysis

All results are expressed on an oven-dry mass basis and reported as mean (±standard deviation) of three field replications. One-way ANOVA was used to determine the significance in the detected differences between soil inorganic N concentrations, nitrification rate, net N mineralization rate, or cumulative mineralized N among the four tree spacing treatments for each soil layer. Significant differences between spacing treatments and measured times were tested using two-way ANOVA (spacing, time frame and the interaction) and Duncan's test with significance at $p = 0.05$. All analyses were performed using SPSS 10.0 statistical software package (SPSS Inc., Chicago, IL, USA).

3. Results

3.1. Nitrogen Mineralization and Nitrification

Seasonal variation in net N mineralization (R_M) and nitrification rate (R_N) in soils under different spacing plantations are shown in Tables 2. Tree spacing treatments and time frames significantly affected R_N and R_M values (Table 3, $p < 0.05$), and a significant interaction between tree spacing treatments and time frames was also detected except for the soil of 0–5 cm depth. Seasonal variation patterns of nitrification and mineralization rates were similar for the three soil layers tested (Table 2), with higher rates detected in 0–5 cm soil layers. The mean net R_N in the surface soil (0–20 cm), where four tree spacings averaged within time frames, followed the order of 0.57 mg·kg^{-1}·day^{-1} in time frame A (from 10 May to 10 August 2012), >0.26 mg·kg^{-1}·day^{-1} in time frame B (from 10 August to 10 November 2012), >0.16 mg·kg^{-1}·day^{-1} in time frame D (from 10 February to 10 May 2013), >0.13 mg·kg^{-1}·day^{-1} time frame C (from 10 November 2012 to 10 February 2013), while the mean net R_M was ranked as time frame A (0.58 mg·kg^{-1}·day^{-1}) > B (0.30 mg·kg^{-1}·day^{-1}) > C (0.12 mg·kg^{-1}·day^{-1}) > D (0.10 mg·kg^{-1}·day^{-1}).

For all of the three soil layers, R_N under 6 × 6 m tree spacing was significantly higher than other three spacing treatments tested at the time frame from 10 May to 10 August 2012, whereas there was no significant difference in R_N values among spacing treatments at other testing periods. For example, in the 5–10 cm layer the R_N under the 6 × 6 m stand achieved 0.24 mg·kg^{-1}·day^{-1}, which was 71.4%, 71.4%, and 50.0% higher than those of 4.5 × 8 m, 5 × 5 m, and 3 × 8 m, respectively. The mean net R_N in the surface soil (0–20 cm), for which the four time frames were averaged within each tree spacing treatment, followed the order of 6 × 6 m spacing (0.38 mg·kg^{-1}·day^{-1}) > 4.5 × 8 m spacing (0.30 mg·kg^{-1}·day^{-1}) > 5 × 5 m spacing (0.22 mg·kg^{-1}·day^{-1}) ≥ 3 × 8 m spacing (0.22 mg·kg^{-1}·day^{-1}).

Table 2. Mean nitrification and mineralization rates measured in four time frames (from 10 May 2012 to 10 May 2013) at three soil layers under different tree spacing treatments (Mean ± SD) *.

Time Frame	Spacing (m)	N Nitrification Rate (mg·kg⁻¹·day⁻¹)			N Mineralization Rate (mg·kg⁻¹·day⁻¹)		
		0–5 cm	5–10 cm	10–20 cm	0–5 cm	5–10 cm	10–20 cm
A	6 × 6	0.40 ± 0.11 a	0.24 ± 0.05 a	0.18 ± 0.08 a	0.40 ± 0.11 a	0.25 ± 0.14 a	0.18 ± 0.08 a
	4.5 × 8	0.31 ± 0.15 a,b	0.14 ± 0.04 b	0.08 ± 0.03 b,c	0.32 ± 0.15 a,b	0.14 ± 0.05 a	0.08 ± 0.04 b
	5 × 5	0.22 ± 0.07 b	0.14 ± 0.08 b	0.11 ± 0.03 b	0.23 ± 0.07 b	0.14 ± 0.08 a	0.11 ± 0.03 a,b
	3 × 8	0.26 ± 0.05 b	0.16 ± 0.05 b	0.05 ± 0.02 c	0.27 ± 0.05 b	0.16 ± 0.05 a	0.04 ± 0.02 b
B	6 × 6	0.14 ± 0.03 a	0.14 ± 0.02 a	0.07 ± 0.09 a	0.16 ± 0.03 a	0.15 ± 0.01 a	0.08 ± 0.05 a,b
	4.5 × 8	0.13 ± 0.02 a	0.11 ± 0.05 a,b	0.09 ± 0.01 a	0.15 ± 0.02 a	0.13 ± 0.07 a,b	0.09 ± 0.01 a
	5 × 5	0.08 ± 0.06 a	0.07 ± 0.06 b	0.02 ± 0.01 b	0.10 ± 0.10 a	0.06 ± 0.04 b	0.03 ± 0.03 b
	3 × 8	0.09 ± 0.03 a	0.08 ± 0.04 a,b	0.03 ± 0.02 b	0.11 ± 0.03 a	0.09 ± 0.05 a,b	0.04 ± 0.03 a,b
C	6 × 6	0.09 ± 0.04 a	0.06 ± 0.06 a	0.04 ± 0.02 a	0.08 ± 0.04 a	0.05 ± 0.05 a	0.02 ± 0.01 a
	4.5 × 8	0.08 ± 0.04 a	0.04 ± 0.02 a	0.02 ± 0.01 a	0.05 ± 0.02 a	0.04 ± 0.01 a	0.003 ± 0.00 a
	5 × 5	0.05 ± 0.02 a	0.04 ± 0.03 a	0.02 ± 0.01 a	0.06 ± 0.01 a	0.05 ± 0.02 a	0.001 ± 0.00 a
	3 × 8	0.05 ± 0.01 a	0.03 ± 0.03 a	0.01 ± 0.01 a	0.05 ± 0.01 a	0.04 ± 0.01 a	0.01 ± 0.01 a
D	6 × 6	0.10 ± 0.06 a	0.04 ± 0.02 a,b	0.02 ± 0.02 a	0.06 ± 0.04 a	0.03 ± 0.02 a	0.01 ± 0.01 a
	4.5 × 8	0.11 ± 0.03 a	0.07 ± 0.02 a	0.02 ± 0.00 a	0.07 ± 0.01 a	0.03 ± 0.01 a	0.01 ± 0.00 a
	5 × 5	0.06 ± 0.03 a	0.05 ± 0.01 a,b	0.02 ± 0.01 a	0.04 ± 0.02 a	0.03 ± 0.02 a	0.02 ± 0.01 a
	3 × 8	0.10 ± 0.03 a	0.03 ± 0.02 b	0.02 ± 0.01 a	0.06 ± 0.04 a	0.02 ± 0.01 a	0.005 ± 0.00 b

* A: 10 May–10 August 2012; B: 10 August–10 November 2012; C: 10 November 2012–10 February 2013; D: 10 February–10 May 2013. Different letters indicate significant differences between spacing treatments of the same soil layer sampled within the same time frame according to Duncan's test ($p < 0.05$).

Table 3. ANOVA for nitrification and mineralization rates in soils under different tree spacing sampled at different time frames.

Soil Layer (cm)	Factor	N Nitrification Rate		N Mineralization Rate	
		F value	p	F value	p
0–5	Spacing	39.04	<0.001	34.87	<0.001
	Time frame	3.87	0.018	3.11	0.040
	Spacing × Time frame	0.95	0.497	0.73	0.681
5–10	Spacing	43.70	<0.001	38.18	<0.001
	Time frame	4.62	0.009	5.85	0.003
	Spacing × Time frame	2.67	0.020	2.41	0.032
10–20	Spacing	58.32	<0.001	41.60	<0.001
	Time frame	13.27	<0.001	10.31	<0.001
	Spacing × Time frame	6.17	<0.001	5.65	<0.001

Similar to soil net R_N, R_M value measured in the 0–5 cm soil layer under 6 × 6 m tree spacing was higher than other spacing treatments tested at time frame A (Table 2), which was also significantly higher than those of the same spacing treatment but at different time frames. The mean net R_M in the surface soil (0–20 cm), for which the four time frames were averaged within each tree spacing treatment, followed the order of 6 × 6 m spacing (0.37 mg·kg^{-1}·day^{-1}) > 4.5 × 8 m spacing (0.29 mg·kg^{-1}·day^{-1}) > 5 × 5 m spacing (0.22 mg·kg^{-1}·day^{-1}) ≥ 3 × 8 m spacing (0.22 mg·kg^{-1}·day^{-1}).

In the time frame A, cumulative N mineralized in the surface soils under 6 × 6 m treatment was significantly higher than the other tested spacing treatments (Figure 1). It was clearly detected that high tree density led to significant reduction in cumulative N mineralization in timeframe B. However, there was no significant difference with respect to N mineralization among spacing treatments within time frames of C and D. Cumulative N mineralization in soils of low density plots was generally higher than those of high density (Figure 1). When summing up data from different tested time frames, total annual cumulative net N mineralization was significantly different in the soils under different tree spacing treatments. The total annual cumulative net N mineralization ranged from 80.3–136.0 mg·kg^{-1} in the surface soils, while total annual cumulative net N mineralization of 6 × 6 m and 4.5 × 8 m spacings was 65% and 24% higher than that of the 5 × 5 m, respectively.

3.2. Soil Inorganic N Concentration

Concentrations of soil NO_3^--N and NH_4^+-N in soils of different layers varied with sampling time and tree spacing treatment (Figure 2). In the 0–5 cm soil layer, temporal variation of NO_3^--N concentrations was similar under different spacing treatments, with the highest concentration detected in August and the lowest in November (Figure 2A). However, there were considerable variations with respect to temporal variation among soils of different depths. In general, mean NO_3^--N concentrations in the samples taken in May and August of 2012 were higher than other sampling times tested. Of spacing treatments tested, variation trend in NO_3^--N concentrations in soils was not consistent across sampling times. In most case, a significant difference in NO_3^--N concentrations was detected among the four tree spacings at the same soil layer and the same sampling time (Figure 2A–C). Temporal variations of soil NH_4^+-N concentrations were somewhat different from those observed for soil NO_3^--N (Figure 1D–F). Overall, concentrations of NH_4^+-N in the three soil layers were by average significantly higher in November 2012 than all other sampling times tested.

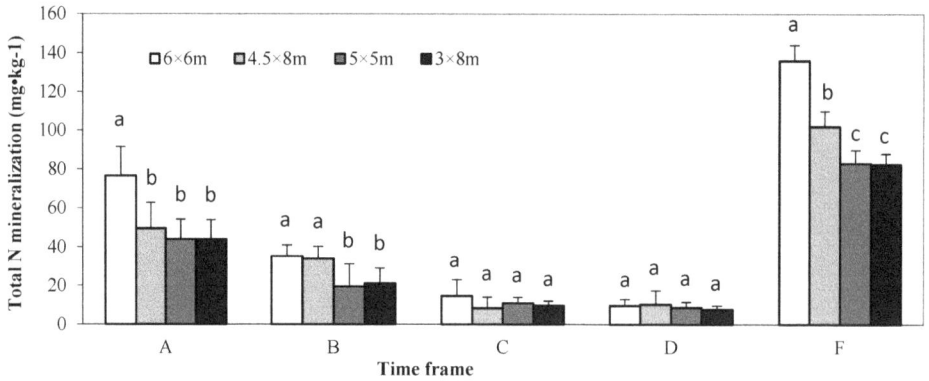

Figure 1. Nitrogen mineralized at the specified time frames (A: 10 May to 10 August 2012; B: 10 August to 10 November 2012; C: 10 November 2012 to 10 February 2013; D: 10 February to 10 May 2013; F: 10 May 2012 to 10 May 2013) in soils under different spacing treatments. Different letters indicate significant differences between four spacing treatments at the same time frame ($p < 0.05$).

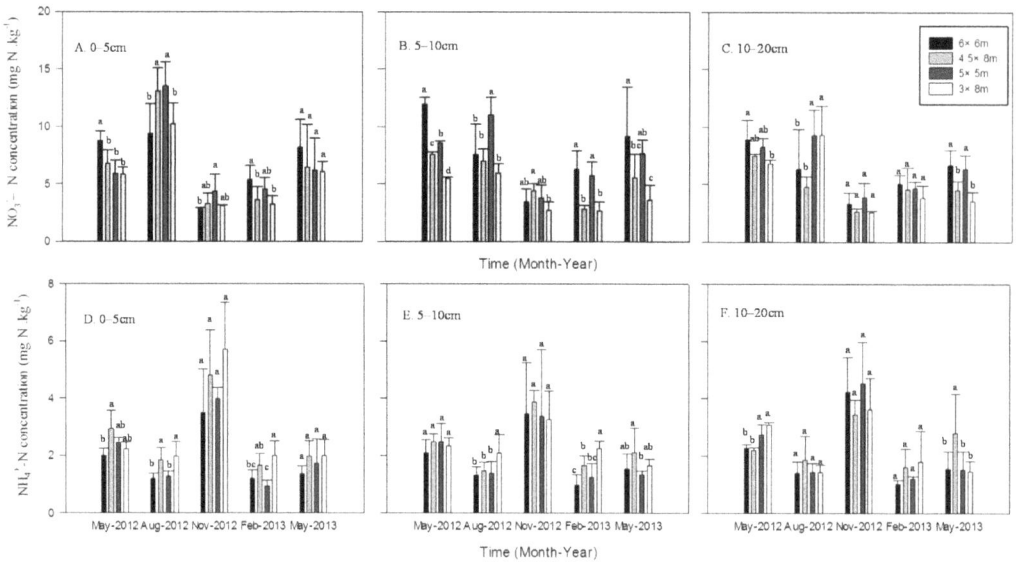

Figure 2. Dynamics of inorganic N concentration in various soil depths (0–5, 5–10, and 10–20 cm) of different tree spacing treatments at different sampling times in poplar plantations. Different letters indicate significant differences between four spacing treatments at the same time ($p < 0.05$).

3.3. Soil Inorganic N Content

Temporal variations of inorganic N content in surface soil (0–20 cm) are shown in Figure 3. Overall, the soil inorganic N content in the growing seasons was higher than in the non-growing seasons, and tree spacing had a significant effect on inorganic N content in the surface soil during the growing season (Figure 3, $p < 0.05$). However, there was a slight difference in this seasonal dynamics among the four tree spacings (Figure 3). For example, the highest content of soil inorganic N appeared in May 2012 for larger spacing plantations (6 × 6 m and 4.5 × 8 m), but occurred in August 2012 for the lower spacings (5 × 5 m and 3 × 8 m). Moreover, the soil inorganic N content in the plantations of square spacings (6 × 6 m and 5 × 5 m) was higher than the rectangular spacings (4.5 × 8 m and 3 × 8 m). During the growing seasons, inorganic N content of the surface soil under the plantation with 6 × 6 m spacing was 12.7% greater than the 4.5 × 8 m spacing, while 19.9% higher in the 5 × 5 m spacing than in the 3 × 8 m spacing.

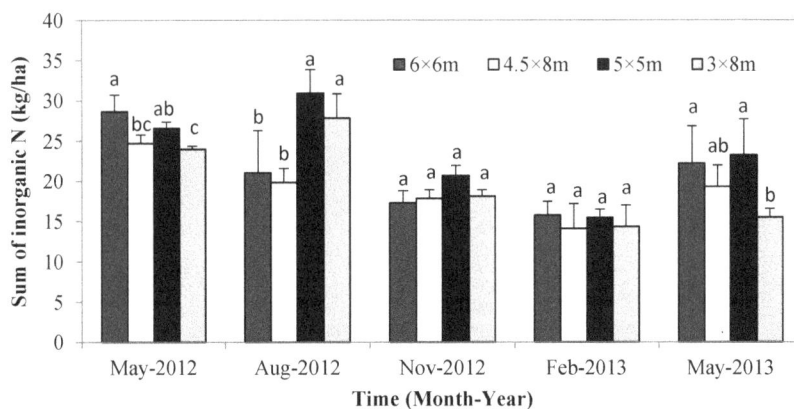

Figure 3. Total inorganic nitrogen content in surface soil (0–20 cm) under different tree spacing treatments at different sampling times.

4. Discussion

4.1. Dynamics of Inorganic N and N Mineralization

Data from this study showed that inorganic N content in the surface soil was higher in the spring and summer seasons and decreased in the later parts of the growing season (in early November 2012), inconsistent with the result of Pajuste and Frey [15], who reported that mineral N in soils of boreal Scots pine and Norway spruce stands was greatest during spring, but decreased as the growing season progressed. Similar temporal variation in mineral N availability in Silver birch stands and grasslands were also reported by Uri *et al.* [26].

Soil N mineralization exhibited spatial and temporal variations, especially seasonal variations [14,27–29]. However, it was not clear whether and how the rate of N mineralization in soil is related to the growing periods of trees above-ground. Most studies suggest that N mineralization was enhanced during the growing season [14,30], which was, in part, attributed to higher temperature coupled with greater soil moisture in the growing season [31,32]. Zhang *et al.* [33] found that 85% of total annual nitrification and 90% of annual N mineralization took place during the growing season. Data from this study also showed that nitrification and N mineralization during the growing season was higher than the dormant period and the peak rates of N mineralization and nitrification were observed during the period of relatively high soil temperature and moisture (during time frame A, Table 2). Overall, the rates of N mineralization and accumulation of NO_3^--N appeared to follow seasonal patterns of temperature and rainfall. However, compared to time frame A, the N mineralization rate in the time frame B decreased significantly possibly in part because of the reduction of rainfall (only half of the time frame A) even if the air temperature was similar between time frames A and B (Figure 4), supporting the fact that the dynamics of N mineralization are also significantly influenced by soil water content [14]. These findings suggest that the temporal pattern of soil N mineralization relates directly to environmental factors and changing ecosystem productivity throughout the year.

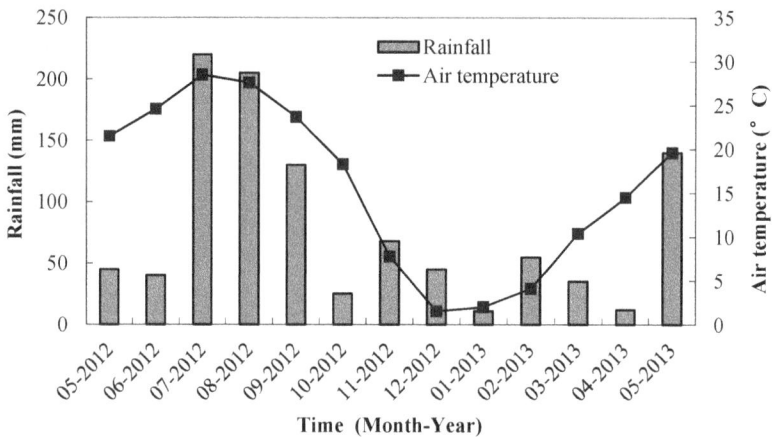

Figure 4. Monthly variation of rainfall and air temperature at the study site during experimental periods.

4.2. Effects of Tree Spacing on Soil N Availability

Results from this study indicated that mean nitrification rate and N mineralization rates in soils under low tree density were higher than those of high tree density (Table 2). The main reason could be the difference in canopy closure among the four tree spacings, which would alter microclimate, understory diversity, litter quality and decomposition rates and finally influence N cycling and availability in the soil [34,35]. However, our results indicated that the NO_3^--N mineralization had a positive relationship with NO_3^--N concentration in the soil, and the values of correlation coefficient R^2 were 0.503 and 0.476 for soils of 0–5 cm and 5–10 cm depth, respectively (Figure 5), suggesting that tree spacing would affect N availability in soil by altering N mineralization rates.

Generally the higher ecosystem activity and greater plant biomass during the middle and later parts of the growing season increase the demands for available N in the soil [36]. In this study, the inorganic N contents in the growing season were higher than the non-growing season (Figure 3), possibly because of the higher microbial activity and nutrient turnover in the growing season [37,38]. However, inorganic N content in soils sampled in August 2012 under lower tree density was much lower than that of higher density treatments in the present study, confirming that available N in forest soils is often strongly influenced by N uptake by forest vegetation rather than microbial N mineralization and immobilization [39]. Our previous study indicated that understory Shannon index, understory biomass, and concentration of N, P, K, and Mg in understory were much greater in low density plantations (6 × 6 m and 4.5 × 8 m) than in high density stands (5 × 5 m and 3 × 8 m) during the growing season [9]. We only investigated inorganic N content in the surface soil (0–20 cm) in the present study, and no consistent tendency in inorganic N content of the soil was observed at the five sampling times for different tree spacings. However, most of the poplar root systems were distributed in soil of 20–40 cm depth [40], therefore, an investigation of deeper soil layer (20–50 cm) is required in order to demonstrate the effects of tree spacings on N availability in the poplar plantations.

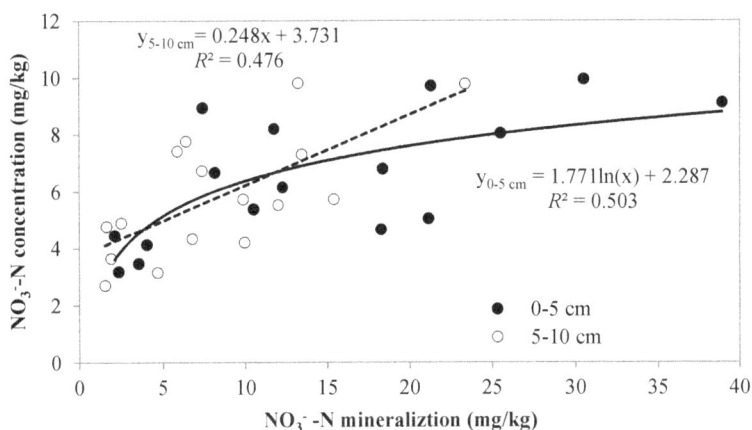

Figure 5. The relationship between soil NO_3^--N concentration (y) and NO_3^--N mineralization (x) in various soil depths ($n = 16$).

5. Conclusions

Seasonal variations of inorganic N content in surface soil (0–20 cm) were detected during the research period, with a higher content in the growing seasons and a lower content in the non-growing seasons. However, a significant variation of inorganic N content in surface soil among the different tree spacings was only observed during the growing season. Tree spacing significantly affected annual net N mineralization, and generally a high annual N mineralization appeared in the soils of low density plantations. However, annual net N mineralization in the soil was higher in square spacing than in rectangular spacing when initial planting density was the same or similar. Seasonal variation patterns of nitrification and mineralization rates were similar in the three soil layers tested, with higher rates detected in the 0–5 cm soil layers. The results obtained suggest that suitable spacing could lead to enhanced N mineralization and soil N availability, but seasonal variation of soil N mineralization may be directly related to plantation productivity.

Acknowledgments

This work was funded by National Basic Research Program of China (973 Program) (2012CB416904) and the Priority Academic Program Development of Jiangsu Higher Education Institutions (PAPD), as well as by the Natural Science fund for colleges and universities in Jiangsu Province (BK2011821). The authors would like to thank Hao Song and Xingjian Dun from Nanjing Forestry University for their contribution and assistance.

Author Contributions

Shengzuo Fang and Luozhong Tang conceived and designed the experiments. Yafei Yan carried out field work, data analysis and prepared the manuscript. Ye Tian and Luozhong Tang contributed to field work and data analysis. Yafei Yan and Dao Ngoc Chuong contributed to the chemical analysis of soil. Shengzuo Fang and Shiping Deng supervised the study, did additional data analysis and reviewed and edited the work.

Conflicts of Interest

The authors declare no conflict of interest.

References

1. Smith, G.; Brennan, P. First thinning in sub-tropical eucalypt plantations grown for high-value solidwood products: A review. *Aust. For.* **2006**, *69*, 305–312.
2. Armstrong, A.; Johns, C.; Tubby, I. Effects of spacing and cutting cycle on the yield of poplar grown as an energy crop. *Biomass Bioenergy* **1999**, *17*, 305–314.
3. Benomar, L.; DesRochers, A.; Larocque, G.R. Comparing growth and fine root distribution in monocultures and mixed plantations of hybrid poplar and spruce. *J. For. Res.* **2013**, *24*, 247–254.
4. Fang, S.Z.; Xu, X.Z.; Lu, S.X.; Tang, L.Z. Growth dynamics and biomass production in short-rotation poplar plantations: 6-Year results for three clones at four spacings. *Biomass Bioenergy* **1999**, *17*, 415–425.
5. Cassidy, M.; Palmer, G.; Smith, R.G.B. The effect of wide initial spacing on wood properties in plantation grown *Eucalyptus pilularis*. *New For.* **2013**, *44*, 919–936.
6. Benomar, L.; DesRochers, A.; Larocque, G.R. The effects of spacing on growth, morphology and biomass production and allocation in two hybrid poplar clones growing in the boreal region of Canada. *Trees* **2012**, *26*, 939–949.

7. Debell, D.S.; Harrington, C.A.; Clendenen, G.W.; Zasada, J.C. Tree growth and stand development of four *Populus* clones in large monoclonal plots. *New For.* **1997**, *14*, 1–18.

8. Kang, K.Y.; Zhang, S.Y.; Mansfield, S.D. The effects of initial spacing on wood density, fibre and pulp properties in jack pine (*Pinus banksiana* Lamb.). *Holzforschung* **2004**, *58*, 455–463.

9. Yan, Y.F.; Fang, S.Z.; Tian, Y.; Song, H.; Dun, X.J. The response of understory plant diversity and nutrient accumulation to stand structure of poplar plantation. *Chin. J. Ecol.* **2014**, *33*, 1170–1177.

10. Oh, T.K.; Cho, M.G.; Chung, J.M.; Jung, H.; Jeon, K.S.; Moon, H.S.; Lee, S.J.; Shinogi, Y. Effect of thinning intensity on soil nitrogen dynamics in *Pinus densiflora* stand. *J. Fac. Agric. Kyushu. Univ.* **2012**, *57*, 473–479.

11. LeBauer, D.S.; Treseder, K.K. Nitrogen limitation of net primary productivity in terrestrial ecosystems is globally distributed. *Ecology* **2008**, *89*, 371–379.

12. Vitousek, P.M.; Howarth, R.W. Nitrogen limitation on land and in the sea: How can it occur? *Biogeochemistry* **1991**, *13*, 87–115.

13. Fang, S.Z.; Xie, B.D.; Liu, D.; Liu, J.J. Effects of mulching materials on nitrogen mineralization, nitrogen availability and poplar growth on degraded agricultural soil. *New For.* **2011**, *41*, 147–162.

14. Gelfand, I.; Yakir, D. Influence of nitrite accumulation in association with seasonal patterns and mineralization of soil nitrogen in a semi-arid pine forest. *Soil Biol. Biochem.* **2008**, *40*, 415–424.

15. Pajuste, K.; Frey, J. Nitrogen mineralization in podzol soils under boreal Scots pine and Norway spruce stands. *Plant Soil* **2003**, *257*, 237–247.

16. Wong, M.T.F.; Nortcliff, S. Seasonal fluctuations of native available N and soil management implications. *Fertil. Res.* **1995**, *42*, 13–26.

17. Booth, M.S.; Stark, J.M.; Rastetter, E. Controls on nitrogen cycling in terrestrial ecosystems: A synthetic analysis of literature data. *Ecol. Monogr.* **2005**, *75*, 139–157.

18. Vitousek, P.M.; Hättenschwiler, S.; Olander, L.; Allison, S. Nitrogen and nature. *Ambio* **2002**, *31*, 97–101.

19. Fang, S.Z.; Zhai, X.C.; Wan, J.; Tang, L.Z. Clonal variation in growth, chemistry and calorific value of new poplar hybrids at nursery stage. *Biomass Bioenergy* **2013**, *54*, 303–311.

20. Thomas, D.; Henson, M.; Joe, B.; Boyton, S.; Dickson, R. Review of growth and wood quality of plantation grown *Eucalyptus dunnii* Maiden. *Aust. For.* **2009**, *72*, 3–11.

21. Sariyildiz, T.; Anderson, J.M.; Kucuk, M. Effects of tree species and topography on soil chemistry, leaf litter quality, and decomposition in Northeast Turkey. *Soil Biol. Biochem.* **2005**, *37*, 1695–1706.

22. Binkley, D.; Aber, J.; Pastor, J.; Nadelhoffer, K. Nitrogen availability in some Wisconsin forests: Comparisons of resin bags and on-site incubations. *Biol. Fertil. Soils* **1986**, *2*, 77–82.

23. Shibata, H.; Urakawa, R.; Toda, H.; Inagaki, Y.; Tateno, R.; Koba, K.; Nakanishi, A.; Fukuzawa, K.; Yamasaki, A. Changes in nitrogen transformation in forest soil representing the climate gradient of the Japanese archipelago. *J. For. Res.* **2011**, *16*, 374–385.

24. Klute, A. *Methods of Soil Analysis. Part 1—Physical Mineralogical Methods*; INC; American Society of Agronomy: Madison, WI, USA, 1986; pp. 363–375.

25. Trindade, H.; Coutinho, J.; Jarvis, S.; Moreira, N. Nitrogen mineralization in sandy loam soils under an intensive doublecropping forage system with dairy-cattle slurry applications. *Eur. J. Agron.* **2001**, *15*, 281–293.

26. Uri, V.; Lohmus, K.; Kund, M.; Tullus, H. The effect of land use type on net nitrogen mineralization on abandoned agricultural land: Silver birch stand *versus* grassland. *For. Ecol. Manag.* **2008**, *255*, 226–233.

27. Pandey, C.B.; Rai, R.B.; Singh, L. Seasonal dynamics of mineral N pools and N mineralization in soils under homegarden trees in South Andaman, India. *Agrofor. Syst.* **2007**, *71*, 57–66.

28. Wei, X.R.; Shao, M.A.; Fu, X.L.; Göran, I.; Yin, X.Q. The effects of land use on soil N mineralization during the growing season on the northern Loess Plateau of China. *Geoedroma* **2011**, *160*, 590–598.

29. Yan, E.R.; Wan, X.H.; Guo, M.; Zhong, Q.; Zhou, W.; Li, Y.F. Temporal patterns of net soil N mineralization and nitrification through secondary succession in the subtropical forests of eastern China. *Plant Soil* **2009**, *320*, 181–194.

30. Zhou, L.S.; Huang, J.H.; Lü, F.M.; Han, X. Effects of prescribed burning and seasonal and interannual climate variation on nitrogen mineralization in a typical steppe in Inner Mongolia. *Soil Biol. Biochem.* **2009**, *41*, 796–803.

31. Ehrenfeld, J.G.; Han, X.G.; Parsons, W.F.J.; Zhu, W.X. On the nature of environmental gradients: Temporal and spatial variability of soils and vegetation in the New Jersey pinelands. *J. Ecol.* **1997**, *85*, 785–798.

32. Zhu, W.X.; Carreiro, M.M. Temporal and spatial variations in nitrogen transformations in deciduous forest ecosystems along an urban-rural gradient. *Soil Biol. Biochem.* **2004**, *36*, 267–278.

33. Zhang, X.L.; Wang, Q.B.; Li, L.H.; Han, X.G. Seasonal variations in nitrogen mineralization under three land use types in a grassland landscape. *Acta Oecol.* **2008**, *3*, 322–330.

34. Tan, X.; Chang, S.X.; Comeau, P.G.; Wang, Y.H. Thinning effects on microbial biomass, N mineralization, and tree growth in a mid-rotation fire-origin lodgepole pine stand in the lower foothills of Alberta, Canada. *For. Sci.* **2008**, *54*, 465–474.

35. Merila, P.; Smolander, A.; Strommer, R. Soil nitrogen transformations along a primary succession transect on the land-uplift coast in Western Finland. *Soil Biol. Biochem.* **2002**, *34*, 373–385.

36. Binkley, D.; Hart, S. The components of nitrogen availability assessments in forest soils. *Adv. Soil Sci.* **1989**, *10*, 57–116.

37. Nadelhoffer, K.J.; Giblin, A.E.; Shaver, G.R.; Laundre, J.A. Effects of temperature and substrate quality on element mineralization in six arctic soils. *Ecology* **1991**, *72*, 242–253.

38. Schmidt, I.K.; Jonasson, S.; Michelsen, A. Mineralization and microbial immobilization of N and P in arctic soils in relation to season, temperature and nutrient amendment. *Appl. Soil Ecol.* **1999**, *11*, 147–160.

39. McKinley, D.C.; Rice, C.W.; Blair, J.M. Conversion of grassland to coniferous woodland has limited effects on soil nitrogen cycle processes. *Soil Biol. Biochem.* **2008**, *40*, 2627–2633.

40. Fang, S.Z.; Xie, B.D.; Liu, J.J. Soil nutrient availability, poplar growth and biomass production on degraded agricultural soil under fresh grass mulch. *For. Ecol. Manag.* **2008**, *255*, 1802–1809.

Seasonal Pattern of Decomposition and N, P, and C Dynamics in Leaf litter in a Mongolian Oak Forest and a Korean Pine Plantation

Jaeeun Sohng, Ah Reum Han, Mi-Ae Jeong, Yunmi Park, Byung Bae Park and Pil Sun Park

Abstract: Distinct seasons and diverse tree species characterize temperate deciduous forests in NE Asia, but large areas of deciduous forests have been converted to conifer plantations. This study was conducted to understand the effects of seasons and tree species on leaf litter decomposition in a temperate forest. Using the litterbag method, the decomposition rate and nitrogen, phosphorous, and carbon dynamics of Mongolian oak (*Quercus mongolica*), Korean pine (*Pinus koraiensis*), and their mixed leaf litter were compared for 24 months in a Mongolian oak stand, an adjacent Korean pine plantation, and a Mongolian oak—Korean pine mixed stand. The decomposition rates of all the leaf litter types followed a pattern of distinct seasonal changes: most leaf litter decomposition occurred during the summer. Tree species was less influential on the leaf litter decomposition. The decomposition rates among different leaf litter types within the same stand were not significantly different, indicating no mixed litter effect. The immobilization of leaf litter N and P lasted for 14 months. Mongolian oak leaf litter and Korean pine leaf litter showed different N and P contents and dynamics during the decomposition, and soil P_2O_5 was highest in the Korean pine plantation, suggesting effects of plantation on soil nutrient budget.

Reprinted from *Forests.* Cite as: Sohng, J.; Han, A.R.; Jeong, M.-A.; Park, Y.; Park, B.B.; Park, P.S. Seasonal Pattern of Decomposition and N, P, and C Dynamics in Leaf litter in a Mongolian Oak Forest and a Korean Pine Plantation. *Forests* **2014**, *5*, 2561-2580.

1. Introduction

Leaf litter decomposition is a critical step in nutrient cycling and providing nutrients to plants [1]. Various factors control leaf litter decomposition, including climate [2], topography [3,4], chemical characteristics of leaf litter [5,6], and terrestrial microbiota [7]. Among those factors, climate, especially temperature and precipitation, is a dominant factor determining leaf litter decomposition patterns in regions experiencing distinct seasonal climate change [8].

Every ecosystem has its own physical characteristics and specific environment that foster the distinct features of its species composition. Each tree species affects leaf litter decomposition differently by providing leaf litters of different quality, which are closely related to the terrestrial microbial community and soil nutrient cycling [9,10]. In temperate natural forests, plant leaf litter usually decomposes in a mixed-species environment. Mixedwood forests are characterized by diverse soil conditions and biological activity within short distances, increasing the importance of small-spatial-scale research on soil nutrient pools, as well as integrating diverse chemical variables of single or mixed species into different temporal and spatial scales [6,11]. However, species composition is one of the factors most altered by human activities as natural mixedwood forests have been converted to single-species plantations. Stand conversion from natural forest to single species plantation changes leaf litter decomposition and eventually alters ecosystem functions [12].

Information on the nutrient cycling within plantations, such as patterns of N, P, and C release, immobilization and mineralization in a decomposing phase, and the planted species' interactions with existing species, is crucial to properly manage plantations and improve their vitality to be similar to natural states [13]. However, limited information is available on the effects of a plantation on nutrient cycling at a site and on plantation interactions with existing ecosystems [14]. Thus, an increasing number of studies investigate the effect of certain species on the decomposition of single or mixed leaf litters for better plantation management or restoration of natural stands [15,16]. Those studies simultaneously integrate features of species, mixed leaf litters, and forest floor conditions by forest type to explain the leaf litter decomposition process [17,18].

Korean pine (*Pinus koraiensis* Siebold et Zucc.) is a representative species in northern temperate old-growth forests and has been widely planted for wood and pine seeds in NE Asia [19,20]. The total plantation area of Korean pine in South Korea is *ca.* 214,357 ha, making Korean pine the third most planted species in Korea [20]. Most Korean pine plantations replace oak-dominant broadleaved stands and are still adjacent to broadleaved forests dominated by oak species such as the oriental cork oak (*Quercus variabilis* Blume), konara oak (*Quercus serrata* Thunb.), and Mongolian oak. The establishment of Korean pine plantations might alter leaf litter decomposition patterns and affect the nutrient budget of an area.

This study compared leaf litter decomposition characteristics and nutrient dynamics between an oak-dominant secondary deciduous stand and a Korean pine plantation to understand the plantation effect on nutrient cycling. We hypothesized that different leaf litter types would show different leaf litter decomposition rates and nitrogen (N), phosphorous (P), and carbon (C) mineralization patterns by forest type. We expected nutrient transport between the Mongolian oak leaf litter and the Korean pine leaf litter to induce complementary interactions, resulting in a synergistic effect that would give the mixed leaf litter the highest decomposition rate.

Efforts to introduce mixedwood forests that are more ecologically sound and resistant to disturbances than monoculture are increasing in NE Asia [21]. Understanding the leaf litter decomposition process in Korean pine plantations would show what happens when a plantation replaces a natural stand and how a plantation interacts with existing ecosystems.

2. Materials and Methods

2.1. Study Sites

The study was conducted at three types of stands in the Seoul National University Forest at Mt. Taehwa (38°17′16″–38°19′26″ N, 127°16′45″ E; 644 m asl; 796 ha), Gyeonggi Province, Korea. The area is a temperate zone with a continental monsoon climate. The annual mean precipitation in this area is 1371 mm with about sixty percent of the annual precipitation is concentrated between June and August [22]. Meteorological data from the weather station of the University Forest during 2009–2011 showed that the annual mean air temperature was 10.9 °C, with a maximum temperature of 33.5 °C in July and a minimum temperature of −21.3 °C in January. It was relatively dry in the winter, with precipitation of 90 mm in the form of snow from December 2010 to February 2011. Snow depth at the study sites ranged from 5 to 20 cm depending on the slope aspect. Soils in the forest are classified as coarse loamy, in the mesic family of Typic Dystrudepts according to the USDA soil classification system [23]. Soil parent material was an eluvium of metamorphic rock, such as biotite and granite gneiss. The depth of the leaf litter layer ranged 3–7 cm in native secondary forests and 2–6 cm in Korean pine plantations in both spring and autumn.

Mt. Taehwa belongs to a broadleaved deciduous forest zone; however, most of the lower slopes are covered by Korean pine and Japanese larch (*Larix kaempferi* (Lamb.) Carrière) plantations. The Korean pine plantations in Mt. Taehwa occupied 136 ha and most of them were established in late 1960s. Because the Korean pine plantations were established in a relatively short period of time and distributed on lower slopes that had similar soil condition, they had a similar stand structure. Secondary natural stands, occupying 497 ha, were distributed on the upper slopes and ridges and were dominated by oaks, such as Mongolian oak, oriental cork oak, and oriental white oak (*Quercus aliena* Blume).

Three types of stands were selected: a secondary natural stand dominated by Mongolian oak (hereafter called oak stand), an adjacent Korean pine plantation (5.5 ha) established in 1969, and a Korean pine—Mongolian oak mixed stand (hereafter called mixed stand). The mixed stand was located between the oak stand and the Korean pine plantation and was established as a Korean pine plantation. However, Mongolian oak from the adjacent natural stands invaded the plantation and replaced part of the Korean pine. The Mongolian oak in the oak stand and the Korean pine in the Korean pine plantation occupied 79.1% and 97.6% of each stand's basal area, respectively. The respective basal areas within the mixed stand were 48.4% Korean pine and 39% Mongolian oak. All study sites were at an altitude of 229–285 m asl, and the slope steepness was 36–67%. The oak stand was on a NW slope and the mixed stand on a NE slope. The Korean pine plantation grew across different oriented slopes from 120–315°; however, it mostly faced south (Table 1).

In each type of stand, three 20 m × 20 m plots were randomly established. All trees ≥2.5 cm in diameter at breast height (DBH; 1.3 m above the ground) were measured within each 20 m × 20 m plot. For each tree, species names, DBH, tree condition (alive, dead, broken top, or leaning), crown width, and proportion of crown to tree height were recorded.

2.2. Soil Measurements, Sampling, and Analysis

A HOBO soil data logger (HOBO H21-002 Micro Station, On-set computer Corporation, USA) was installed in the center of each study site. The soil temperature (°C) and soil moisture content ($m^3 m^{-3}$) were measured from 28 March 2010 to 31 March 2011 at 30-minute intervals at 10 cm soil depth. The daily mean soil temperature and daily mean soil water content were calculated.

Table 1. Selected characteristics of the three study sites in the Seoul National University Forest at Mt. Taehwa, Gyeonggi Province.

Stand types	Oak stand	Mixed stand	Korean pine plantation
Altitude (m)	284–285	258–270	229–239
Slope steepness (%)	43–67	36–47	42–55
Aspect (°)	315–335	36–47	120–315
Soil texture	Loam	Loam	Loam
Sand (%)	38	36	36
Silt (%)	39	41	41
Clay (%)	23	23	23
Species (density (number of stems ha^{-1}), relative basal area (%))	*Quercus mongolica* (567, 79.1), *Pinus koraiensis* (308, 7.9), *Kalopanax septemlobus* (17, 2.9), Others (767, 10.1)	*P. koraiensis* (425, 48.4), *Q. mongolica* (256, 39), *Q. variabilis* (25, 3.7), Others (127, 8.9)	*P. koraiensis* (725, 97.6), *Q. variabilis* (17, 1.0), *Styrax obassia* (42, 0.7), Others (50, 0.7)

Soil samples were collected from the soil surface to a depth of 10 cm with a soil sampling container (100 cm^3) from all the plots in March, June, and October 2010. At each plot, four sampling points were randomly selected 1.5–2 m from the base of a tree trunk to minimize the direct physical and chemical influences of roots [4]. A total of 36 soil samples at a time were collected from the three stand types. The soils were oven-dried at 105 °C for 24 hours and the soil water content (%) was measured on a gravimetric basis. A particle-size analysis was conducted with the pipette method [24], and soil texture (standard of US Department of Agriculture) was determined. We measured soil organic matter contents by combusting it at 600 °C for 4 hours in a muffle furnace and weighing the residue [25]. For chemical analyses, the sampled soils were air-dried and sieved (<2 mm). We measured soil pH (soil:water = 1:5), total nitrogen content (TN) [25], cation exchange capacity (CEC) [26], and total phosphorus (TP; inductively coupled plasma ICPS-1000IV, Shimazu, Japan). Available soil phosphorous (P$_2$O$_5$) was determined by analyzing the Bray No. 1 extracts with an ICP (inductively coupled plasma) optical emission spectrometer (ICP-730 ES, Varian, USA). We used an elemental analyzer to measure the C:N ratio (Flash EA 1112, Thermo Electron Corporation, USA).

2.3. Leaf litter Decomposition Experiment

Leaf litter fall of Mongolian oak and Korean pine (5–6 kg per each species) was collected from forest floor of the Mongolian oak stand and Korean pine plantation, respectively, in the study sites in March 2009. Our study focused on in-situ condition and year-round changes in leaf litter on the forest floor. Leaf litter fall of Mongolian oak occurred in autumn and Korean pine was a conifer with leaf litter fall distributed over the whole year, so we collected forest floor litter in March before active decomposition began instead of autumn to avoid the combination of very fresh Mongolian oak litter fall and relatively less fresh Korean pine litter fall. We were careful to collect intact leaves without clear sign of decomposition from the top layer. The leaf litter sample was oven-dried at 80 °C for 48 hours. The oven-dried leaf litter was weighed into 10 g samples and put into litterbags, either pure or in a 1:1 ratio of the two leaf litters by mass [27,28]. Each litterbag was 20 cm × 10 cm and was made of nylon mesh (1 mm × 1.2 mm mesh size).

The litterbags were placed on topsoil parallel to the forest floor in each study plot in April 2009. A total of 972 litterbags were deployed at the study sites: 3 stand types × 3 plots at each stand type × 3 leaf litter types × 3 replicates × 12 retrieving dates from April 2009 to March 2011. Each litterbag was at least 15 cm away from other litterbags, and all litterbags at each site were tied together with a nylon string to minimize unnecessary movements. Every two months, 27 litterbags were retrieved from each study site. The retrieved leaf litter was oven-dried at 80 °C for 48 hours and brushed carefully to minimize soil contamination [29]. Oven-dried litter was weighed to the nearest 0.01 g to determine its mass loss, and milled for chemical analyses [30]. Total nitrogen and C concentrations (mg g^{-1}) were measured using a dry combustion method with a C-N analyzer (Yanaco MT-500, Kyoto, Japan) [28]. Total P in the leaf litter was determined with an automated ion analyzer (QuikChem AE, Lachat Inc., Loveland, Colorado, USA). The amount of nutrients in the leaf litter was calculated as follows:

$$N_t = \frac{C_t}{100} \times L_t \tag{1}$$

Where N_t is the amount of nutrient (g) at time t, C_t is the percent concentration of nutrient at time t, and L_t is the leaf litter remaining mass (g) at time t.

2.4. Data Analysis

The decomposition rate constant (k) was derived from the negative exponential decomposition equation proposed by Olson [31] as follows:

$$-k = \ln\left(\frac{X}{X_0}\right)t \tag{2}$$

Where X_0 is the initial mass, and X is the remaining mass at time t.

The predicted remaining mass for the mixed leaf litter was estimated using equation (3) based on the observed remaining mass of the single species to verify a mixed effect on leaf litter decomposition [29]. The predicted results were compared with the observed remaining mass of the mixed leaf litter.

$$E_{ML} = \sum_{i=1}^{S} O_{MLi} \times P_{IMi} \tag{3}$$

Where E_{ML} is the predicted remaining mass of mixed litter (%); S is the species; O_{MLi} is the observed leaf litter mass of the single species (%); and P_{IMi} is the initial proportion of species i.

When the observed remaining mass of mixed leaf litter is not significantly different from the predicted results, mixing has an additive effect; otherwise, it has a non-additive effect. The non-additive effect is differentiated into a synergistic effect (decomposition rates are more than predicted values) or an antagonistic effect (decomposition rates are less than predicted values) [16]. The predicted and observed remaining mass were compared using a paired t-test to determine possible mixed leaf litter effects (*i.e.*, additive or non-additive).

One-way ANOVA was used to compare the decomposition rate constant (k) and the changes in leaf litter N, P, and C concentrations among the three stand types. A simple linear regression was used to assess the relationship between mass loss (%) and total C in decomposing leaf litters. The relationship between the total N, P, and C in each leaf litter type over 24 months was estimated using polynomial regression analysis.

A repeated-measures ANOVA was used to compare the soil physical and chemical characteristics in different seasons (March, June, and October 2010) and identify whether significant differences occurred in mass loss and changes in leaf litter N and P every two months by leaf litter type in a single stand or among the stand types. Tukey's test was applied for post-hoc comparisons, with significance set at $p < 0.05$. The SPSS statistical package program Version 18.0 was used for all statistical analyses (SPSS Software 2009, IBM SPSS Inc., Chicago, IL, USA).

3. Results

3.1. Soil Characteristics

All study sites had similar soil temperatures. The maximum soil temperature reached at 24 °C in August, and it stayed near 0 °C in January and February (Figure 1). The Korean pine plantation had a daily mean soil moisture content lower than the other two stand types in June, October, and November (Figure 1), and soil moisture content measurements lower than the other two stand types in March and October 2010 (Table 2; $p < 0.001$). All study sites showed similar daily soil moisture content in the summer.

The three stand types showed significant differences in soil TN concentrations, which were highest in the oak stand and lowest in the Korean pine plantation in spring and autumn ($p = 0.03$). However, soil TN did not significantly differ among stand types in the summer. The oak stand had the highest organic matter content (19.7%) in March 2010 ($p < 0.05$), but we found no differences in organic matter content in the summer and autumn (Table 2).

The soil P_2O_5 content was highest in the Korean pine plantation with 126.7 mg L^{-1}; was more than twice the 49 mg L^{-1} observed in the oak stand in March 2010. The largest soil N:P ratio was therefore in the oak stand (Figure 2), which had the lowest soil P_2O_5 and substantial seasonal differences in soil P_2O_5 and soil N:P ($p = 0.015$) between March and October, 2010, though few seasonal changes in soil TN were observed. Due to the high variability of soil P_2O_5 within the stands, soil P_2O_5 contents among the stands were not significantly different. Soil pH, TP, and CEC were not significantly different among the stand types.

Figure 1. Daily mean soil temperatures (St, °C) and daily mean soil water contents (Sw, $m^3 m^{-3}$) of the Mongolian oak stand, the mixed stand, and the Korean pine plantation from 28 March 2010 to 31 March 2011.

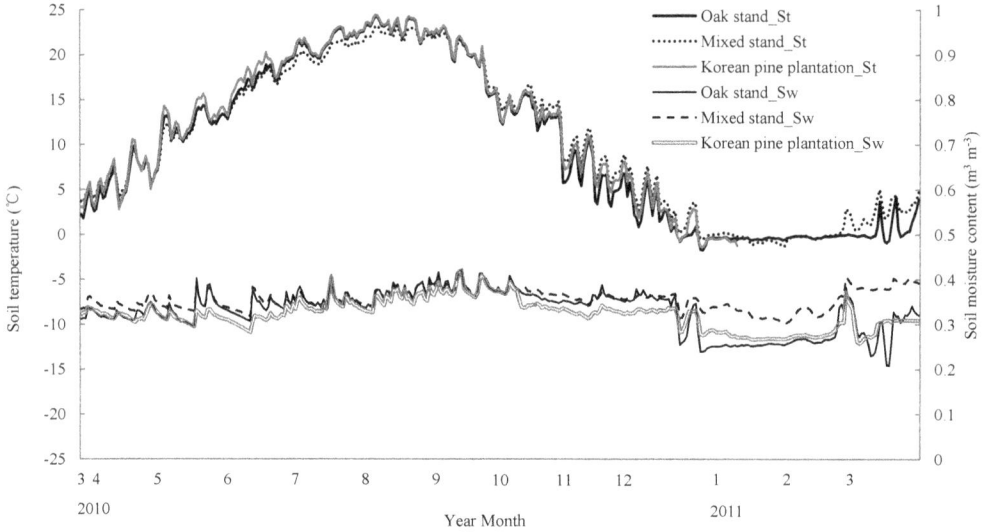

Figure 2. Seasonal changes in the soil N:P ratio in soil (0–10 cm) at the study sites. Data are means ± SE ($n = 9$). Data in the same stand followed by the different letters are significantly different at $p < 0.05$ according to t-tests.

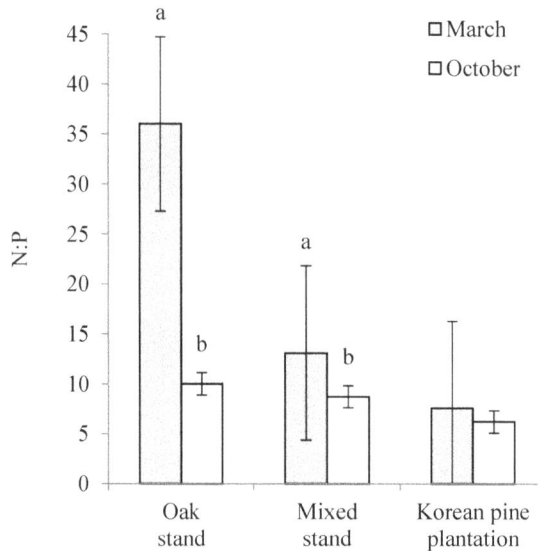

Table 2. Soil chemical characteristics of the three study sites in the Seoul National University Forest at Mt. Taehwa, Gyeonggi Province in March, June, and October 2010.

Month	Stand type	Soil water content (%)	Organic matter (%)	pH	Total N (%)	Total P (mg L⁻¹)	P₂O₅ (mg L⁻¹)	CEC (cmol kg⁻¹)	C:N ratio
March	Oak stand	67.3 ± 11.5 [a]	19.7 ± 2.8 [a]	4.0 ± 0.2	0.60 ± 0.00 [a]	301.2 ± 12.8	49.0 ± 7.7	23.0 ± 0.8	14.6 ± 0.2 [ab]
	Mixed stand	53.4 ± 6.4 [b]	16.2 ± 2.1 [b]	4.0 ± 0.1	0.43 ± 0.02 [b]	308.5 ± 15.6	97.9 ± 20.1	21.4 ± 1.4	14.0 ± 0.0 [b]
	Korean pine plantation	42.7 ± 5.8 [c]	13.3 ± 2.7 [c]	4.3 ± 0.9	0.30 ± 0.03 [c]	278.1 ± 43.7	126.7 ± 34.2	18.6 ± 4.2	15.0 ± 0.3 [a]
June	Oak stand	-	-	-	0.44 ± 0.04	248.5 ± 16.8	60.2 ± 5.2	-	15.0 ± 0.6 [ab]
	Mixed stand	-	-	-	0.36 ± 0.02	269.1 ± 21.0	79.8 ± 12.1	-	13.8 ± 0.3 [b]
	Korean pine plantation	-	-	-	0.30 ± 0.05	238.3 ± 45.1	73.0 ± 20.0	-	16.1 ± 0.5 [a]
October	Oak stand	60.4 ± 7.1 [a]	16.6 ± 3.0	4.1 ± 0.3	0.50 ± 0.03 [a]	315.1 ± 19.3	129.9 ± 14.2	-	14.0 ± 0.2 [b]
	Mixed stand	49.3 ± 8.1 [b]	14.4 ± 2.0	4.0 ± 0.1	0.47 ± 0.02 [a]	337.3 ± 18.7	129.4 ± 7.1	-	15.4 ± 0.4 [a]
	Korean pine plantation	39.6 ± 7.0 [c]	13.8 ± 4.4	4.1 ± 0.1	0.30 ± 0.05 [b]	296.1 ± 53.0	143.4 ± 99.6	-	15.8 ± 0.3 [a]

Data in the same month followed by the same letter are not significantly different at $p < 0.05$ according to Tukey's multiple range tests; Values are means $\pm$ SE.

3.2. Leaf litter Decomposition and Mixed Leaf litter Effects

All leaf litter showed distinct seasonal changes in decomposition rates. About 20% of mass loss occurred in all study sites during the initial six months from April to October 2010 regardless of leaf litter type. Little change in the remaining mass was detected in the subsequent six months (during winter and spring). During the second summer, fourteen months after the experiment started, all types of leaf litter in each stand lost 14%–29% of the original mass (Figure 3).

The leaf litter decomposition rate in the oak stand was significantly slower than those in the Korean pine plantation and the mixed stand until the second summer ($p < 0.001$). The leaf litter decomposition rate was highest in the Korean pine plantation and lowest in the oak stand for 17 months. Although the leaf litter mass loss in the oak stand was less than that in the other stands prior to the second summer, the remaining leaf litter mass continuously decreased in the oak stand during the second autumn, whereas further decomposition did not progress in the other stands.

As a result, after 24 months the remaining leaf litter mass was similar among the stand types, ranging from approximately 41–47%. No significant difference was found in the decomposition rate among the leaf litter types in the same stand.

Figure 3. Percent remaining leaf litter mass for Mongolian oak leaf litter (black solid line), Korean pine leaf litter (gray solid line) and mixed leaf litter (dashed line) in (**A**) the oak stand, (**B**) the mixed stand, and (**C**) the Korean pine plantation. Data are means ± SE ($n = 9$).

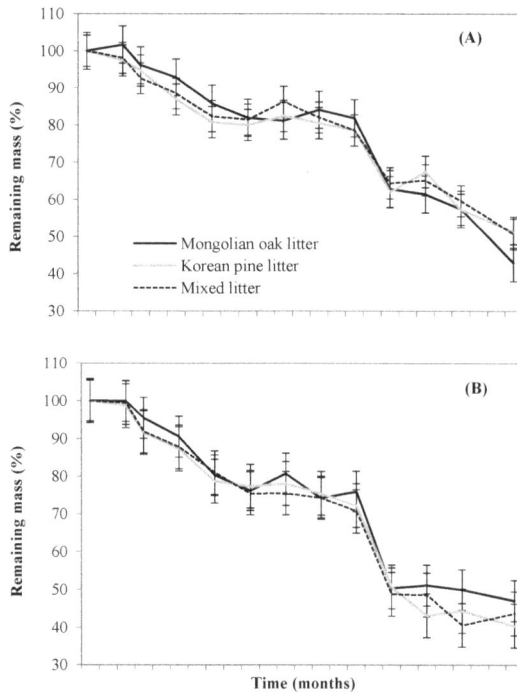

Figure 3. *Cont.*

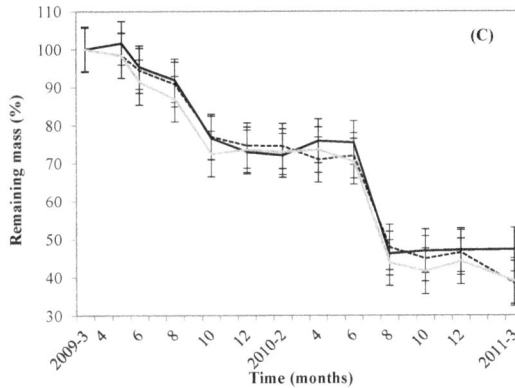

Observed and predicted decay constants in each stand were not significantly different, indicating that the leaf litter mixing of Korean pine and Mongolian oak had an additive effect on leaf litter decomposition. We found no significant synergistic or antagonistic interactions between those two species in leaf litter decomposition over 24 months (Figure 4).

Figure 4. Comparisons of predicted (white bars) and observed (grey bars) decay constant k at each site. Predicted values calculated from the corresponding monoculture litterbags (paired t- test). Data are means $\pm$ SE ($n = 9$).

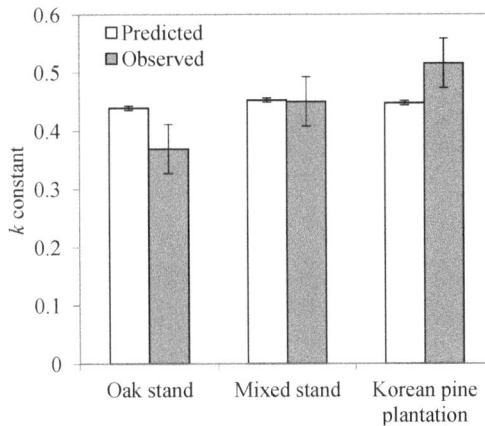

3.3. Leaf litter N, P, and C Changes and Relationships

Changes in the decomposition constant (k) for 24 months corresponded to changes in the amount of C in decomposing leaf litters, and those changes were highly correlated ($p < 0.001$; Table 3). The amount of leaf litter C decreased in a pattern similar to leaf litter mass loss (Figure 3). Consequently, the changes in leaf litter C mass did not vary among leaf litter types, but they were significantly different among stands: the oak stand retained significantly more leaf litter C than the other stands ($p < 0.001$).

Table 3. Leaf litter decomposition constant (k) and C percentage after 24 months relative to the initial leaf litter C ($n = 9$), and R^2-values of simple linear regression for the relationship between mass loss (%) and leaf litter C in the decomposing phase ($p < 0.05$).

Forest type	Leaf litter type	k	C (%)	Regression	R^2
Oak stand ***	Mongolian oak	0.4588	34.0	C = −0.0935 x + 7.8699	0.86
	Mixed leaf litter	0.3694	32.5	C = −0.0934 x + 7.9997	0.72
	Korean pine	0.3774	33.1	C = −0.0927 x + 8.1664	0.87
Mixed stand	Mongolian oak	0.4035	29.7	C = −0.0921 x + 7.9344	0.88
	Mixed leaf litter	0.4504	28.3	C = −0.0902 x + 8.1363	0.98
	Korean pine	0.4886	25.2	C = −0.1045 x + 8.4264	0.93
Korean pine	Mongolian oak	0.4056	25.1	C = −0.0894 x + 7.9403	0.97
plantation	Mixed leaf litter	0.5159	23.1	C = −0.0875 x + 7.9612	0.92
	Korean pine	0.5048	22.7	C = −0.0947 x + 8.3453	0.93

The notation *** indicates that the leaf litter C percentage among the forest types after 24 months is significantly different at $p < 0.001$.

Percentage changes in leaf litter N and P showed that N and P were immobilized during the first year in all study sites. The immobilization was sustained until the beginning of the second summer, June 2010. Mineralization of N and P was observed after 14 months (Figure 5).

Figure 5. Percent of original N ((**A**), (**B**) and (**C**)) and P ((**D**), (**E**) and (**F**)) remaining for Mongolian oak leaf litter (black solid line), Korean pine leaf litter (gray solid line) and mixed leaf litter (dashed line) during the decomposition in (**A**), (**D**) the oak stand, (**B**), (**E**) the mixed stand, and (**C**), (**F**) the Korean pine plantation after 24 months. Data are means ± SE ($n = 9$).

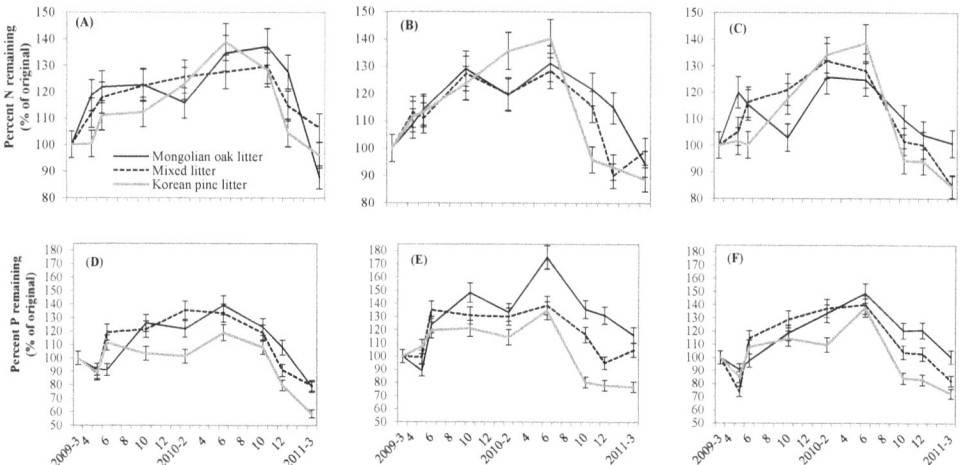

The shift from N immobilization to N mineralization in the mixed stand and the Korean pine plantation occurred in the 16th month, June 2010, whereas the process of N immobilization in the oak stand was sustained until October 2010. Although the oak stand had a longer N immobilization phase, it showed a higher rate of N mineralization in the second autumn, and the percent of N remaining in the leaf litter in the oak stand was similar to that in the other stands in the 24th month. Changes in the percent of N remaining in the oak leaf litter was significantly different from that of Korean pine leaf litter in the oak stand ($p < 0.05$), but the other stands did not show significant differences in the percent of N remaining by leaf litter type (Figure 5A–C).

Temporary leaching of P was shown at the beginning of the study, and then a process of P immobilization started within two months and continued until the 16th month in all study sites (Figure 5D–F). The highest P immobilization rate (175%) was observed in the oak leaf litter at the mixed stand (Figure 5E). The Korean pine leaf litter had the lowest P immobilization percentage in all stand types. Korean pine leaf litter showed the lowest percent of P remaining among leaf litter types in all study sites, indicating the greatest leaf litter P decomposition, and oak leaf litter showed the highest percent of P remaining ($p < 0.05$). Percent of P remaining in the leaf litter in the 24th month was 59–79% in the oak stand, 77–116% in the mixed stand, and 73–100% in the Korean pine plantation. The oak stand showed the greatest P mineralization among the stand types. The percent of P remaining in the Korean pine leaf litter was lower in the oak stand than in the other stands ($p = 0.018$).

The changes in leaf litter N and P had a polynomial correlation with leaf litter C (Figure 6, Table 4). The relationship among leaf litter N, P, and C differed among the leaf litter types. The leaf litter N amount relative to leaf litter C loss was significantly different in the order Korean pine leaf litter[b] < mixed leaf litter[ab] < Mongolian oak leaf litter[a] (Figure 6A–C; $p < 0.001$). The inflection points between N immobilization and mineralization differed by stand type ($p < 0.05$). The maximum amount of leaf litter N peaked when the total amount of C ranged from 5 g to 6 g, which was interpreted to be 50–60% of the initial leaf litter C amount, in the mixed stand and the Korean pine plantation (Figure 5B,C) and 4 g of C in the oak stand (Figure 6A). After 24 months of the experiment, leaf litter C:N ranged from 25 to 46. Leaf litter C:N was not significantly different among stand types, but differed among the leaf litter types as Korean pine leaf litter had significantly higher C:N than the other litter types ($p < 0.001$).

Korean pine leaf litter had the greatest leaf litter P content among leaf litter types. The amount of leaf litter P in the Mongolian oak leaf litter was significantly smaller than that in the mixed leaf litter and Korean pine leaf litter until the remaining C reached 4 g in all the stand types. The amount of leaf litter P became similar among all leaf litter types in the mixed stand and the Korean pine plantation after remaining C was less than 4 g (Figure 6E,F). Similar amounts of C (4–5 g) were observed at the inflection point of P changes in all stand types (Figure 6D–F; $p < 0.05$). The Mongolian oak stand had significantly higher leaf litter C:P of 476 than the Korean pine plantation of 383 after 24 months ($p = 0.019$). The Mongolian oak stand had significantly higher leaf litter N:P of 14 than the other two sites of *ca.* 13 ($p < 0.05$). Three litter types showed significant difference in leaf litter N:P in 24 months in the order Korean pine leaf litter[c] < mixed leaf litter[b] < Mongolian oak leaf litter[a] ($p < 0.05$).

Table 4. The relationship between amount of C, N and P in each decomposing leaf litter type as described by polynomial regression.

Forest type	Leaf litter type	Regression	R^2	P-value
Oak stand	Mongolian oak	$N = -0.0006C^2 + 0.0034C + 0.0924$	0.6046	0.156
		$P = -0.0001C^2 + 0.0006C + 0.0010$	0.7871	0.045
	Mixed leaf litter	$N = -0.0011C^2 + 0.009C + 0.0643$	0.9198	0.006
		$P = -0.0001C^2 + 0.0009C + 0.0009$	0.8509	0.022
	Korean pine	$N = -0.0012C^2 + 0.0104C + 0.0503$	0.7944	0.042
		$P = -0.00003C^2 + 0.0003C + 0.0024$	0.3239	0.457
Mixed stand	Mongolian oak	$N = -0.0016C^2 + 0.0145C + 0.0606$	0.9046	0.009
		$P = -0.0001C^2 + 0.001C + 0.0007$	0.8408	0.025
	Mixed leaf litter	$N = -0.0012C^2 + 0.012C + 0.0508$	0.8081	0.037
		$P = -0.0001C^2 + 0.0008C + 0.0012$	0.8119	0.035
	Korean pine	$N = -0.0019C^2 + 0.0203C + 0.017$	0.8809	0.014
		$P = -0.0001C^2 + 0.001C + 0.0004$	0.8722	0.016
Korean pine plantation	Mongolian oak	$N = -0.0011C^2 + 0.0108C + 0.0603$	0.5020	0.248
		$P = -0.0001C^2 + 0.0005C + 0.0014$	0.7598	0.058
	Mixed leaf litter	$N = -0.0017C^2 + 0.017C + 0.0368$	0.9258	0.006
		$P = -0.0001C^2 + 0.0009C + 0.001$	0.7812	0.048
	Korean pine	$N = -0.0017C^2 + 0.018C + 0.0233$	0.7957	0.042
		$P = -0.0001C^2 + 0.0009C + 0.0009$	0.6936	0.094

Figure 6. Changes in the total N ((**A**), (**B**), (**C**)) and P ((**D**), (**E**), (**F**)) relative to total C for Mongolian oak leaf litter, Korean pine leaf litter and mixed leaf litter in (**A**), (**D**) the oak stand, (**B**), (**E**) the mixed stand, and (**C**), (**F**) the Korean pine plantation.

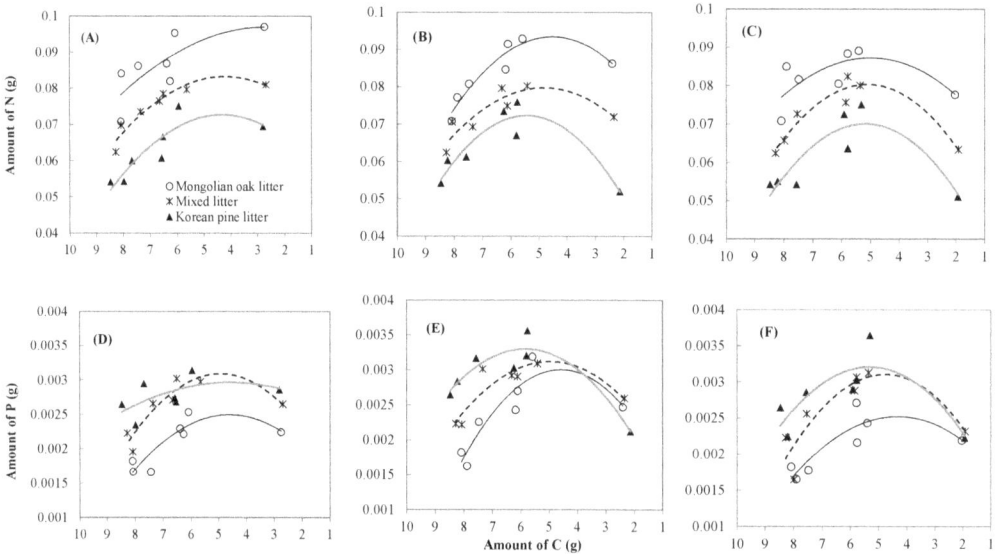

4. Discussion

4.1. Seasonal Leaf litter Decomposition Patterns

Seasonal weather changes have a large effect on leaf litter breakdown in temperate deciduous forests, especially when abundant rainfall and high summer temperatures contribute to a substantial leaf litter mass loss [32]. The majority of leaf litter decomposition in all study sites occurred from June to August, although the decomposition rates differed among the stand types. The seasonal effect on the decomposition rate of *Quercus* leaf litters in a temperate deciduous forest was also reported in Lee *et al.* [33]: the majority of leaf litter mass loss occurred in the summer, and little mass loss was detected from February to April.

In this study, the percent mass of the Korean pine leaf litter remaining after 24 months was 40–50%, which was similar to the 51.6% in 26 months reported in the Gwangneung Experimental Forest, in the same province as our study sites [30]. On Mt. Gyebang, which is north of Mt. Taehwa and has lower temperatures, the percent mass of leaf litter remaining after 24 months was *ca.* 52% [34]. The percent average leaf litter mass remaining after 24 months was 55–60% for Korean pine leaf litter in Jilin Province, northeast China, which has a higher latitude than Mt. Taehwa [19]. About 50% of the leaf litter was decomposed during the initial two years, and the decomposition rate lowered slightly as the latitude increased, showing that temperature was a determining factor for leaf litter decomposition rate: a hotter humid summer corresponds to faster leaf litter decomposition in temperate deciduous forests in NE Asia [35].

All study sites had the same soil texture (loam) and similar soil temperatures. Soil moisture contents were similar in the summer and different in spring and autumn. The oak stand and mixed stand were on a north slope, and the Korean pine plantation was mostly on a south slope, which explained why the lowest soil moisture content was found in the Korean pine plantation in the dry seasons of spring and autumn [36,37]. Leaf litter decomposition rate generally increases with soil moisture content [38]. However, our study sites had similar decomposition rates after 24 months, indicating that the soil moisture content in summer was the key determinant of leaf litter decomposition and confirming the importance of summer weather in leaf litter decomposition [39].

4.2. Mixed Leaf Litter Effect

The leaf litter decomposition rate of broadleaved species is generally higher than that of conifers [40,41]. However, the decomposition rates of Mongolian oak leaves and Korean pine needles in the same stand were not different in this study. Mongolian oak leaves have thick hard blades and low nutrient content compared with other oak species, retarding the leaf litter decomposition rate [34] and contributing to the similar decomposition rates of the two species.

The mixed leaf litter of Korean pine and Mongolian oak showed no mixed litter effect on leaf litter decomposition, indicating that the interaction strength between the leaf litters of the two species was weak. The leaf litter decomposition rate of the mixed leaf litter could therefore be predicted from the decomposition rate of each individual species [15]. Mixed leaf litter could have a higher decomposition rate than a single-species leaf litter because nutrient exchange among leaf litters of different species through microbes could enhance the nutrient condition of low-quality leaf litters, accelerating the decomposition rate [27,42]. At the beginning, Mongolian oak leaves contained more N, and Korean pine needles had more P (Figure 6). We could observe more physical breakdown of the Mongolian oak leaf litter in the mixed leaf litter. However, contrary to our expectations that transport of nutrients between the Mongolian oak leaf litter and the Korean pine leaf litter would induce complementary interactions and a synergistic effect, an actual mixed effect on the remaining mass was not detected within 24 months, suggesting that synergistic effects of mixed leaf litters of contrasting qualities is questionable [29,43]. Moreover, the seasonal climate was a primary control on the leaf litter decomposition pattern, making species difference less powerful in leaf litter decomposition in a temperate deciduous forest [44]. However, the possibility that an additive effect is temporary and switches to a non-additive effect as decomposition progresses still exists because the non-additive effect of mixed leaf litters for mass loss could be time dependent [45].

4.3. Leaf Litter Nutrient Dynamics

After two years of experiments, only 10–20% of net N mineralization was detected due to the immobilization process in the initial decomposition period, which lasted longer than one year. Nutrients in decaying foliar leaf litter temporarily increase due to nutrient immobilization by microbes in the initial mass loss process. The leaf litter N and P immobilization phase prolonged more than 14 months in this study, which was similar to the N and P immobilization phase of more than 15 months in the foliar leaf litter decomposition of oak and Korean pine in a broadleaf-Korean pine mixed forest in NE China [19].

Different leaf litter types had a similar period for the immobilization and mineralization of N and P to diminishing C as well as similar leaf litter decomposition rates at the same stand. Net N mineralization rates of leaf litter are greatly influenced by the site conditions rather than leaf litter species, causing the same species to show different mineralization rates depending on the site [46].

Although soil N is an important determinant in leaf litter decay, soil P becomes a stronger control for leaf litter decay when it is limited under abundant soil N [47]. Seasonal P dynamics are more variable than N in leaf litter decomposition [48]. High precipitation in the summer promotes the movement of phosphate in the soil, inducing seasonal changes in the distribution of soil phosphate [33]. Thus, the proportions of P with other nutrients, such as C:P and N:P, as well as the concentration of P itself, were major regulators that could inhibit or further decay rates by providing a balanced nutritional pool for decomposers [28,49]. Seasonal changes in soil N:P in the oak stand were related to changes in available P content. The lowest soil P_2O_5 and spatial and seasonal variances of soil P_2O_5 (Table 2) and N:P (Figure 2) might contribute to a fluctuation of nutrient supply that could explain the lowest leaf litter decomposition rate prior to the second summer and the different patterns of immobilization and mineralization of N and P in the oak stand compared with the other stands. Therefore, leaf litter nutrient dynamics were more affected by site conditions than by leaf litter types, suggesting that leaf litter decomposition and leaf litter nutrient dynamics patterns between a secondary forest and an adjacent plantation would not be much different if the site condition and stand age were similar, due to the overwhelming seasonal climatic effect on leaf litter decomposition in temperate deciduous forests.

Oak leaf litter showed the smallest P content among leaf litter types, which might be related to the occurrence of least soil P_2O_5 in the oak stand. The Korean pine plantation had the lowest soil TP, but it had the highest soil P_2O_5 among the stand types, and the Korean pine leaf litter had the greatest leaf litter P content and mineralization rate, suggesting that after 40 years of occupation of a Korean pine plantation might increase soil P_2O_5 [50]. Lee *et al.* [33] reported that rapid P cycling in black locust communities resulted in increased membrane phosphate in soil compared with adjacent Mongolian oak communities. Although the effects of species composition might be less powerful than climate on leaf litter decomposition in temperate deciduous forests, effects of species change on detailed leaf litter decomposition process, such as leaf litter nutrient dynamics and soil nutrients, can be expected [10]. We could suggest different leaf litter decomposition patterns relevant to leaf litter N and P dynamics and soil properties between the Korean pine plantation and the oak stand. We expect that further studies on decomposer communities would clarify the decomposition mechanism in temperate deciduous forests.

5. Conclusions

This study compared the leaf litter decomposition patterns and nutrient dynamics between a Korean pine plantation and an adjacent native Mongolian oak–dominant broadleaved temperate forest. Seasonal climate, especially hot and humid summer weather, was a more powerful determinant in leaf litter decomposition patterns than leaf litter types, resulting in no mixed leaf litter effect on leaf litter decomposition. However, we did find effects of the changes in dominant species on leaf litter nutrient dynamics and soil nutrients. The conversion of natural forests into plantations may have long-term effects on nutrients available in the soil, eventually modifying soil nutrient budgets.

Acknowledgments

This study was supported by the Korea Forest Service (Project No. S111214L020130 and S211314L020100). We thank all the participants in the field survey and data processing. We thank the Seoul National University Forest for providing access to the study sites and weather data. We acknowledge the Research Institute of Agriculture and Life Sciences, Seoul National University for grammar assistance and anonymous reviewers for their valuable comments.

Author Contributions

Jaeeun Sohng and Pil Sun Park conceived and designed the experiments. Jaeeun Sohng carried out field work, data analysis and prepared the manuscript. Ah Reum Han, Mi-Ae Jeong, and Yunmi Park contributed to field work and data analysis. Byung Bae Park contributed to the chemical analysis of soil and leaf litter. Pil Sun Park supervised the study, did additional data analysis and reviewed and edited the work.

Conflicts of Interest

The authors declare no conflict of interest.

References

1. Gosz, J. Organic matter and nutrient dynamics of the forest and forest floor in the Hubbard Brook Forest. *Oecologia* **1976**, *22*, 305–320.
2. Meentemeyer, V. An approach to the biometereology of decomposer organisms. *Int. J. Biometeorol.* **1978**, *22*, 94–111.
3. Mudrick, D.A.; Hoosein, M.; Ray, R.H.J.; Townsend, E.C. Decomposition of leaf litter in an Appalachian forest: Effects of leaf species, aspect, slope position and time. *For. Ecol. Manage.* **1994**, *68*, 231–250.
4. Sariyildiz, T.; Anderson, J.M.; Kucuk, M. Effects of tree species and topography on soil chemistry, leaf litter quality, and decomposition in Northeast Turkey. *Soil Biol. Biochem.* **2005**, *37*, 1695–1706.
5. Fogel, R.; Cromack, K.J. Effect of habitat and substrate quality on Douglas-fir leaf litter decomposition in western Oregon. *Can. J. Bot.* **1977**, *55*, 1632–1640.
6. Valachovic, Y.S.; Caldwell, B.A.; Cromack, K. Jr.; Griffiths, R.P. Leaf litter chemistry controls on decomposition of Pacific Northwest trees and woody shrubs. *Can. J. For. Res.* **2004**, *34*, 2131–2147.
7. Prescott, C.E. Influence of forest floor type on rates of leaf litter decomposition in microcosms. *Soil Biol. Biochem.* **1996**, *28*, 1319–1325.
8. Zhou, G.; Guan, L.; Wei, X.; Tang, X.; Liu, S.; Liu, J.; Zhang, D.; Yan, J. Factors influencing leaf litter decomposition: An intersite decomposition experiment across China. *Plant Soil* **2008**, *311*, 61–72.
9. Rouifed, S.; Handa, I.T.; David, J.-F.; Hättenschwiler, S. The importance of biotic factors in predicting global change effects on decomposition of temperate forest leaf litter. *Oecologia* **2010**, *163*, 247–256.
10. Aponte, C.; García, L.V.; Marañón, T. Tree species effect on leaf litter decomposition and nutrient release in Mediterranean oak forests changes over time. *Ecosystems* **2012**, *15*, 1204–1218.
11. Madritch, M.D.; Cardinale, B.J. Impacts of tree species diversity on leaf litter decomposition in northern temperate forests of Wisconsin, USA: A multi-site experiment along a latitudinal gradient *Plant Soil* **2007**, *292*, 147–159.
12. Wang, Q.; Wang, S.; Xu, G.; Fan, B. Conversion of secondary broadleaved forest into Chinese fir plantation alters leaf litter production and potential nutrient returns. *Plant Ecol.* **2010**, *209*, 269–278.
13. Lian, Y.; Zhang, Q. Conversion of a natural broad-leafed evergreen forest into pure and mixed plantation forests in a subtropical area: Effects on nutrient cycling. *Can. J. For. Res.* **1998**, *28*, 1518–1529.

14. Rothe, A.; Binkley, D. Nutritional interactions in mixed species forests: A synthesis. *Can. J. For. Res.* **2001**, *31*, 1855–1870.

15. Hoorens, B.; Aerts, R.; Stroetenga, M. Does initial leaf litter chemistry explain leaf litter mixture effects on decomposition? *Oecologia* **2003**, *137*, 578–586.

16. Gartner, T.B.; Cardon, Z.G. Decomposition dynamics in mixed-species leaf litter. *Oikos* **2004**, *104*, 230–246.

17. Gholz, H.; Wedin, D.; Smitherman, S.; Harmon, M.; Parton, W. Long-term dynamics of pine and hardwood leaf litter in contrasting environments: Toward a global model of decomposition. *Glob. Change Biol.* **2000**, *6*, 751–765.

18. Laganiere, J.; Pare, D.; Bradley, R.L. How does a tree species influence leaf litter decomposition? Separating the relative contribution of leaf litter quality, leaf litter mixing, and forest floor conditions. *Can. J. For. Res.* **2010**, 40, 465–475.

19. Li, X.; Han, S.; Zhang, Y. Foliar decomposition in a broadleaf-mixed Korean pine, *Pinus koraiensis* Sieb. et Zucc. plantation forest, the impact of initial leaf litter quality and the decomposition of three kinds of organic matter fraction on mass loss and nutrient release rates. *Plant Soil* **2007**, *295*, 151–167.

20. Korea Forest Service. *Statistical Yearbook of Forestry*; Korea Forest Service: Daejeon, Korea, 2011.

21. Zhang, P.; Shao, G.; Zhao, G.; Le Master, D.C.; Parker, G.R.; Dunning, J.B. Jr.; Li, Q. China's forest policy for the 21st century. *Science* **2000**, *288*, 2135–2136.

22. Weather information. Available online: http://www.kma.go.kr/weather/climate/average_30years.jsp (accessed on 3 October 2013).

23. Buol, S.W.; Southar, R.J.; Graham, R.C.; McDaniel, P.A. *Soil Genesis and Classification*; 5th ed.; Iowa State Press: Iowa, IA, USA, 2003.

24. Miller, W.P.; Miller, D.M. A micro‑pipette method for soil mechanical analysis. *Commun. Soil Sci. Plan.* **1987**, *18*, 1–15.

25. Konen, M.E.; Jacobs, P.M.; Burras, C.L.; Talaga, B.J.; Mason, J.A. Equations for predicting soil organic carbon using Loss-on-Ignition for North Central U.S. soils. *Soil Sci. Soc. Am. J.* **2002**, *66*, 1878–1881.

26. Sumner, M.E.; Miller, W.P. Cation exchange capacity, and exchange coefficients. In *Methods of Soil Analysis, Part 3. Chemical methods. Soil Science of Society of America Book Series no. 5*; 3rd ed.; Sparks, D.L., Ed.; Soil Science Society of America: Madison, WI, USA, 1996; pp. 1201–1229.

27. Wardle, D.A.; Bonner, K.I.; Nicholson, K.S. Biodiversity and plant leaf litter: Experimental evidence which does not support the view that enhanced species richness improves ecosystem function. *Oikos* **1997**, *79*, 247–258.

28. Xu, X.; Hirata, E. Decomposition patterns of leaf litter of seven common canopy species in a subtropical forest: N and P dynamics. *Plant Soil* **2005**, *273*, 279–289.

29. Bonanomi, G.; Incerti, G.; Antignani, V.; Capodilupo, M.; Mazzoleni, S. Decomposition and nutrient dynamics in mixed leaf litter of Mediterranean species. *Plant Soil* **2010**, *331*, 481–496.

30. You, Y.H.; Namgung, J.; Lee, Y.Y.; Kim, J.H.; Lee, J.Y.; Mun, H.T. Mass loss and nutrients dynamics during the leaf litter decomposition in Kwangnung experimental forest. *Jour. Korean For. Soc.* **2000**, *89*, 41–48.

31. Olson, J.S. Energy stores and the balance of producers and decomposers in ecological systems. *Ecology* **1963**, *44*, 322–331.

32. Anaya, C.A.; Jaramillo, V.J.; Martínez-yrízar, A.; García-oliva, F. Large rainfall pulses control leaf litter decomposition in a tropical dry forest: Evidence from an 8-year study. *Ecosystems* **2012**, *15*; 652–663.

33. Lee, Y.C.; Nam, J.M.; Kim, J.G. The influence of black locust, *Robinia pseudoacacia.* flower and leaf fall on soil phosphate. *Plant Soil* **2011**, *341*, 269–277.

34. Lee, I.K.; Lim, J.H.; Kim, C.S.; Kim, Y.K. Nutrient dynamics in decomposing leaf litter and leaf litter production at the long-term ecological research site in Mt. Gyebangsan. *J. Ecol. Field Biol.* **2006**, *29*, 585–591.

35. Zhang, D.; Hui, D.; Luo, Y.; Zhou, G. Rates of leaf litter decomposition in terrestrial ecosystems: Global patterns and controlling factors. *J. Plant Ecol.* **2008**, *1*, 85–93.

36. Han, A.R.; Lee, S.K.; Suh, G.U.; Park, Y.; Park, P.S. Wind and topography influence the crown growth of *Picea jezoensis* in a subalpine forest on Mt. Deogyu, Korea. Agric. *For. Meteorol.* **2012**, *166–167*, 207–214.

37. Hinckley, E-L. S.; Ebel, B.A.; Barnes, R.T.; Anderson, R.S.; Williams, M.W.; Anderson, S.P. Aspect control of water movement on hillslopes near the rain-snow transition of the Colorado Front Range. *Hydrolo. Proc.* **2014**, *28*, 74–85.

38. Wang, S.; Ruan, H.; Han, Y. Effects of microclimate, leaf litter type, and mesh size on leaf litter decomposition along an elevation gradient in the Wuyi Mountains, China. *Ecol. Res.* **2010**, *25*, 1113–1120.

39. Austin, A.T.; Vitousek, P.M. Precipitation, decomposition and leaf litter decomposability of *Metrosideros polymorpha* in native forests on Hawaii. *J. Ecol.* **2000**, *88*, 129–138.

40. Mun, H.T.; Joo, H.T. Leaf litter production and decomposition in the *Quercus acutissima* and *Pinus rigida* forests. *Korean J. Ecol.* **1994**, *17*, 345–353.

41. Hisabae, M.; Sone, S.; Inoue, M. Breakdown and macrointervebrate colonization of needle and leaf litter in conifer plantation streams in Shikoku, southwestern Japan. *J. For. Res.* **2011**, *16*, 108–115.

42. Bardgett, R.D.; Shine, A. Linkages between plant leaf litter diversity, soil microbial biomass and ecosystem function in temperate grasslands. *Soil Biol Biochem* **1999**, *31*, 317–321.

43. De Marco, A.; Meola, A.; Maisto, G.; Giordano, M.; De Santo, A.V. Non-additive effects of leaf litter mixtures on decomposition of leaf litters in a Mediterranean maquis. *Plant Soil* **2011**, *344*, 305–317.

44. García-Palacios, P.; Maestre, F.T.; Kattge, J.; Wall, D.H. Climate and leaf litter quality differently modulate the effects of soil fauna on leaf litter decomposition across biomes. *Ecol. Lett.* **2013**, *16*, 1045–1053.

45. Chapman, S.K.; Newman, G.S.; Hart, S.C.; Schweitzer, J.A.; Koch, G.W. Leaf litter mixtures alter microbial community development: mechanisms for non-additive effects in leaf litter decomposition. *PLoS One* **2013**, *8*, doi:10.1371/journal.pone.0062671.

46. Prescott, C.E.; Vesterdal, L.; Pratt, J.; Venner, K.H.; Montigny, L.M.D. Trofymow, J.A. Nutrient concentrations and nitrogen mineralization in forest floors of single species conifer plantations in coastal British Columbia. *Can. J. For. Res.* **2000**, *30*, 1341–1352.

47. Hobbie, S.E.; Vitousek, P.M. Nutrient limitation of decomposition in Hawaiian forests. *Ecology* **2000**, *81*, 1867–1877.

48. Aerts, R.; Callaghan, T.V.; Dorrepaal, E.; van Logtestijn, R.S.P.; Cornelissen, J.H.C. Seasonal climate manipulations have only minor effects on leaf litter decomposition rates and N dynamics but strong effects on leaf litter P dynamics of sub-arctic bog species. *Oecologia* **2012**, *170*, 809–819.

49. Moore, T.R.; Trofymow, J.A.; Prescott, C.E.; Titus, B.D.; Group, C.W. Nature and nurture in the dynamics of C, N and P during leaf litter decomposition in Canadian forests. *Plant Soil* **2011**, *339*, 163–175.

50. Celi, L.; Cerli, C.; Turner, B.L.; Santoni, S.; Bonifacio, E. Biogeochemical cycling of soil phosphorus during natural revegetation of *Pinus sylvestris* on disused sand quarries in Northwestern Russia. *Plant Soil* **2013**, *367*, 121–134.

Increased Biomass of Nursery-Grown Douglas-Fir Seedlings upon Inoculation with Diazotrophic Endophytic Consortia

Zareen Khan, Shyam L. Kandel, Daniela N. Ramos, Gregory J. Ettl, Soo-Hyung Kim and Sharon L. Doty

Abstract: Douglas-fir (*Pseudotsuga menziesii*) seedlings are periodically challenged by biotic and abiotic stresses. The ability of endophytes to colonize the interior of plants could confer benefits to host plants that may play an important role in plant adaptation to environmental changes. In this greenhouse study, nursery-grown Douglas-fir seedlings were inoculated with diazotrophic endophytes previously isolated from poplar and willow trees and grown for fifteen months in nutrient-poor conditions. Inoculated seedlings had significant increases in biomass (48%), root length (13%) and shoot height (16%) compared to the control seedlings. Characterization of these endophytes for symbiotic traits in addition to nitrogen fixation revealed that they can also solubilize phosphate and produce siderophores. Colonization was observed through fluorescent microscopy in seedlings inoculated with *gfp*- and *mkate*-tagged strains. Inoculation with beneficial endophytes could prove to be valuable for increasing the production of planting stocks in forest nurseries.

Reprinted from *Forests*. Cite as: Khan, Z.; Kandel, S.L.; Ramos, D.N.; Ettl, G.J.; Kim, S.-H.; Doty, S.L. Increased Biomass of Nursery-Grown Douglas-Fir Seedlings upon Inoculation with Diazotrophic Endophytic Consortia. *Forests* **2015**, *6*, 3582-3593

1. Introduction

Millions of Douglas-fir (*Pseudotsuga menziesii*) seedlings are grown in nurseries for reforestation purposes in North America, Europe and elsewhere [1]. The majority of seedlings are grown for more than a year as bare root seedlings, although some are also grown in containers. Environmental disturbances and stresses contribute to seedling failure when transplanted in the field, and in order to maximize production, nurseries apply intensive cultural practices, fertilization being one of the most common [1,2]. This can result in additional costs to nursery operators and could also have negative environmental repercussions if nitrogen leaches from nurseries situated in areas near lakes and rivers and/or where ground water reservoirs are found. An inexpensive and environmentally-benign alternative for enhancing the productivity of newly-established forest plantations involves nursery inoculation of seedlings with plant growth-promoting microorganisms. In forestry, this has traditionally been restricted to inoculation with mycorrhizal fungi [3,4]; however, it is reported that in many cases, mycorrhizal inoculation resulted in low to no measurable benefit in outplanting success. Recently, root-associated bacteria have been shown to stimulate tree seedling growth in addition to improving mycorrhizal colonization and may be valuable in current reforestation efforts [5,6].

Some species of conifers are able to inhabit harsh and nutrient-poor subalpine sites where few other plants grow. A key component of their ecological success may be attributed to the presence of endophytes. Endophytes are microbes (bacteria and fungi) that reside inside plants and do not cause disease [7–10], and there is increasing evidence of their profound impact on plant development, physiology, evolution and adaptation [11]. An important trait of some endophytes is the ability to supply nitrogen to their host plant through biological nitrogen fixation. Nitrogen-fixing endophytes have been isolated from a variety of species, such as sugarcane [12], wild rice [13], corn [14,15], African sweet potato [16], kallar grass [17], coffee [18], cactus [19] and woody plants, including poplar, willow and coniferous trees [20–23]. It has been demonstrated that N fixed by diazotrophic bacteria can be utilized by plants [24]. Endophytic bacteria have been found in different parts of tree tissues, such as roots [25], stems [26], shoots [27], leaves [28] and buds [29]. More recently, diazotrophic endophytes were isolated from conifers growing in nitrogen-poor soils. Inoculation of lodgepole pine and western red cedar seedlings with the isolate *Paenibacillus polymyxa* P2b2R resulted in significant foliar ^{15}N dilution, indicative of biological nitrogen fixation [30,31]. The importance of endophytes to the establishment and persistence of long-lived western conifers is unknown; however, endophytes may allow establishment and growth on nutrient-limited and environmentally-stressed sites; for example, similar endophytes were found in subalpine conifers from distant sites in California and Colorado [32].

Phosphorous is another essential macronutrient promoting plant growth and development. Many woody plants are dependent on ectomycorrhizal fungi for their growth and survival. Mycorrhizae inoculation of Douglas-fir seedlings is a common practice in commercial nurseries [33,34]. However, commercial nursery treatments, including frequent addition of fertilizer and water, may not be favorable for mycorrhizae colonization of container-grown seedlings [35]. The importance of phosphate-solubilizing endophytes in increasing the phosphorous availability by solubilization of inorganic phosphate and mineralization of organic phosphate has recently been demonstrated as one of the important mechanisms of increased plant growth [36]. Siderophores are low molecular weight ferric iron-specific chelating agents produced by bacteria and fungi [37]. Microbial siderophores have been reported to have a positive correlation with plant growth promotion, and the production of siderophores is being considered as one of the key traits for primary screening of beneficial endophytes [38].

In this preliminary study, we chose a consortia of diazotrophic endophytes, originally isolated from poplar and willow trees [20–22], because they had shown robust growth-promoting activity with agricultural crops, grasses and poplar plants grown under abiotic stress [22,39–41]. Our objective was to determine if these endophytes enhance the growth of Douglas-fir seedlings under nutrient-poor conditions.

2. Experimental Section

2.1. Plant Material and Endophytes

Douglas-fir container (1 + 0) (4 cu. in ; Ray Leach "cone-tainers") seedlings from a local western WA provenance obtained from Silvaseed Company Foresters (Roy, WA, USA) were individually planted into plastic pots containing Sunshine Mix #2 (Steubers, WA, USA). The propagation protocol for seedling production is described elsewhere [42]. Seven bacterial strains and one yeast strain (Table 1) were chosen based on strong growth in nitrogen-free media, the presence of the nitrogen fixation gene (*nifH*) [20,21], production of phytohormones [43] and/or their robust plant growth-promoting abilities.

Table 1. Poplar and willow endophytes used in this study and their phosphate solubilization index (PSI) in the plate assay [20,21].

Endophyte	Closest 16SrDNA Match	PSI
WP1	*Rhodotorula graminis*	1.05
WP5	*Rahnella* sp.	1.64
WP9	*Burkholderia* sp.	1.50
WP19	*Acinetobacter calcoaceticus*	3.10
PTD1	*Rhizobium tropici* bv *populus*	1.40
WW5	*Sphingomonas yanoikuyae*	1.10
WW6	*Pseudomonas putida*	1.64
WW7	*Sphingomonas* sp.	1.82

2.2. Methods

Inoculation suspensions were prepared by first growing the individual endophyte strains on nutrient-rich media from −80 °C glycerol stocks. Bacteria were grown on mannitol glutamate/Luria–Bertani (MG/L) medium [44]. The yeast strain was grown on yeast extract peptone dextrose (YPD) medium. Single colonies were inoculated into liquid broth and grown overnight on a rotary shaker (150 rpm) and agitated for 24 h at 30 °C. Bacteria were harvested by centrifugation at 8000 rpm for 10 min, resuspended in nitrogen-free Murashige and Skoog (NFMS) (Caisson Labs, USA) broth and washed three times in NFMS. The endophyte consortia treatment was prepared by mixing equal concentrations of each inoculation suspension as determined by measuring the optical density at 600 nm (OD_{600}) of the individual strains and adjusting to a final OD_{600} of 0.1. Using a sterile conical tube, a 50-mL sample of the endophyte consortia was then delivered 5–10 cm below the surface and in close proximity to the roots. Mock-inoculated control seedlings received the same volume of sterile liquid broth. A total of 24 pots, 12 replications of inoculated or mock-inoculated control plants, were arranged in a randomized fashion. Seedlings were kept in an environment with an 18h photoperiod, a 20/14 °C day/night temperature cycle and a relative humidity of 70%–85%. Drip trays were placed under each individual pot to prevent cross-contamination through run-off. Each pot was individually irrigated with tap water. Irrigation frequency and duration were adjusted, such that the pots were not allowed to flood nor dry out between watering. Additionally, each pot was fertigated with received 100 mL of 1/2 strength, 1/8 nitrogen modified Hoagland's medium containing (g L^{-1}): $CaCl_2·2H_2O$, 0.22; K_2SO_4, 0.17; $MgSO_4·7H_2O$, 0.26; KH_2PO_4, 0.136; NaFeEDTA(10% Fe), 0.015; with 1 mL·L^{-1} micronutrient solution containing (g·L^{-1}): H_3BO_3, 0.773; $MnSO_4$, 0.169; $ZnSO_4·7H_2O$, 0.288; $CuSO_4·5H_2O$, 0.062; H_2MoO_4 (83% MoO_3), 0.04; weekly for the first two months post-inoculation, then 250 mL twice monthly until the end of the experiment.

Height measurements were recorded after inoculation and every month thereafter. After 15 months, the plants were removed from the soil, and roots and shoots were gently washed to remove dirt and debris. Roots and stems were then blotted dry, cut into root and shoot sections, and the lengths and wet weights were determined. Foliar samples were taken from each plant and oven dried at 65 °C, ground to a fine powder and analyzed for nitrogen content using a PE 2400 series II CHN elemental analyzer (Perkin-Elmer, MA, USA) at the University of Washington's School of Environmental and Forest Sciences Analytical Soils laboratory.

Potential plant growth-promoting traits, such as phosphate solubilization and siderophore production, were determined for the selected endophytes. The ability of the isolates to solubilize tricalcium orthophosphate (TCP) was tested on the National Botanical Research Institute's growth medium [45] containing $g \cdot L^{-1}$: glucose, 10; $Ca_3(PO_4)_2$, 5; $MgCl_2 \cdot 6H_2O$, 5; $MgSO_4 \cdot 7H_2O$, 0.25; KCl, 0.2; $(NH_4)_2SO_4$, 0.1. The inoculum was added in quadruplicate on the medium in Petri dishes. The halo and colony diameters were measured after 14 days of incubation of the plates at 25 °C. The ability of the microbes to solubilize the insoluble phosphate is described by the solubilization index (the ratio of the total diameter (halo + colony diameter (mm) to the colony diameter (mm)). Analysis of siderophore production by the endophyte isolates was performed following the chrome azurol S (CAS) method of Alexander and Zuberer [46]. For each isolate, 10 µL of inoculum were placed in quadruplicate on the medium. After incubation at 25 °C for 3 days, the discoloration of the medium (blue to orange) indicated siderophore-producing endophytes.

To evaluate endophytic colonization by some of the isolates, we used strains that were labeled with fluorescent markers with a broad host range plasmid. The *mkate* and *gfp* plasmids were introduced into *Acinetobacter calcoaceticus* strain WP19 and *Rahnella* sp. strain WP5, respectively, via triparental mating using *E. coli* DH5α as the donor and *E. coli* HB101 (pRK 2073) as the helper strain [47]. The inoculations were done as described above, and the colonization of root tissue and needles was verified through fluorescent microscopy using a Zeiss Imager M2 equipped with an AxioCam MRM and recorded with Zeiss AxioVision software (Karl Zeiss, LLC, Thornwood, NY, USA).

2.3. Statistical Analysis

One-way ANOVA (analysis of variance) and Tukey's honestly significant difference (HSD) test were used to identify significant differences between inoculated and mock-inoculated control seedlings.

3. Results and Discussion

Differences in plant heights and the health of inoculated and mock-inoculated control plants were apparent after fifteen months of growth in the greenhouse under N-limited conditions. As shown in Figure 1, the mock-inoculated control plants were stunted, with some being chlorotic and showing signs of nutrient deficiency, whereas the inoculated plants appeared taller, more robust and greener than the controls. The inoculated plants were 16% ($p = 0.008$) taller, had 13% longer roots ($p = 0.024$) (Figure 2) and a significant 48% ($p = 0.019$) increase in biomass (Figures 2 and 3) when compared to their mock-inoculated control plants. The concentration of N in plant tissues was higher in the inoculated plants compared to the controls, although it was not statistically significant ($p = 0.156$) (Figure 4).

Figure 1. Representative photograph of mock-inoculated (left) and inoculated (right) Douglas-fir seedling after 15 months of growth in nutrient-poor soil.

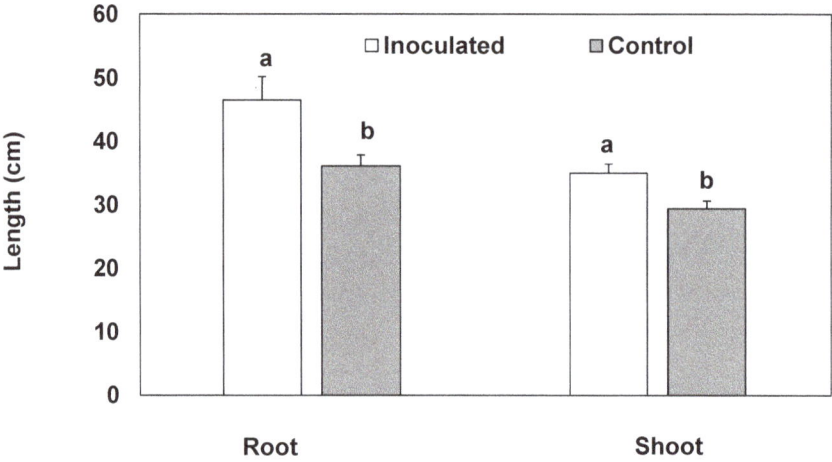

Figure 2. Root and shoot lengths of inoculated and control Douglas-fir seedlings at harvest. Data are the mean of 12 replicates. Error bars show the standard deviation. Means designated with different letters are significantly different.

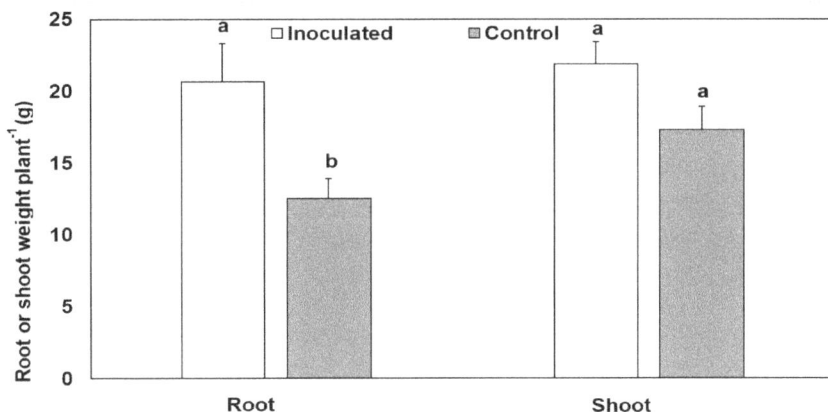

Figure 3. Root and shoot weights (g) of inoculated and control Douglas-fir seedlings at harvest. Data are the mean of 12 replicates. Error bars show the standard deviation. Means designated with different letters are significantly different.

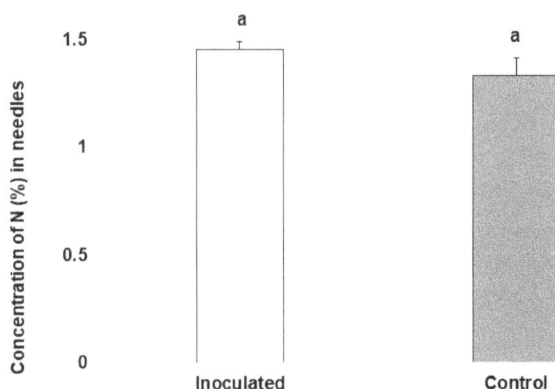

Figure 4. Total N concentration in the inoculated and control Douglas-fir seedlings needles at harvest. Data are the mean of 12 replicates. Error bars show the standard deviation. ($p = 0.156$).

Our results demonstrate that inoculation with the poplar and willow endophyte consortia stimulated growth of Douglas-fir seedlings under low nutrient conditions. It was previously concluded that these endophytes are not host-specific colonizers [39]. Poplar and willow plants are angiosperms, which diverged from gymnosperms more than 100 million years ago [48]. Remarkably, the endophytes from poplar and willow plants colonized Douglas-fir, a gymnosperm. It is possible that the required plant-microbe communication may be ancient, pre-dating the divergence of angiosperms and gymnosperms. Interestingly, the response of Douglas-fir to inoculation with the endophytes took more than a year to develop, which suggests that this plant species may require a period of growth under controlled conditions to facilitate the establishment of the symbiotic relationship with these endophytes. Parker and Dangerfield [49] also reported a delayed response to microbial inoculation in Douglas-fir. In studies done by Chanway *et al.* [50], rhizospheric bacteria isolated from perennial rye grass and white clover stimulated the growth of container Douglas-fir seedlings after 12 weeks of inoculation. More recently, Anand *et al.* [31] showed that seedling growth enhancement in other conifers, lodgepole pine and western red cedar occurred only after 13 months.

To assess other potential mechanisms in addition to N fixation for the improved growth of the inoculated plants, we investigated the potential of phosphate solubilization by the endophytes. The majority of the isolates were able to produce a clear zone in minimal medium containing insoluble phosphorous. As shown in Table 1, the PSI of isolate WP19 was the highest, indicating the high solubilization capacity of this endophyte compared to the rest of the endophytes. There have been a number of reports on plant growth promotion by bacteria that have the ability to convert the insoluble inorganic forms of P into soluble forms through acidification, secretion of organic acids or protons and exchange reactions [36]. Many symbiotic phosphate-solubilizing microorganisms belonging to species of the genera *Bacillus, Pseudomonas*, *Rhizobium, Enterobacter* and *Burkholderia* and certain fungi produce phytase to mineralize organic phosphate [51]. In our study, phosphate solubilization may be one of the important mechanism through which these endophytes promoted growth of Douglas-fir seedlings; however, additional studies including phosphorous quantification and mutant analysis will be necessary to determine what symbiotic factors contribute the most to growth enhancement. The role of mycorrhizal symbiosis in increasing phosphorous uptake has been studied in a number of conifer species, with recent studies showing how mycorrhizal effects on seedling establishment change with soil conditions [52]. In dry Douglas-fir forests, reduced mycorrhizal diversity and abundance have been associated with reduced survival and growth of newly-planted conifer seedlings [53,54]. Inoculating nursery seedlings with beneficial endophytes may help offset some of these disadvantages. Furthermore, endophytes may increase mycorrhizal development by affecting root colonization, as well as by enhancing N and P uptake.

Rungin *et al.* [55] recently demonstrated plant growth-enhancing effects by a siderophore-producing endophytic streptomycete isolated from Thai jasmine rice plants. Significant increases in total biomass of rice and mung bean plants were observed in the inoculated controls compared to the untreated controls and siderophore-deficient mutant treatment. Most of the endophytes in our study produced siderophores, which may have been beneficial in plant growth promotion. However, this is largely speculative and warrants further research. Siderophore-producing microbes may also have significant activity against plant pathogens [56,57]. Interestingly, we found that Douglas-fir seeds inoculated with these endophytes were significantly less affected by Fusarium, a common Douglas-fir pathogen [58].Therefore, these endophytes merit further investigation, as reducing mortality by inoculations could prove valuable to tree nurseries.

To evaluate endophytic colonization, Douglas-fir seedlings were inoculated with fluorescent-labelled strains of WP5 and WP19. As seen from the fluorescent microscope images, *mkate*-tagged cells of *Acinetobacter* strain WP19 were localized in the intercellular spaces of root tissue of the inoculated seedlings (Figure 5A). In addition, GFP-tagged cells of *Rahnella* sp. strain WP5 colonized the needles (Figure 5B), suggesting that this endophyte is able to move up the plant into aerial parts after inoculation. Similar colonization zones have been reported for other endophytes [30].

The enhanced growth in inoculated seedlings seen in this preliminary study merits further research in larger greenhouse trials and field trials to assess the long-term benefits of endophyte inoculations.

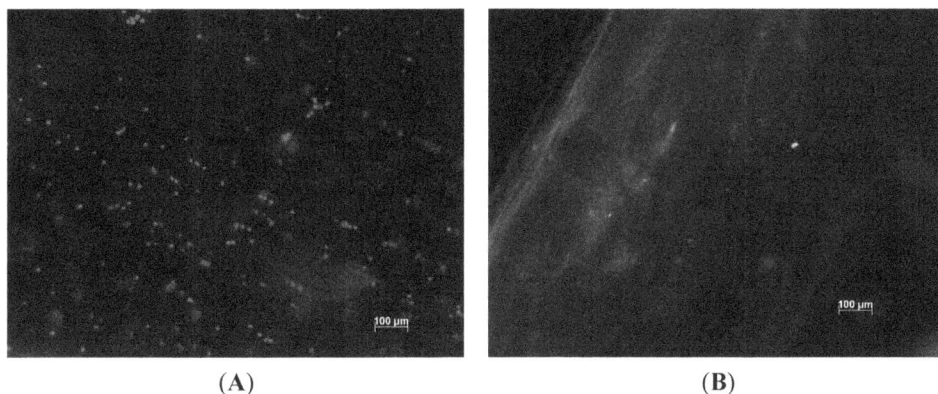

(A)   (B)

Figure 5. Fluorescence micrographs showing intercellular colonization of *mkate*-tagged strain WP19 in the root tissue of the inoculated seedling of Douglas-fir (**A**) and GFP-tagged strain WP5 in the needle of the inoculated Douglas-fir (**B**). Samples were surface-sterilized and visualized (1000×) three weeks after inoculation.

4. Conclusions

The results of the present study indicate that inoculation with a diazotrophic endophytic consortia significantly increased Douglas-fir seedling growth under nutrient-limited conditions. Most commercial nursery treatments add high rates of fertilizers to maximize seedling shoot and total growth to increase establishment and success in the field [59] which is not only expensive, but also suppresses mycorrhiza development of container seedlings [60]. Therefore, endophyte inoculation of seedlings may be considered as a cost-effective technique to increase growth in the early stage for the nursery stock, which can lead to later stage successes in transplanting and establishment after transplanting.

Acknowledgments

This work was funded by the Mc Intire-Stennis and USDA National Institute of Food and Agriculture grants (# 2012-00931). The authors would like to thank Mike Gerdes and David Gerdes from Silvaseed Company Foresters for providing the seedlings and Sam Miller (University of Washington, Microbiology) for providing the plasmids.

Author Contributions

Zareen Khan carried out the experiments and wrote the manuscript. Shyam L Kandel and Daniela N Ramos contributed to the experiments and data analysis. Doty served as the Principal Investigator (Mc Intire-Stennis and NIFA grants) and directed the project. Ettl and Kim served as co-PI's for the project. Doty, Ettl and Kim edited the manuscript.

236

Conflicts of Interest

The authors declare no conflict of interest.

References

1. Van den Driessche, R. Growth, survival and physiology of Douglas-fir seedlings following root wrenching and fertilization. *Can. J. For. Res.* **1983**, *13*, 270–278.
2. Aldhous, I. Nursery Practice. *For. Comm. Bull.* **1972**, *43*, 184.
3. Cram, M.M.; Dumroese, K.R. Mycorhizae in Forest Tree nurseries. In *Forest Nursery Pests. Agriculture Handbook*; Cram, M.M.F., Mallams, K.M., Eds.; Forest Service: Washington, DC, USA, 2012; pp. 20–25.
4. Kropp, B.R.; Langlois, C.G. Ectomycorrhizae in reforestation. *Can. J. For. Res.* **1990**, *20*, 438–451.
5. Tien, T.M.; Gaskins, M.H.; Hubbell, D.H. Plant growth substances produced by *Azospirillum brasilense* and their effect on the growth of Pearl Millet. *Appl. Environ. Microbiol.* **1979**, *37*, 1016–1024.
6. Chanway, C.P.; Holl, F.B.; Rurkington, R. Genotypic coadaptation in growth promotion of forage species by *Bacillus polymyxa*. *Plant Soil* **1988**, *106*, 281–284.
7. Ryan, R.P.; Germaine, K.; Franks, A.; Ryan, D.J.; Dowling, D.N. Bacterial endophytes: Recent developments and applications. *FEMS. Microbiol. Lett.* **2008**, *278*, 1–9.
8. Hardoim, P.R.; van Overbeek, L.S.; van Elsas, J.D. Properties of bacterial endophytes and their proposed role in plant growth. *Trends Microbiol.* **2008**, *16*, 463–471.
9. Gaiero, J.R.; McCall, C.A.; Thompson, K.; Day, N.J.; Best, A.S.; Dunfield, K.E. Inside the root microbiome: Bacterial root endophytes and plant growth promotion. *Am. J. Bot.* **2013**, *100*, 1738–1750.
10. *Symbiotic Endophytes*; Springer-Verlag Berlin Heidelberg: New York, NY, USA, 2014.
11. Compant, S.; van der Heijden, M.G.A.; Sessitch, A. Climate change on beneficial plant microorganism interactions. *FEMS Microbiol. Ecol.* **2010**, *73*, 197–214.
12. Welbaum, G.E.; Sturz, A.V.; Dong, Z.; Nowak, J. Endophytic nitrogen fixation in sugarcane: Present knowledge and future applications. *Plant Soil* **2003**, *252*, 139–149.
13. Peng, G.; Zhang, W.; Luo, H.; Xie, H.; Lai, W.; Tan, Z. Enterobacter oryzae sp. nov., a nitrogen-fixing bacterium isolated from the wild rice species *Oryza latifolia*. *Int. J. Syst. Evol. Microbiol.* **2009**, *59*, 1650–1655.
14. Caballero-Mellado, J.; Martinez-Aguilar, L.; Paredes-Valdez, G.; Estrada-de Los Santos, P. *Burkholderia unamae* sp. nov., an N_2-fixing rhizospheric and endophytic species. *Int. J. Syst. Evol. Microbiol.* **2004**, *54*, 1165–1172.
15. Montanez, A.; Abreu, C.; Gill, P.R.; Hardarson, G.; Sicardi, M. Biological nitrogen fixation in maize (*Zea mays* L.) by N-15 isotope-dilution and identification of associated culturable diazotrophs. *Biol. Fert. Soils* **2009**, *45*, 253–263.
16. Reiter, B.; Burgmann, H.; Burg, K.; Sessitsch, A. Endophytic nifH gene diversity in African sweet potato. *Can. J. Microbiol.* **2003**, *49*, 549–555.

17. Reinhold-Hurek, B.; Hurek, T. Life in grasses: Diazotrophic endophytes. *Trends Microbiol.* **1998**, *6*, 139–144.

18. Jimenez-Salgado, T.; Fuentes-Ramirez, L.F.; Tapia-Hernandez, A.; Mascarua-Esparza, M.A.; Martinez-Romero, E.; Caballero-Mellado, J. *Coffea arabica* L., A new host plant for *Acetobacter diazotrophicus*, and isolation of other nitrogen-fixing acetobacteria. *Appl. Environ. Microbiol.* **1997**, *63*, 3676–3683.

19. Lopez, B.R.; Bashan, Y.; Bacilio, M. Endophytic bacteria of Mammillaria fraileana, an endemic rock-colonizing cactus of the southern Sonoran Desert. *Arch. Microbiol.* **2011**, *193*, 527–541.

20. Doty, S.L.; Dosher, M.R.; Singleton, G.L.; Moore, A.L.; Aken, B.; van Stettler, R.F.; Gordon, M.P. Identification of an endophytic Rhizobium in stems of Populus. *Symbiosis* **2005**, *39*, 27–35.

21. Doty, S.L.; Oakley, B.; Xin, G.; Kang, J.W.; Singleton, G.; Khan, Z.; Vajzovic, A.; Staley, J.T. Diazotrophic endophytes of native black cottonwood and willow. *Symbiosis* **2009**, *47*, 23–33.

22. Xin, G.; Zhang, G.; Kang, J.W.; Staley, J.T.; Doty, S.L. A diazotrophic, indole-3-acetic acid-producing endophyte from wild cottonwood. *Biol. Fert. Soils* **2009**, *45*, 669–674.

23. Bormann, B.T.; Bormann, W.B.; Bowden, R.S.; Piece, S.P.; Hamburg, D.; Wang, M.; Snyder, C.; Ingersoll, R.C. Rapid N_2 Fixation in Pines, Alder, and Locust: Evidence From the Sandbox Ecosystems Study. *Ecosystems* **1993**, *74*, 583–598.

24. Pankievicz, V.C.S.; do Amaral, F.P.; Santos, K.F.D.N.; Agtuca, B.; Xu, Y.; Schueller, M.J.; Arisi, A.C.M.; Steffens, M.B.R.; de Souza, E.M.; Pedrosa, F.O.; *et al.* Robust biological nitrogen fixation in a model grass-bacterial association. *Plant J.* **2015**, *81*, 907–919.

25. Mocali, S.; Bertelli, E.; Dicello, F.; Mengoni, A.; Falanga, A.; Villani, F.; Caciotti, A.; Tegli, S.; Surico, G.; Fani, R. Fluctuation of bacteria isolated from elm tissues during different seasons and from different plant organs. *Res. Micribiol.* **2003**, *154*, 105–114.

26. Bal, A.S.; Anand, R.; Berge, O.; Chanway, C.P. Isolation and identification of diazotrophic bacteria from internal tissues of *Pinus contorta* and *Thuja plicata*. *Can. J. For. Res.* **2012**, *42*, 807–813.

27. Izumi, H.; Anderson, I.C.; Killham, K.; Moore, E.R. Diversity of predominant endophytic bacteria in European deciduous and coniferous trees. *Can. J. Microbiol.* **2008**, *54*, 173–179.

28. Ulrich, K.; Ulrich, A.; Ewald, D. Diversity of endophytic bacterial communities in poplar grown under field conditions. *FEMS Microbiol.* **2008**, *63*, 169–180.

29. Pirttila, A.M.; Laukkanen, H.; Pospiech, H.; Myllyla, R.; Hohtola, A. Detection of intracellular bacteria in the buds of scotch pine (*Pinus sylvestris* L.) by insitu hybridization. *Appl. Environ. Microbiol.* **2000**, *66*, 3073–3077.

30. Bal, A.S.; Chanway, C.P. Evidence of nitrogen fixation in lodgepole pine inoculated with diazotrophic *Paenibacillus polymyxa*. *Botany* **2012**, *90*, 891–896.

31. Anand, R.; Grayston, S.; Chanway, C.P. N_2-fixation and seedling growth promotion of lodgepole pine by endophytic *Paenibacillus polymyxa*. *Microb. Ecol.* **2013**, *66*, 369–374.

32. Carrell, A.A.; Frank, A.C. *Pinus flexilis* and *Picea engelmannii* share a simple and consistent needle endophyte microbiota with a potential role in nitrogen fixation. *Front. Microbiol.* **2014**, doi:10.3389/fmicb.2014.00333.

33. Stack, R.W.; Sinclair, W.A. Protection of Douglas-fir seedlings against *Fusarium* root rot by a mycorrhizal fungus in the absence of mycorrhiza formation. *Phytopathology* **1975**, *65*, 468–472.

34. Horton, T.R.; Bruns, T.D.; Parker, V.T. Ectomycorrhizal fungi associated with *Arctostaphylos* contribute to *Pseudotsuga menziesii* establishment. *Can. J. Bot.* **1999**, *77*, 93–102.

35. Nara, K. Ectomycorhizal networks and seedling establishment during early primary succession. *New Phytol.* **2006**, *169*, 169–178.

36. Rodríguez, H.; Fraga, R. Phosphate solubilizing bacteria and their role in plant growth promotion. *Biotechnol. Adv.* **1999**, *17*, 319–339.

37. Neilands, J.B. Siderophores: Structure and function of microbial iron transport compounds. *J. Biol. Chem.* **1995**, *270*, 26723–26726.

38. Glick, B.R. Plant Growth-Promoting Bacteria: Mechanisms and Applications. *Scientifica* **2012**, *2012*, doi:10.6064/2012/963401.

39. Khan, Z.; Guelich, G.; Phan, H.; Redman, R.; Doty, S. Bacterial and Yeast Endophytes from Poplar and Willow Promote Growth in Crop Plants and Grasses. *ISRN Agric.* **2012**, *2012*, 890280:1–890280:11.

40. Knoth, J.L.; Kim, S.H.; Ettl, G.J.; Doty, S.L. Effects of cross host species inoculation of nitrogen-fixing endophytes on growth and leaf physiology of maize. *GCB Bioenergy* **2013**, *5*, 408–418.

41. Knoth, J.L.; Kim, S.H.; Ettl, G.J.; Doty, S.L. Biological nitrogen fixation and biomass accumulation within poplar clones as a result of inoculations with diazotrophic endophyte consortia. *New Phytol.* **2014**, *201*, 599–609.

42. Dumroese, K.R.; Wenny, L.D. Propagation protocol for production of Container (plug) *Pseudotsuga menziesii* Franco plants 66 mL (4cu.in) Ray Leach "Cone-tainers". 2009. Available online: http://www.NativePlantNetwork.org (accessed on 28 September 2015).

43. Khan, Z.; Rho, H.; Firrincieli, A.; Hung, S.H.; Luna, V.; Masciarelli, O.; Kim, S.H.; Doty, S.L. Growth improvement and drought tolerance of hybrid poplar by inoculation with endophyte consoria. **2015**, Submitted.

44. Cangelosi, G.A.; Best, E.A.; Martinetti, G.; Nester, E.W. Genetic analysis of *Agrobacterium*. *Meth. Enzymol.* **1991**, *204*, 384–397.

45. Nautiyal, C.S. An efficient microbiological growth medium for screening phosphate solubilizing microorganisms. *FEMS Microbiol. Lett.* **1999**, *170*, 265–270.

46. Alexander, D.B.; Zuberer, D.A. Use of chrome azurol S reagent to evaluate siderohore production by rhizosphere bacteria. *Biol. Fert. Soils* **1991**, *12*, 39–45.

47. Doty, S.; Chang, M.; Nester, E.W. The chromosomal virulence gene. ChvE of *Agrobacterium tumafaciens* is regulated by a LysR family member. *J. Bacteriol.* **1993**, *175*, 7880–7886.

48. Martin, W.; Lydiate, D.; Brinkmann, H.; Forkmann, G.; Saedle, R.H.; Cerff, R. Molecular phylogenies in angiosperm evolution. *Mol. Biol. Evol.* **1993**, *10*, 140–162.

49. Parker, A.K.; Dangerfield, J.A. Influence of bacterial inoculations on growth of containerized Douglas-fir seedlings. *Can. J. For. Res.* **1975**, *31*, 14–15.

50. Chanway, C.P.; Radley, R.A.; Holl, F.B. Bacterial Inoculation of Lodgepole Pine, White Spruce. In Proceedings of the Intermountain Forest Nursery Association Annual Meeting, Bismarck, ND, USA, 14–18 August 1989; Landis, T.D., Ed.; USDA Forest Service: Fort Collins, CO, USA, 1989.

51. Singh, P.; Kumar, V.; Agrawal, S. Evaluation of phytase producing bacteria for their plant growth promoting activities. *Int. J. Microbiol.* **2014**, *2014*, 7.

52. Horton, T.R.; Cazaras, E.; Bruns, T.D. Ectomycorrhizal, vesicular-arbuscular and dark septate fungal colonization of bishop pine (*Pinus muricata*) seedlings in the first 5 months of growth after wildfire. *Mycorrhiza* **1998**, *7*, 11–18.

53. Kazantseva, O.; Bingham, M.; Simard, S.W.; Berch, S.M. Effects of growth medium, nutrients, water, and aeration on mycorhization and biomass allocation of greenhouse-grown Douglas-fir seedlings. *Mycorrhiza* **2009**, *20*, 51–66.

54. Teste, F.P.; Schmidt, M.G.; Berch, S.M.; Bulmer, C.; Egger, K.N. Effects of ectomycorrhizal inoculants on survival and growth of interior Douglas-fir seedlings on reforestation sites and partially rehabilitated landings. *Can. J. For. Res.* **2004**, *4*, 2074–2088.

55. Rungin, S.; Indananda, C.; Suttiviriya, P.; Kruasuwan, W.; Jaemsaeng, R.; Thamchaipenet, A. Plant growth enhancing effects by a siderophore producing endophytic streptomycete isolated from a Thai jasmine rice plant (*Oryza sativa* L. KDML105). *Antonie Leeuwenhoek* **2012**, *102*, 463–472.

56. Thomashow, L.S. Biological control of plant root pathogens. *Curr. Opin. Biotech.* **1996**, *7*, 343–347.

57. Duijiff, B.J.; Alabouvette, C.; Lemanceau, P. Involvement of pseudobactin 358 in the inhibition of *Fusarium oxysporum* f.sp.lini in flax rhizosphere by the combination of *Pseudomas putida* WCS358 and nonpathogenic *Fusarium oxysporum* F047 and *Pseudomas putida* WCS358. *Phytopathology* **1997**, *89*, 1073–1079.

58. Khan, Z.; Ramos, D.; Matthew, A.; Ettl, J.G.; Doty, S.L. Impact of endophyte inoculation on health of Douglas-fir seedlings. Unpublished work, 2015.

59. Hunt, G.A. Effects of Mycorrhizal Fungi on Quality of Nursery Stock and Plantation Performance in the Southern Interior of British Columbia; FRDA Report ISSN 0835-0752; Natural Resources Canada: Victoria, BC, Canada, 1992; p. 185.

60. Marx, D.H.; Hatch, A.B.; Mendicino, J.F. High soil fertility decreases sucrose content and susceptibility of loblolly pine to ectomycorrhizal infection by *Pisolithis tinctorius. Can. J. Bot.* **1977**, *55*, 1569–1574.

MDPI AG

Klybeckstrasse 64

4057 Basel, Switzerland

Tel. +41 61 683 77 34

Fax +41 61 302 89 18

http://www.mdpi.com/

Forests Editorial Office

E-mail: forests@mdpi.com

http://www.mdpi.com/journal/forests